AF819150

Unmanned Aircraft Systems

This book provides comprehensive details on unmanned aerial vehicle applications and allied technologies to enrich the readers with the endless potential of the unmanned aerial vehicles ecosystem. It also presents various international regulatory compliances to enable the readers with informed decisions.

Key Features

- Thoroughly examines the core principles that form the foundation of unmanned aerial vehicles technology, highlighting the critical significance of sensors in unmanned aerial vehicles operations.
- Presents a step-by-step approach for designing custom security protocols for secure unmanned aerial vehicles communications.
- Discusses topics including robot operating systems, communication protocols, and swarm intelligence.
- Showcases practical case studies to bridge the gap between theory and real-world implementation.
- Illustrates designing and formulating security protocols for secure drone communication on practical platforms.

The text is primarily written for senior undergraduates, graduate students, and academic researchers in electrical engineering, electronics and communications engineering, computer science and engineering, aerospace engineering, and information technology.

Unmanned Aircraft Systems

Concepts, Simulation, Communication, and Applications

Edited by
Kalpana Gupta, Ummer Iqbal, Kriti Saroha,
Rahul Johari, and Kiran Walia

CRC Press
Taylor & Francis Group
Boca Raton London New York

CRC Press is an imprint of the
Taylor & Francis Group, an **informa** business

Cover image: Getty Images

First edition published 2027
by CRC Press
2385 NW Executive Center Drive, Suite 320, Boca Raton FL 33431

and by CRC Press
4 Park Square, Milton Park, Abingdon, Oxon, OX14 4RN

CRC Press is an imprint of Taylor & Francis Group, LLC

© 2027 selection and editorial matter, Kalpana Gupta, Ummer Iqbal, Kriti Saroha, Rahul Johari and Kiran Walia; individual chapters, the contributors

Reasonable efforts have been made to publish reliable data and information, but the author and publisher cannot assume responsibility for the validity of all materials or the consequences of their use. The authors and publishers have attempted to trace the copyright holders of all material reproduced in this publication and apologize to copyright holders if permission to publish in this form has not been obtained. If any copyright material has not been acknowledged please write and let us know so we may rectify in any future reprint.

Except as permitted under U.S. Copyright Law, no part of this book may be reprinted, reproduced, transmitted, or utilized in any form by any electronic, mechanical, or other means, now known or hereafter invented, including photocopying, microfilming, and recording, or in any information storage or retrieval system, without written permission from the publishers.

For permission to photocopy or use material electronically from this work, access www.copyright.com or contact the Copyright Clearance Center, Inc. (CCC), 222 Rosewood Drive, Danvers, MA 01923, 978-750-8400. For works that are not available on CCC please contact mpkbookspermissions@tandf.co.uk

For Product Safety Concerns and Information please contact our EU representative GPSR@taylorandfrancis.com. Taylor & Francis Verlag GmbH, Kaufingerstraße 24, 80331 München, Germany.

Trademark notice: Product or corporate names may be trademarks or registered trademarks and are used only for identification and explanation without intent to infringe.

ISBN: 9781032983868 (hbk)
ISBN: 9781032983875 (pbk)
ISBN: 9781003598350 (ebk)

DOI: 10.1201/9781003598350

Typeset in Sabon
by codeMantra

Contents

Preface viii
Editor Biographies ix
List of Contributors xi

PART 1
Foundations of unmanned aerial vehicles

1 Introduction to drone technology 3
DINESH KUMAR RAJENDRAN, Y.G. BALA, AND SUMAYA AKHTAR

2 Seeing beyond: sensor technologies in drone operations 29
PRERNA AJMANI, AYUSH ARYA, ASHISH KHANNA, PANKAJ KUMAR VARSHNEY, AND BAL VIRDEE

3 Navigating the skies: UAVs and the emerging global frameworks 52
PRASAD GAYATHRI, ANJU MOHAN, AND MURUGAN RAMU

PART 2
Unmanned aerial vehicles simulation and modeling

4 Simulators for UAVs 79
NARESH TRIPATHI, KALPANA JOHARI, NEELIMA GUPTA, AND SANDHYA ANEJA

5 Flight controller selection: an investigation for the development of drones for specific applications 96
N.C. AJAY VISHWATH, ARVIND R. YADAV, ROSHAN KUMAR, AND UDAYKUMAR V. JOSHI

6 Introduction to the Robot Operating System (ROS) 128

SARLA MORE, PRERNA AJMANI, DURGESH MISHRA, AND EBRAHIM A. MATTAR

7 Image-processing applications in modern UAV systems 156

PRANAV PARANJPE AND RAJESH BODADE

PART 3
Communication and security

8 Mist computing in distributed UAV swarm control 171

ROLANDO HERRERO AND MARTIN CADIROLA

9 MAVLink communication framework: from basic operations to secure deployment 197

IRAQ AHMAD RESHI, SUMAYA AKHTAR, AND HRISIKESH ROY

10 Secure horizons: future trends in drone communication 217

SUMAYA AKHTAR, IRAQ AHMAD RESHI, UMMER IQBAL KHAN, PRERNA AJMANI, AND J. JOSHUA THOMAS

PART 4
Application and allied technologies for unmanned aerial vehicles

11 Global awareness and adoption of UAVs in agriculture: a case study of urban India 237

IPSEETA SATPATHY, B. CHANDRA MOHAN PATNAIK, VISHAL JAIN, AND ANISH PATNAIK

12 Sensor-driven UAV technologies for disaster management: a comprehensive framework for real-time response and recovery 265

BHAVYA, ARPIT MISHRA, UJJWAL SINGH KATIYAR, AND RAHUL JOHARI

13 Drones in real estate: applications, challenges, and opportunities 304

VIJAY PATEL, ARVIND R. YADAV, ROSHAN KUMAR, JAYENDRA KUMAR, AND UDAYKUMAR V. JOSHI

14 Application of UAVs in industrial manufacturing and transportation 333

UJJWAL SINGH KATIYAR, ARPIT MISHRA, BHAVYA, AND RAHUL JOHARI

Index 383

Preface

We are pleased to present *Unmanned Aircraft Systems: Concepts, Simulation, Communication, and Applications*, a curated volume that brings together contemporary research, practical insights, and emerging innovations in the unmanned aerial vehicle (UAV) domain. The chapters featured in this book reflect the collective expertise of contributors from academia, research institutions, and industry, providing an interdisciplinary perspective on the evolving landscape of unmanned aerial systems. This book has been structured to guide the readers from fundamental concepts to advanced implementations. The initial chapters lay a strong foundation by discussing drone technology, sensing systems, and global regulatory frameworks. Subsequent chapters delve into essential technical elements such as UAV simulation tools, flight controller evaluation, ROS-based development, and image-processing techniques—topics that are critical for both learners and practitioners building autonomous and intelligent UAV systems.

A dedicated section explores communication and security aspects, including LoRa-based mist computing for UAV swarms, the MAVLink communication framework, and future trends in secure UAV networking. These discussions highlight the growing need for resilience, robustness, and secure communication in modern drone systems.

We express our sincere gratitude to all contributing authors and reviewers for their dedication and timely cooperation. Their efforts have ensured that this volume maintains high academic quality and practical relevance. We hope this book serves as a valuable resource for students, researchers, engineers, and professionals working in the UAV domain.

Editor Biographies

Ummer Iqbal is currently serving as a Scientist 'D' at the National Institute of Electronics and Information Technology J & K, Ministry of Electronics and Information Technology, Government of India. He earned his B.E. in Computer Science and Engineering from Dayananda Sagar College (VTU, Karnataka) in 2007 and completed his M.Tech. and Ph.D. in Computer Science and Engineering at NIT Srinagar in 2015 and 2022, respectively. He joined the National Institute of Electronics and Information Technology (NIELIT) as a Scientist 'B' in March 2010 and has over 15 years of experience in teaching and research. His research interests include Internet of Things security, drone security, and cybersecurity, and he has authored many publications in reputed international journals and conferences. He has also served as Chief Investigator and Co-Chief Investigator for several R&D and capacity-building projects.

Kalpana Gupta is presently working as Associate Director (Scientist F) at the Centre for Development of Advanced Computing. She holds an M.Tech and PhD from Guru Gobind Singh Indraprastha University, Delhi. She has around 23 years of research and teaching experience. Her job profile includes managing government funded research projects and teaching at postgraduate level. Her area of interest includes artificial intelligence and algorithms.

Kriti Saroha is presently working as Joint Director (Scientist E) at the Centre for Development of Advanced Computing, Noida. She has received her PhD from the Guru Gobind Singh Indraprastha University, Delhi and MS from BITS, Pilani. She has around 24 years of teaching and research experience. Her research interest includes data warehousing and data mining, artificial intelligence, machine learning, and computer architecture. She has published a book on computer organization and several research papers in national/international conferences and journals.

Rahul Johari is an Associate Professor at the University School of Automation and Robotics, Guru Gobind Singh Indraprastha University, East Delhi Campus, Delhi, India. He has completed his postdoctoral research under the supervision of Prof. Deo Prakash Vidyarthi at the School of Computer and System Sciences (SC&SS), JNU and his PhD under the supervision of Prof. Neelima Gupta and Dr. Sandhya Aneja in the Department of Computer Science at the University of Delhi. Presently he is PhD Program Coordinator and B.Tech. (Industrial Internet of Things—IIoT) Program Coordinator in USAR and is Associate Director of Students' Welfare of East Delhi Campus, Guru Gobind Singh Indraprastha University. He is the Head and Founder of SWINGER (Security, Wireless, IoT Network Group of Engineering and Research). He is a Microsoft/Google-Certified Professional and an ODEM-certified BlockChain Faculty. He is the recipient of the 'Outstanding Researcher Award' for the academic year 2023–2024, Best Teacher/Researcher Award for the academic years 2013–2014 and 2014–2015, and Vice Chancellor's Certificate of Appreciation in 2020–2021 and 2022–2023 at GGSIP University. His academic and research profile has been accredited as the Best Profile of the GGSIP University by IRINS (Indian Research Information Network System), a web-based Research Information Management (RIM) service developed by INFLIBNET Centre under the National Mission on Education through ICT (NME-ICT) under the Ministry of Education, Government of India. He has been selected as ACM Eminent Speaker (ESP) by ACM India Council in 2021. He was conferred with the 'Best Faculty Award' by the PHD Chamber of Commerce and Industry and SEMS, Inc. in April 2022. With 128 SCOPUS-indexed publications (including 25+ book chapters) in leading publishers and ACM-indexed international conferences and journals of repute, his current H-index is 18.

Kiran Walia is currently a Project Engineer at the Centre for Development of Advanced Computing. She has completed her M.Tech from the Electronic Science Department at Kurukshetra University. With approximately 14 years of teaching experience, she has a keen interest in VLSI design and fabrication, microelectromechanical systems, and the Internet of Things. Her job profile includes teaching at the postgraduate level and actively participating in research projects.

List of Contributors

Prerna Ajmani
School of Information Technology
Vivekananda Institute of Professional Studies—Technical Campus
New Delhi, Delhi, India

Sumaya Akhtar
National Institute of Electronics and Information Technology
Srinagar, Jammu & Kashmir, India

Sandhya Aneja
School of Computer Science and Mathematics
Marist University
Poughkeepsie, New York

Ayush Arya
Department of Computer Applications
St. Andrews Institute of Technology & Management
Gurugram, Haryana, India

Y.G. Bala
Department of Mechanical Engineering
MNNIT Allahabad
Prayagraj, Uttar Pradesh, India

Bhavya
University School of Automation and Robotics
GGSIPU EDC
New Delhi, Delhi, India

Rajesh Bodade
Department of ECE
Military College of Telecommunication Engineering (MCTE)
Indore, Madhya Pradesh, India

Martin Cadirola
Ecotronics Ventures LLC
Baltimore, Maryland

Prasad Gayathri
Department of Law
Saveetha School of Law
Saveetha Institute of Medical and Technical Sciences
Chennai, Tamil Nadu, India

Neelima Gupta
Department of Computer Science
University of Delhi
Delhi, India

Rolando Herrero
College of Engineering
Northeastern University
Boston, Massachusetts

Vishal Jain
Department of Computer Science and Engineering, School of Engineering and Technology,
Sharda University, Greater Noida India

Kalpana Johari
Centre for Development of Advanced Computing
Noida, Uttar Pradesh, India

Rahul Johari
University School of Automation and Robotics
GGSIPU EDC
New Delhi, Delhi, India

Udaykumar V. Joshi
Department of Mechanical Engineering
Government Engineering College Bharuch
Bharuch, Gujarat, India

Ujjwal Singh Katiyar
University School of Automation and Robotics
GGSIPU EDC
New Delhi, Delhi, India

Ummer Iqbal Khan
National Institute of Electronics and Information Technology (NIELIT)
Srinagar, Jammu & Kashmir, India

Ashish Khanna
Department of Computer Science & Engineering
Maharaja Agrasen Institute of Technology
New Delhi, Delhi, India

Jayendra Kumar
Department of Electronics and Communication
NIT Jamshedpur
Jamshedpur, Jharkhand, India

Roshan Kumar
Department of Electronic and Information Technology
Miami College of Henan University
Kaifeng, China

Ebrahim A. Mattar
College of Engineering
University of Bahrain
Zallaq, Bahrain

Arpit Mishra
University School of Automation and Robotics
GGSIPU EDC
New Delhi, Delhi, India

Durgesh Mishra
School of Computer Science and Information Technology
Symbiosis University of Applied Sciences
Indore, Madhya Pradesh, India

Anju Mohan
Department of Law
Saveetha School of Law
Saveetha Institute of Medical and Technical Sciences
Chennai, Tamil Nadu, India

Sarla More
School of Computer Science and Information Technology
Symbiosis University of Applied Sciences
Indore, Madhya Pradesh, India

Pranav Paranjpe
Department of ECE
Military College of Telecommunication Engineering
Indore, Madhya Pradesh, India

Vijay Patel
Mechatronics
School of Science, Manufacturing and Infrastructure
Kaushalya—The Skill University
Ahmedabad, Gujarat, India

Anish Patnaik
KIIT School of Management
KIIT University
Bhubaneswar, India

B. Chandra Mohan Patnaik
KIIT School of Management
KIIT University
Bhubaneswar, India

Dinesh Kumar Rajendran
Department of Mechanical Engineering
NIT Srinagar
Srinagar, India

Murugan Ramu
Department of Law
Saveetha School of Law
Saveetha Institute of Medical and Technical Sciences
Chennai, India

Iraq Ahmad Reshi
Research Associate
NIELIT Srinagar
Srinagar, Jammu & Kashmir, India

Hrisikesh Roy
Research Intern
NIELIT Srinagar
Srinagar, Jammu & Kashmir, India

Ipseeta Satpathy
Senior Professor, KIIT School of Management, KIIT University
India

J. Joshua Thomas
Department of Computing
School of Engineering, Computing, and Built Environment
UOW Malaysia
KDU Penang University College
Penang, Malaysia

Naresh Tripathi
University of Delhi
New Delhi, Delhi, India

Pankaj Kumar Varshney
Department of Computer Science
Institute of Information Technology & Management
GGSIPU
New Delhi, Delhi, India

Bal Virdee
Communications Technology Research Centre
London Metropolitan University
London, United Kingdom

N.C. Ajay Vishwath
Maintenance, repair and operations & R&D
Skye Air Mobility Pvt. Ltd.
Gurugram, Haryana, India

Arvind R. Yadav
Electronics & Instrumentation Engineering Department
Institute of Technology
Nirma University
Ahmedabad, Gujarat, India

Part I

Foundations of unmanned aerial vehicles

Chapter 1

Introduction to drone technology

Dinesh Kumar Rajendran, Y.G. Bala, and Sumaya Akhtar

1.1 INTRODUCTION

A **drone**—technically known as an unmanned aerial vehicle (UAV)—is an aircraft operated without a human pilot onboard. Commonly, the term unmanned aircraft system (UAS) encompasses the vehicle, ground station, and communication system, while remotely piloted aircraft system (RPAS) emphasizes remote pilot involvement. Thus, "drone" in this context covers small quadcopters used recreationally and large, fully autonomous systems for commercial or governmental operations [1–5]. The nomenclature is shown in Figure 1.1 and the drone market size is shown in Figure 1.2.

Drones are a disruptive technology with applications in precision agriculture, delivery logistics, infrastructure inspection, disaster relief, and more [6–8]. Their ability to collect data from inaccessible locations, automate repetitive or dangerous tasks, and provide aerial perspectives redefines traditional workflows across numerous industries [10].

The UAV market has rapidly grown since 2020, supported by substantial advances in autonomy, payloads, and battery technologies, with slow but steady assimilation of regulations. Advancements in artificial intelligence (AI) with machine learning algorithms have improved the ability of UAVs to complete tasking with little human intervention, and the development of advanced payload and delivery systems such as LiDAR, hyperspectral sensors, and delivery modules has enhanced capabilities beyond the traditional hobbyist market. Furthermore, advances in energy systems, including high-energy-density lithium-ion and lithium-polymer batteries, as well as potential hydrogen fuel cell and solar-powered options, will enable UAVs to remain aloft longer, increase their range, and make them more viable on an industrial scale and for beyond visual line of site (BVLOS) operations. Furthermore, evolving regulatory initiatives adopt remote ID, geo-fencing, and organized operator classifications, allowing safer integration of UAVs into the national airspace and thus widen their application. The revenue generated by the global drone market exceeded \$30 billion by 2023 and is predicted to exceed \$80 billion by 2030 [9–12]. This rapid economic growth has been mirrored by the exponential growth of academic research

DOI: 10.1201/9781003598350-2

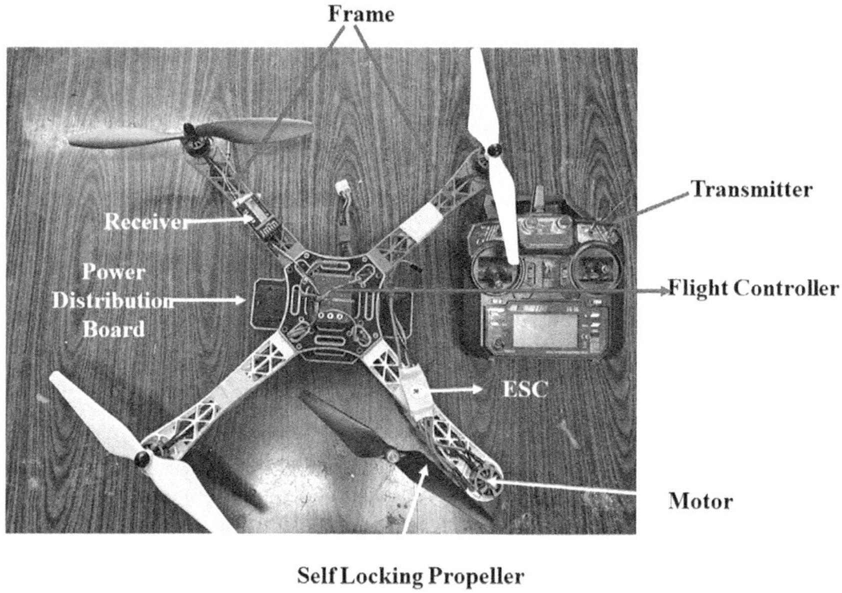

Figure 1.1 Types of UAV.

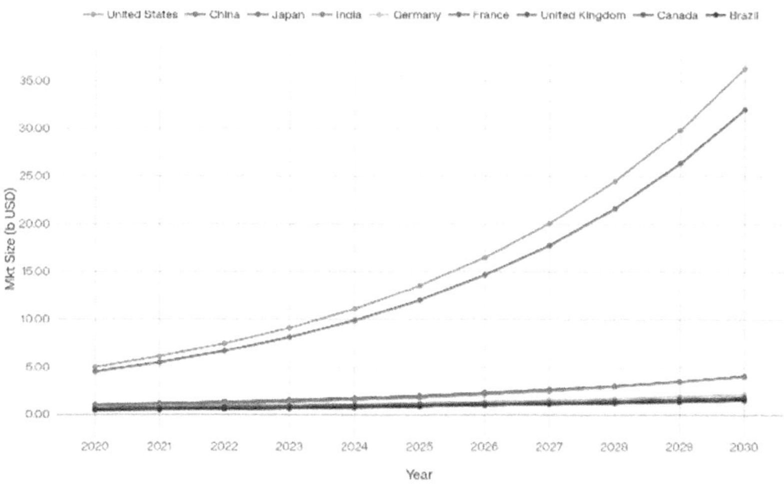

Figure 1.2 Drone market size 2020–2030.

output, the growth of patents on both hardware and software, and the emergence of many startups that are constantly evolving drone design, navigation, data analytics, and application-specific solutions [13]. All these trends demonstrate that UAVs have evolved beyond niche tools to become mainstream technologies that are transforming industries, driving digital transformation, and establishing themselves as one of the fastest-growing sectors of the global technology marketplace.

This chapter provides foundational insight into drone technologies, tracing their evolution, classification schemes, architectures, practical applications, and regulatory and ethical frameworks. It sets the stage for deeper technical and sectorial dives in subsequent chapters.

1.2 DRONE EVOLUTION: HISTORICAL AND TECHNOLOGICAL MILESTONES

1.2.1 Early developments

Drone technology originated with military applications during World Wars I and II, evolving from *target drones* to more complex reconnaissance and surveillance platforms [14]. Military R&D throughout the latter 20th century accelerated advances in remote control, sensor integration, and autonomous flight. Early drones have a long and fascinating history, beginning with early concepts of flying machines such as gliders and ornithopters. Leonardo da Vinci's designs, though never built or flown, laid an important foundation for the future development of unmanned flight. However, the true advancement of drone technology happened during World Wars I and II in the names of "Aerial Targets" by the British military to confuse and divert enemy defenses, "Radioplane OQ-2" by the US military as the target for anti-aircraft gunnery practice, "V-1" flying bomb by the German military launched from ground and guided to its target, the Ryan Firebee [15–20], developed for the U.S. military, was an important early unmanned aircraft. Initially designed as a target drone, it was later adapted for high-altitude reconnaissance and aerial surveillance, helping shape the development of modern military UAVs [21]. The Timeline of major UAV milestones, from 1800 to 2025 is shown in Figure 1.3.

1.2.2 Civil and commercial drone expansion

Technological democratization—miniaturization of sensors, inexpensive GPS modules, and open-source autopilot, among others—propelled the drone technology to the foreground of civilian and commercial usage in the beginning of the 21st century, moving drone technology away from military installations and onto the civilian and commercial landscape. The emergence of consumer quadcopters, specifically the DJI Phantom in 2012, accelerated adaptation of drone use and eventually made them to be used

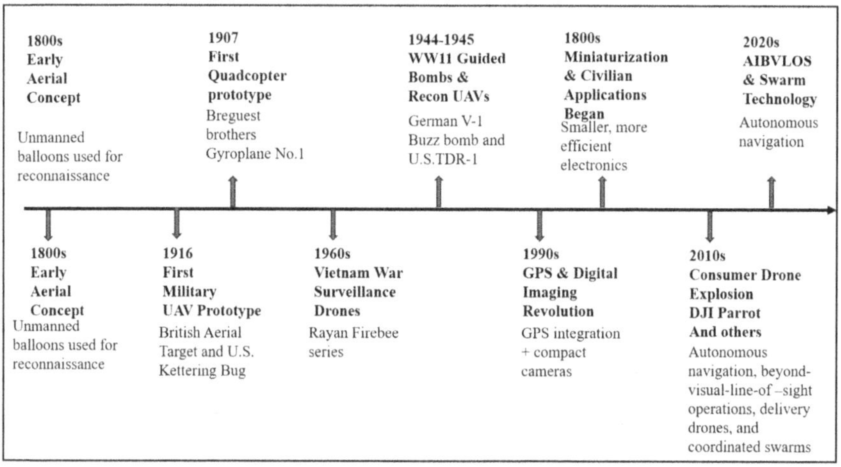

Figure 1.3 Timeline of major UAV milestones, from 1800 to 2025.

by all. The evolution of consumer drones provided access to advanced aerial imagery and navigation systems for more users rapidly, making the drone technology part of normal recreational practices for hobbyists, photographers, surveyors, and small businesses. Consumer drones are designed for recreational purposes in broad range, allowing individuals to capture aerial footages, experience new different views, and enjoy the thrill of flying. The easy access to aerial imagery from a high-resolution camera, in combination with an electric motor, gimbals, and stabilization systems, has made aerial photography or videography simple and inexpensive. Commercial drones can be used in businesses such as real estate, events/pop-virtual coverage, journalism, and tourism, thus adding more values to these businesses. Not only did this democratization of available technology reduce barriers to entry, but it also accelerated innovation in the UAV industry, as firms sought to improve flight times, stability, and image quality, at competitive prices. Drone technology has become widely established as a creative and practical technology for individuals and professionals alike, paving the way for more developments in fields such as agriculture, surveying, logistics, and infrastructure inspection [22–25].

1.2.3 Contemporary trends (2020–2025)

In the past few years, there were many transformative improvements in several aspects of drone technology, especially in autonomy, communications, and environmental versatility. Breakthroughs in AI and machine learning have allowed drones to execute new levels of autonomy, including real-time object detection, obstacle avoidance, and adaptive mission planning without continuous human oversight. Vision-based navigation systems, such as simultaneous localization and mapping (SLAM) methods, have improved operational confidence through their ability to operate in environments where GPS is unavailable, such as indoors, in congested urban environments, or through forest canopies. At the same time, better communication infrastructure has increased the types of operations possible with drones. The introduction of 5G has enabled ultra-low-latency data transfers, allowing to receive real-time video, carry out cloud-based processing, and coordinate many UAVs in a shared air space simultaneously. Mesh networking allows UAVs to form communication links in a resilient and decentralized way, which opens the possibilities for swarming techniques and enhanced range missions. Also noteworthy are developments in environmental adaptability. The advent of ruggedized drones that can operate in severe weather and submersible, or amphibious, platforms expands the use of UAVs to disaster response, maritime watching, and environmental monitoring where the terrain is demanding. This trend reflects a broader move toward more versatile, reliable, and mission-critical UAVs that can perform in an extensive range of operational environments [26]. The future of drone technology is shown in Figure 1.4

1.2.3.1 Integration of AI and machine learning

Drones are being evolved into smart platforms, making them capable of taking real-time decisions, thanks to enhancements stemming from the application of AI. Today's UAVs have the capability to accurately identify, classify,

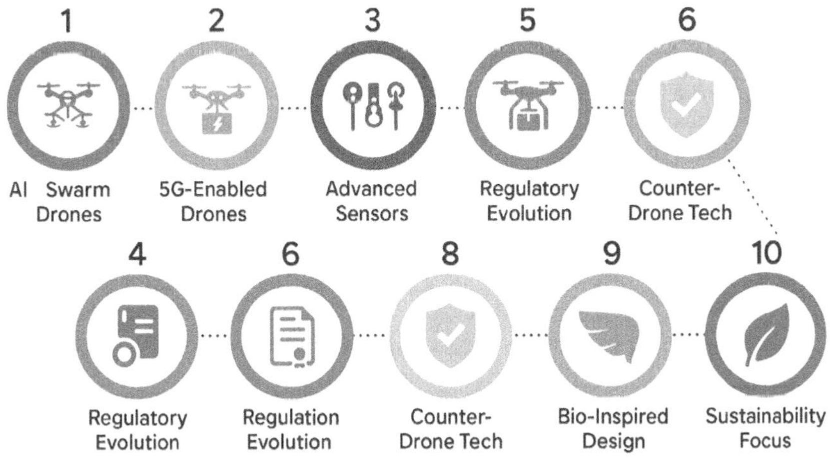

Figure 1.4 Future of DRONE Technology.

and track objects while on the wing, enabling applications like surveillance, traffic monitoring, wildlife tracking, and security. This capability relies on the advances in computer vision and machine learning algorithms, allowing drones to autonomously adapt to changes in the environment—for example, unexpected changes in weather—without constant human intervention. In this regard, a further advancement in the drone industry is the introduction of visual SLAM which enables drones to operate in environments where GPS is not available, such as warehouses and disaster areas. Ultimately, these advancements have made UAV mobility possible not only in the traditional open-air flight environment but also in indoor and underground environments for carrying out complex missions. Also, the integration of AI has further improved applications in specific industries. In emergency response situations, drones equipped with AI can quickly and effectively map the disaster area, find survivors, and communicate important information back to rescue teams on the ground. In construction sites, drones assist in automatic site inspections and maintain records of work progress. Moreover, in agriculture, drones help manage crop stress, diseases, and pests at surgical levels of precision for more effective resource management. Additionally, predictive analytics integrated into UAV systems can recognize early signs of wear and tear on specific components of equipment, inform wear properties, and improve maintenance scheduling, thus increasing the uptime of equipment. Collectively, developments in drone application have enabled them to perform as autonomous, adaptable, and mission-critical devices [27].

1.2.3.2 Swarm drones

Swarm drone technology is one of the most transformative innovations in UAV technologies, which allows dozens or even hundreds of drones to fly in cooperation to achieve multiple formations. The swarm approach to drone technology with distributed intelligence and strong communication protocols enables a group of drones to work together and complete tasks that are tedious or slow with only one UAV. In precision agriculture, swarm drones are a key enabler for precision farming as they use multiple UAVs to monitor large fields of little plots at the same time, apply fertilizers, and perform pollination studies. In disaster response and search and rescue operations, swarm drone technology enables rapid scanning that covers large areas to identify survivors and share the information in real time to significantly reduce response time. In the defense world, swarm technology is used for perimeter security, area surveillance, and tactical operations without the need to deploy a large number of personnel. Within the swarm communication links, drones can share sensor information, organize mapping activities, and even divide up their payloads through cooperative transport. Apart from operational applications, drone swarms have also demonstrated their potential in synchronized aerial displays, highlighting the use of swarm technology for entertainment, large-scale spectacles, and public engagement. Another strength of swarm technologies is the ability to operate autonomously, which reduces operators' decision-making overhead as the swarm drones can self-organize, can adapt to different changing environments, and can be deployed on complex missions with fast response times. With collective intelligence as a core component, drone swarms can create measurable and scalable impact across society, industry, security, and emergency response by addressing complex problems more effectively [24–28].

1.2.3.3 Long-endurance and BVLOS flights

Flight endurance has been the central focus of UAV development programs in the last few years, and new technologies developing energy systems are dramatically increasing service durations. Solar-powered designs (autonomous and non-autonomous) now allow to use renewable energy to continuously fly for hours to days. Hydrogen fuel cells are utilized to provide energy-dense power levels with lower environmental impacts in comparison to conventional batteries. At the same time, the chemistry of lithium-ion and lithium-polymer batteries has improved—contributing to longer flight times and higher safety standards—to make drones much more reliable for various missions. Another major enabler for long-range missions is the regulatory acceptance of BVLOSwhich mandates to bring drones far beyond the direct view of the pilot. This is especially important for large-scale environmental monitoring projects, such as monitoring pipelines and powerlines, border surveillance, or any delivery services where large areas need to be covered. Many of these use cases of drones are replacing manned aircraft operational methodologies due to their lower operational costs, increased safety, and higher flexibility [27–29]. In addition, VTOL (Vertical Take-Off and Landing) drones, such as hybrid fixed-wing–rotor designs, blend the efficiency of fixed-wing flying with the flexibility of multirotors. These types of drones can both take off and land without runways, while cruising over long endurance. They are apt for performing the same inspection task repeatedly over time and similarly fit into longer-range exploits in logistics. The engagement of energy optimization software optimizes the power drain of drones intelligently to accomplish longer missions.

1.2.3.4 5G-Enabled drones

With incredible ultra-low-latency connections, drones will soon be connected to 5G networks to provide seamless real-time control and high-definition video. Significantly increased bandwidth permits drones to transmit extremely large datasets (e.g., multispectral images, LiDAR scans, or live video) across cloud servers almost instantaneously. These capabilities support use cases ranging from precision agriculture to urban traffic circulation, where time-sensitive information is critical for decision makers.

Similarly, as edge computing begins to gain traction, drones can ingest and process data in the field, thereby not only limiting reliance on an always-available network, but also improving response times on mission-critical operations. Most importantly, 5G-enabled systems are able to control many drones, all at once, in crowded airspace while ensuring that drones do not interfere with each other's control, and this has ramifications for the possibility of system-wide coordination, rather than drone-system-wide coordination (i.e., scale). As discussed earlier, drone swarms deployed across different sectors and arranged as coordinated fleets have the potential to transform applications such as environmental monitoring, logistics and drone delivery, and defense operations, where improved situational awareness can help save lives. Their significant commercial applications can also be found today in sports, where live images from drone streams are used to provide immersive experiences; infrastructure inspection, where fusing real-time analytics to data helps improve safety; and disaster response, where UAVs can write a map of an area and share actionable intelligence rapidly with ground teams [30]. In the context of Industry 5.0, drones with 5G capabilities significantly contribute to enabling human–machine collaboration. Drones help form smarter, more sustainable industrial ecosystems through real-time data sharing, automation, and human supervision, thus facilitating production of personalized services, resilient supply chains, and enhanced workplace safety.

1.2.3.5 Drone delivery networks

Drone-based logistics systems are developed by business concerns such as Wing, Zipline, and Amazon. Drones can deliver items in a matter of minutes by avoiding traffic in urban areas. Drones are utilized to deliver medical supplies to far-flung sites in rural communities. Drones' take-off and landing are controlled by automated distribution centers without the need for human pilots. Developments in regulations are driving full-scale integration into current supply chains [31].

1.2.3.6 Counter-drone technology

Counter-drone systems have developed into an important layer of security against the use of unauthorized UAVs in restricted or sensitive environments. The threat of rogue drones creating safety, privacy, and security risks to critical infrastructure has led to the development of counter-drone systems designed to detect, track, and, if necessary, defeat any rogue UAVs. Advances in technologies such as AI-assisted radar, electro-optical sensors, and radio frequency (RF) detection enable the identification of drones even when they are flying low, hidden, or in a way that avoids conventional surveillance. Drones with entangling nets, GPS spoofing to disrupt navigation, and RF jamming to disrupt communications can be effective methods of neutralization. Counter-drone systems are becoming common at military bases, airports, government facilities, and large public venues, like stadiums, that are prone to threats. Based on the increasing threat level and restrictive regulations around their use, investment in counter-UAV technology continues to accelerate [32].

1.2.3.7 Bio-inspired designs

Recent developments in the bio-inspired drone design are providing exciting new avenues of research, by way of leveraging characteristics from insects, birds, and bats, to create micro-drones that fly highly agile and energy-efficiently. These flapping-wing drones are gaining interest by providing added flight dynamics that traditional UAVs with rotors cannot match. More importantly, they consume lesser energy compared to even the most advanced rotor UAVs. These design characteristics allow drones to perform complex aerial maneuvers that are ideally suited to environments/operations characterized by confined or cluttered spaces, where traditional drones are incapable of doing so. Additionally, bio-inspired drones can perch on surfaces such as trees, walls, and ledges to complete "long-duration" monitoring tasks without needing to sustain flight. This characteristic is most revealing in areas such as surveillance, environmental observation, and wildlife monitoring, where not having to continuously fly can produce useful data when an observation period is a longer before it can be modified. Moreover, aerodynamics taken from nature provides better performance and functionality from visualized urban spaces we're used to, where pilots of UAVs need to observe movement and avoid collisions in a fast-paced and dynamic environment. The reduced noise levels of flapping-wing drones, similar to the wings of birds and insects, make them ideal to stealthy operations in defense, security, and research/testing applications. By utilizing biological principles in the design of UAVs in conjunction with the advancements in materials and miniaturized sensors, it is anticipated that bio-inspired designs can enhance the capability of next-generation UAVs, which can be used in more responsive and sustainable ways than conventional UAVs [33].

1.3 CLASSIFICATION OF DRONES

1.3.1 Type and size

Drones are classified based upon the number of wings, structure, and weight, as these elements are significant in determining flight characteristics and applications. The number of wings or rotors affects how a drone creates lift and stability. Single-rotor drones are helicopter-like, creating large amounts of lift for carrying heavier weights, while multirotor drones, like quadcopters, hexacopters, and octocopters, have precise steering and stability and offer the ability to hover, which is useful for surveillance, filming, or short-range delivery. Conversely, multirotor drones can't fly as far or as fast as fixed-wing drones (with airplane-type wings), which uses forward motion for lift create. Fixed-wing types are best for long distances and very useful for mapping or surveying agriculture or very large areas. Structural classification takes in to account shapes, for example, tilt-rotor, ducted-fan drones, and blended-wing body (BWB) drones. Structural differences will account for differences in performance, such as aerodynamic advantages, stability in wind, and payload, which all apply to increasingly specialized applications such as long-range environmental monitoring, industrial inspection, and cargo transport. Weight-based classification categorizes drones as small, medium, and large. Small drones are generally intended to be lightweight and portable, and generally used for recreational purposes, indoor inspection, or short-range monitoring. Medium drones have a larger payload capacity, longer flying times, and mid-range durations, which may allow for applications in areas like precision agriculture or logistics. Large drones, typically with hybrid or combustion engines, can be used for heavy cargo transportation purposes, extended endurance missions, or defense applications. Ultimately, considering wings, structure, and weight will offer a broad awareness of limits and capabilities of drone types so that we may use and learn about the right type of drone for given activities specific to agriculture, logistics, defense, or environmental monitoring.

The details are as shown in Figure 1.5 and Table 1.1 [34].

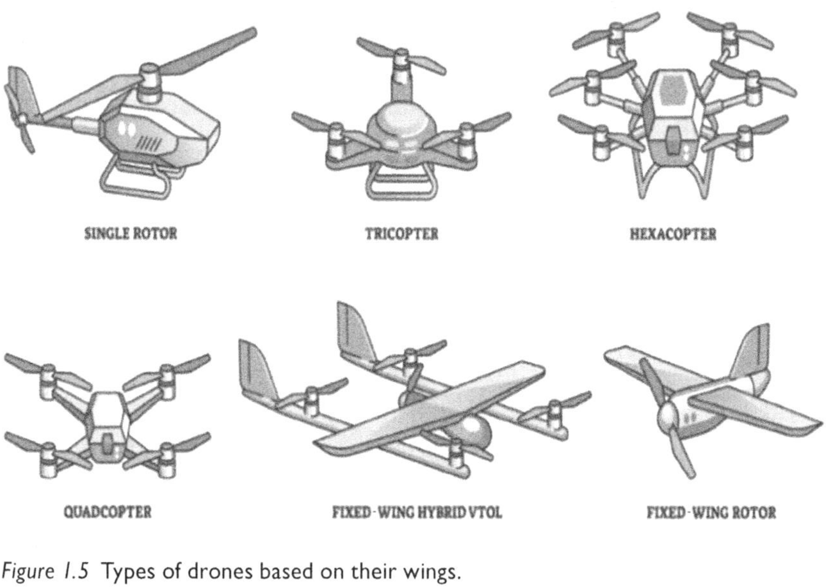

Figure 1.5 Types of drones based on their wings.

Table 1.1 Comparing battery chemistries and typical flight durations

Battery chemistry	*Typical flight duration*	*Key characteristics*
Lithium-ion (Li-ion)	20–50 minutes (for consumer/prosumer drones), up to 2 hours for industrial drones	High energy density, relatively lightweight, good cycle life
Lithium polymer (LiPo)	Varies widely, depends on size and application, can be similar to Li-ion	High power output, can be lighter than Li-ion for same capacity, but more susceptible to damage and thermal issues
Hybrid (fuel-electric)	Many hours or even days	Efficient, long flight times, but more complex system

1.3.2 Autonomy level [30–34]

- Manual: These UAVs are purely human-operated, and the pilot is responsible for navigation, speed, altitude, and safety—either Visual Line of Sight (VLOS) or BVLOS. There is no autonomous decision-making here as the drone's performance is solely based on the skill of the operator.
- Semi-autonomous: These drones can perform some tasks automatically following a preprogrammed flight path, maintaining stability, or fusing data from different sensors. However, the human pilot can also intervene to control and/or override the system. The semi-autonomous mode reduces the workload of pilots, while providing both safety and flexibility.
- Fully autonomous: These UAVs operate autonomously using AI, which performs mission planning and execution and adapts to mission goals in real time. UAVs operating fully autonomously can identify obstacles, be adaptive to changes in the environment, and optimize its path without any human intervention. Human involvement is minimal, and in most cases, it is only necessary for mission setup or, of course, in any emergency situation. The Flowchart of autonomy levels, from manual (radio controller) to fully AI-based operations is shown in Figure 1.6.

1.3.3 Function

Depending on their capabilities and functions, drones can come in different shapes and forms. Surveillance and security drones are used in many sectors such as defense, enforcement, and critical infrastructure monitoring These drone applications are highly varied and demonstrate the capability of drones to provide real-time, forward-looking intelligence and rapid response support in dynamic operational environments. In filmmaking, journalism, and entertainment, drones enable aerial cinematography, live broadcasting, and event coverage from perspectives that were previously difficult, expensive, or impossible to achieve. In cargo and logistics, drones

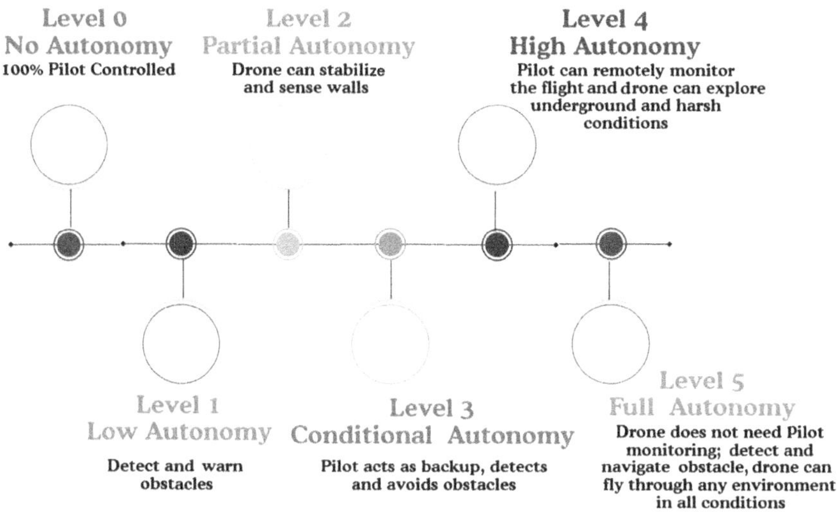

Figure 1.6 Flowchart of autonomy levels, from manual (radio controller) to fully AI-based operations.

are frequently used for transportation and delivery services. Drones provide fast, efficient, pandemic-friendly, and contactless delivery in both urban and remote settings. In agriculture, drones help in crop monitoring, judicious spraying, and health assessments, utilizing multispectral sensors and vegetation indices, to enhance production levels and planning for resource allocations. Depending on the desired outcome area, mapping and surveying drones are frequently used in a range of industries for industrial planning, city planning, and infrastructure site inspection, as they provide high-resolution topographical data. Drones are often employed for search and rescue operations, damage assessments, relief services during disaster events, and assessing areas that humans cannot reach. Drones for environmental monitoring track and monitor organisms or wildlife, assess pollution readings and levels, and monitor forests and waterways, thus contributing to conservation efforts. Drones are increasingly used in diverse entertainment applications, including coordinated and choreographed drone light shows that create visually engaging experiences for recreation, celebrations, and public events. All of these considerations represent different drone applications which extend the value of drones in agriculture, logistics, industry, defense, environment, disaster management, and recreation.

1.3.4 Design

Drones can be classified not only by their purpose but also by their physical configurations and flight characteristics that affect their performance, stability, payload capacity, and versatility of operation. Single-rotor helicopters have a single large main rotor with a small tail rotor operating on a vertical backward-angled axis, which allows them to carry heavier payloads and have longer flight time. The dual-rotor, single-rotor helicopter drone design is useful for industrial inspection, agricultural chemical spraying, military reconnaissance, and other industries where flight time and lift capacity are critical. Multirotor drones, quadcopters, hexacopters, and octocopters can be controlled and maneuvered while flying, which makes them more useful for surveillance, aerial photography, filming, and short-distance delivery packages compared to fixed-wing drones. Multirotors have less energy efficiency than helicopters and fixed wings, which limits their flight time. Fixed-wing drones have airplane-like wings and rely on continuous forward motion to produce lift. They are extremely efficient when covering large swaths of land quickly, which is important for applications like mapping, surveying, environmental monitoring, and agricultural applications. Tilt-rotor drones represent a combination of VTOL capabilities and efficient flying capability. These drones allow for both the best hovering ability provided by a multirotor and the best flight duration provided by a fixed-wing drone. Tilt-rotor drones lend themselves to unique designs with specialized attributes that would be beneficial for logistics and transport of goods, services, emergencies, and land/water craft solutions in industrial and military operations. Ducted-fan drones have propellers that offer better safety, more aerodynamic efficiency, and smaller size—ideal for indoor inspection and for operating in tight spaces of urban environments. Lastly, BWB drones have wings and a fuselage in a single aerodynamic body shape and have improved lift-to-drag ratio, better fuel use, and enhanced stability. As a result, BWB designs are increasingly evaluated in the commercial sector for long-distance cargo transportation, extensive environmental monitoring, and special military operations.

1.4 PROPERTIES OF DRONES

Power Sources: Drones utilize various power sources for its applications based on their size and duration. Batteries are used in smaller UAVs, and for other system of drones, alternative energy sources such as solar power, fuel cells, gasoline engines, and hybrid system can be utilized [35].

Landing Capability: Vertical landing capability can be achieved in some specific drones such as rotary-wing drones of helicopter, quadcopter, and VTOL drones [36].

Autonomous Operation: Utilizing systems like GPS navigation for obstacle avoidance and flight path automation, modern drones can function independent of human control.

Automated Hovering: Automated hovering is possible in quadcopters and helicopters through rotary wings, but it is not possible in fixed-wing UAVs.

Payload Flexibility: UAVs can carry various tools like sensors, cameras, and other deliver packages for a variety of applications ranging from aerial photography to logistics [34–36].

1.5 DRONE SYSTEM ARCHITECTURES

1.5.1 The flight controller

The **flight controller** is a UAV's "brain," integrating multiple sensors (IMU, GPS, barometer, compass) with algorithms for stability, path planning, and fail-safe routines. Recent flight controllers embed neural processors for edge AI and vision-based tasks. The Block diagram of a modern flight controller and sensor integration is shown in Figure 1.7 [35–38].

1.5.2 Propulsion and energy systems

1.5.2.1 Motors and propellers

The propulsion systems of drones vary widely depending on their size, mission, and performance requirements, and they are related to aircraft dynamics and aerodynamics. Small UAVs such as consumer quadcopters or

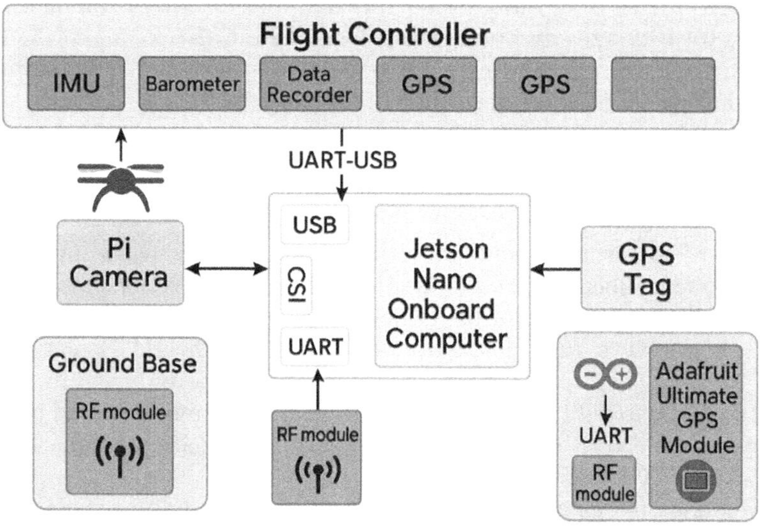

Figure 1.7 Block diagram of a modern flight controller and sensor integration.

surveillance UAVs, which include Class 1 UAVs, typically use brushless DC (BLDC) motors, found in drones that use fixed or variable pitch propellers. For stability across a range of operating conditions (i.e., hover, take-off, stowing, and agile transit), these motors must be lightweight and efficient and allow precise control of speed. Variable-pitch propellers, although less common than fixed-pitch propellers in commercial quadcopters, allow the blade angle to be adjusted according to flight conditions. This enables better control of thrust and lift during acceleration, deceleration, and hovering, thereby improving aerodynamic efficiency and overall flight performance. The thrust produced can be controlled to gain accurate pitch, roll, and yaw dynamics and to ensure precise navigation in tight, yaw, or turbulent spaces. Hybrid drones and larger UAVs use combustion engines or hydrogen fuel cells to gain more thrust and longer endurance. Because of larger mass and higher speeds, it is more important to get aerodynamic performance right; everything related to drones' aerodynamic design such as wing structure, fuselage shape/type, and rotor configuration should be optimized to reduce drag, improve lift-to-drag ratio, and maintain a state of flight despite changing wind conditions. Their propulsion systems will need to create sufficient force to overcome gravity and provide forward motion, efficient turns, and smooth altitude change by integrating aspects of flight dynamics including momentum, lift, drag, and torque. By aligning the propulsion system with aerodynamic and dynamic requirements, drones can be effectively designed for mission-specific applications, ranging from agile indoor flight to long-distance cargo transport, precision agriculture, and surveillance operations.

1.5.2.2 Battery and power management

- Small- to mid-scale UAVs commonly use **lithium polymer batteries** due to their high discharge rate, lightweight structure, and suitability for compact aerial platforms. Larger and endurance-focused UAVs increasingly adopt **advanced lithium-ion batteries** and **hydrogen fuel cells** to achieve longer flight duration and improved energy efficiency. Since 2020, major innovations in UAV energy systems have focused on higher energy-density batteries, intelligent battery management systems, and solar-assisted power augmentation [38–42].

1.5.3 Structural materials

Composite frames of drones, usually comprised of carbon fiber and engineered plastics, are designed to combine a number of performance properties concurrently. Carbon fiber adds a great deal of strength with lightweight. It provides high level of rigidity which makes drones resilient to forces of impact, vibrations, aerodynamic stresses, and loading during high-speed flight in turbulent conditions. Low weight supports battery efficiency and flight endurance, enabling the longest possible mission with the most payload weight. The range of engineered plastics offers flexibility, corrosion resistance, variety in mold and shape of parts, and aerodynamic efficiency. An additional benefit of composite materials is their ability to provide electromagnetic shielding to onboard electronics, sensors, and communication devices from interferences caused by the drone's motors, radio signals, or external electromagnetic sources while conducting a mission. Stability of control, accuracy of sensor readings, and reliable transmission of data are critical to missions involving surveillance, mapping, or industrial inspection. Composite frames are tough and lightweight, and hence provide electromagnetic protection to achieve performance, durability, and reliability in urban protected or remote hostile environments.

1.6 COMMUNICATION AND CONTROL

The communication and control mainly depend on two major parameters, namely, range and security protocol. The combined version is shown in the Figure 1.8 [40–42].

1.6.1 LOS and BVLOS operation

LOS (Line of Sight): LOS refers to drone operations where the pilot or operator can see the drone without the aid of technology (i.e., through naked eye) or with limited visual assistance (e.g., using binoculars). LOS provides more situational awareness, lower latency for

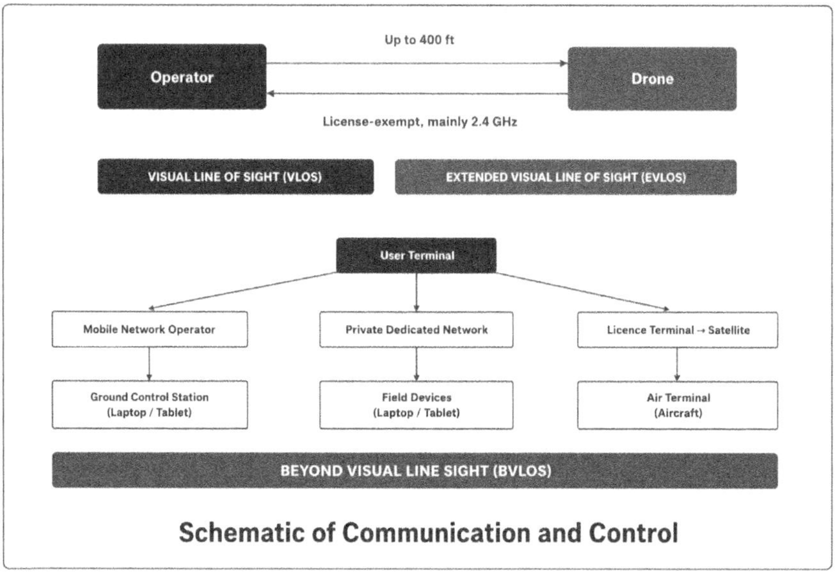

Figure 1.8 Schematic of communication and control.

manual control, and less regulatory hurdles. LOS is typically used for short-distance missions (aerial photography, inspection, or recreational) where the pilot can maintain visual awareness of the drone as this provides a greater degree of safety and is practical.

BVLOS: In this mode, the pilot or operator has no visual contact with the drone and is typically operated over much longer distances, or in environments where visual contact is impossible due to obstacles in the pilot's view. BVLOS offers safety and reliability in drone operations by using advanced communication systems such as encrypted radio link, 5G mesh networks, or communication from satellites. These communication systems allow for real-time video streams, remote real-time monitoring, and the ability to control the drones from a command center thousands of miles away. BVLOS is almost a necessity for many large-scale applications including, but not limited to, pipeline monitoring, precision agriculture, delivery service, disaster management, and military surveillance. However, to ensure safe and effective operations, BVLOS requires strict regulatory approvals and advanced fail-safe mechanisms (collision avoidance, autonomous navigation, and redundant communication links).

1.6.2 Redundancy and secure protocols

Drones utilize various tactics to maintain their communication links despite possible failures or attacks. One tactic is redundancy in a communication link where a drone has multiple or back channels covering primary and auxiliary links like radio, LTE/5G, and satellite. If any interference or failure occurs with one channel, the other begins the switch automatically without interruption. Another method is usually referred to as frequency hopping. Frequency hopping requires a rapid change of transmission frequency to prevent any attacker to jam the signal and to reduce the possibility of other devices' intelligence gathering across the communications bandwidth. When a link is established, secure handshake protocols are used as cryptographic authentication steps to verify both ends: the drone and the ground station remote control. These authentication stages prevent unauthorized access and hijacking efforts before a proper link is secured. These made framework resiliency and security important for essential operations of drones such as BVLOS operations, military surveillance operations, and commercial intelligence gathering from high-value targets.

1.7 REPRESENTATIVE APPLICATION DOMAINS

Agricultural UAVs are changing the way farmers operate by adding efficiency, accuracy, and automation to standard farming practices. Agricultural UAVs are outfitted with various sensors and imaging technology that enable precision spraying. Instead of wasting chemicals, fertilizers, pesticides, and herbicides, UAVs can be used to provide chemical input directly where needed. UAVs also provide mapping of the health of plants through thermal and multispectral imaging, which helps farmers to recognize if plants are stressed, diseased, or deficient of nutrients. Some UAVs are also developed as potential solution to natural pollinator declines through autonomous pollination trials. Another important function of agricultural UAVs are their ability to provide real-time analytics about fields by collecting and sending necessary data to farmers. Farmers can use the information generated from UAVs for crop yield predictions, irrigation plans, and resource optimization. In an integrated way, they assist the growth of precision agriculture by providing all these functions, which improves agricultural productivity while allowing for more sustainable practices.

Drones enabled NDVI-based fertilizer allocation, resulting in yield increases and input savings. In **infrastructure inspection**, robotic drones are used to inspect bridges, power lines, pipelines, and large industrial assets, thus avoiding hazardous manned interventions and reducing downtime. **In journalism and filmmaking**, aerial imaging and real-time broadcast have transformed event coverage, creative cinematography, and investigative journalism [40–43]. In **public safety, disaster relief, and health**, for example, COVID-19 pandemic, drones were rapidly adopted for medical delivery, social monitoring, and crisis mapping in both urban and rural areas. In **environmental research**, wildlife monitoring, environmental sampling, and atmospheric data collection increasingly depend on UAV-based sensing. In **new entertainment and art forms**, large-scale light shows and installation art now use programmable drone swarms [43–45].

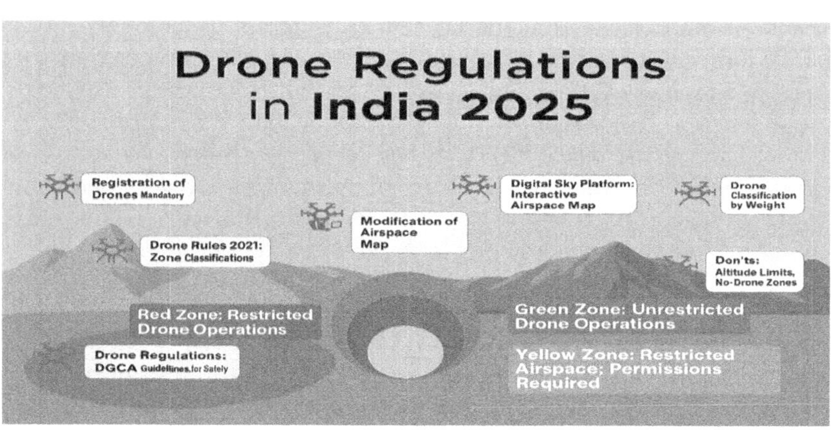

Figure 1.9 Map showing global regulatory regimes in 2025.

1.8 REGULATORY ENVIRONMENT

1.8.1 Major international frameworks

- FAA PART 107 (US): Registration, pilot certification, operational limits
- EASA (EU): Risk-based operation classes, digital U-space
- DGCA Digital Sky (India): Online permissions, geo-fencing, operator classes The Map showing global regulatory regimes in 2025 is shown in Figure 1.9.

1.8.2 Key global developments (2020–2025)

- Mandated *Remote ID*
- Automated air-traffic management (U space)
- Privacy and data protection laws for aerial sensors

1.9 BENEFITS OF DRONES

Various aspects of the lives and industries are influenced by drones. In **military transformation**, UAVs replace the traditional jet aircraft and redefine the dynamics of warfare with cost-effective and risk-free operations with lower fuel consumption, acquisition, and maintenance cost. In **operational efficiency**, enhanced continuous task execution by drones allows stress-free operations, ensuring high safety during prolonged flight duration. In **security advancement**, autonomous drones are ideal for border patrol due to their superior speed, safety, and cost-effectiveness. In **hazardous navigation**, by performing precise and repetitive tasks during extended flight duration, UAVs excel in navigating dangerous environments and thus reduce the operator's risk [46,47]. In **repetitive task**, drones can perform invaluable repetitive tasks such as raster scanning, ranging from geological survey to advanced imaging and communication measurements UAVs are now used across diverse industries, including agriculture, construction, filmmaking, and law enforcement. They are also expanding possibilities in recreational racing, education, training, and research applications. In **education and cinema**, drones play a pivotal role, fostering STEM learning and contributing to film production. As technology advances, the potential for drone use in passenger and cargo transportation expands.

1.10 EMERGING TECHNOLOGIES AND TRENDS

1.10.1 Autonomy and AI

Onboard neural networks enable navigation, target recognition, and environmental adaptation—real-time inference at the network edge. The increasing capabilities of advanced drone technology, combined with modern sensors, artificial intelligence, and machine learning techniques, enable drones to operate more efficiently and versatilely across various applications with minimal human intervention.

1.10.2 Swarm intelligence

Swarm-enabled platforms perform large-scale agricultural, mapping, and entertainment tasks through distributed coordination. Swarm drone technology allows multiple drones to work together in coordinated missions, helping in environmental monitoring, military operations, and rescue missions.

1.10.3 Extended flight and energy sources

- Extended flight endurance is a major requirement for advanced UAV applications. Modern drones are increasingly adopting improved energy sources and intelligent power-management strategies to support longer missions, wider coverage, and reduced operational interruptions. Hydrogen fuel cell drones are being explored for long-endurance missions due to their higher energy density and longer flight duration compared to conventional battery-powered systems. Similarly, solar-powered UAVs are emerging as a promising option for continuous or long-duration operation, especially in environmental monitoring, communication relay, and surveillance applications. In addition, energy-aware route planning optimizes flight paths by considering battery status, wind conditions, payload weight, mission distance, and return-to-home requirements, thereby improving mission efficiency and operational safety.

1.10.4 Smart communication and Unmanned Aircraft System Traffic Management (UTM)

UTM, coupled with 5G and edge computing, represents an area of intensive research focused on handling dense aerial operations in urban air mobility. The ultra-low latency and high bandwidth characteristics of 5G networks allow for real-time interaction between aerial vehicles and ground infrastructure management systems through commercial and secure network data communication, preserving safety and enabling a coordinated outline of flight paths. Edge computing strengthens the UAV ecosystem by processing critical data close to its source, reducing latency and enabling faster decision-making during dynamic conditions such as weather changes, emergency route adjustments, software-defined airspace corridors, and aerial congestion. When integrated with 5G networks and UTM frameworks, edge computing supports scalable traffic management, collision avoidance, and airspace optimization. These capabilities are essential for the safe operation of large numbers of passenger drones, cargo UAVs, and other aerial platforms in complex urban environments.

1.10.5 Integrated Internet of Things drones

Drone capabilities in data collection, monitoring, and surveillance can be further enhanced through integration with the Internet of Things (IoT). By connecting with other devices, sensors, and systems in real time, drones can support more coordinated, synchronized, and intelligent operations [49].

1.10.6 Delivery and logistics

Drone services are efficiently used by delivery and logistics companies for more efficient package delivery. The widespread adoption of drone delivery is being driven by advances in battery technology, flight range, and obstacle-avoidance systems. However, regulatory challenges, safety considerations, and the smooth integration of drones into existing delivery processes remain important factors for large-scale deployment.

1.11 ETHICAL, SOCIAL, AND ENVIRONMENTAL CONSIDERATIONS

1.11.1 Privacy and security

Concerns about aerial surveillance have prompted updated privacy laws and technical measures such as geo-fencing and image masking. Drones can capture and record high-resolution images and videos from significant heights, raising serious concerns related to privacy, personal space, individual consent, and the protection of sensitive national or public infrastructure. Modern drones use redundancy and encryption to counter rising risks from jamming, hacking, or signal spoofing [45–50].

1.11.2 Issues on airspace

Drones operate within shared airspace and, in some cases, can reach altitudes that may interfere with manned aircraft operations. This raises concerns related to technical faults, malfunctioning, accidents, disaster response, and emergency airspace management.

1.11.3 Environmental impact

Drones may reduce carbon footprints compared to traditional methods but introduce new challenges such as e-waste (batteries, electronics) and wildlife disturbance. The climatic change in the environment has to be taken in to consideration since drones may worsen climatic conditions if they utilize fossil fuels as their power source.

1.11.4 Ethical and legal regulatory

Autonomous drones in conflict zones can perform precise strikes and thus reduce the risk to military personnel. Though they provide tactical advantages, it may lead to the discussion of ethics of remote warfare due to no human involvement in the battle. When the advancement of technology surpassed the legal and regulatory frameworks, there is always the absence of clear guidelines always about transparency and accountability.

1.12 FUTURE PROSPECTS

Deployments of urban air mobility and passenger drones are happening in a few select cities and are indicative of the enormous potential to enhance urban transportation by alleviating road congestion, decreasing travel time, and providing greener alternatives. Continuously evolving technology will ultimately make air taxis to become fully automatic without pilots, thus delivering safer and more reliable transportation on a larger scale than the urban air mobility can provide now. The success of air taxis will depend on their integration into smart city ecosystems, which connect drones with urban infrastructure, air-traffic management, and mobility platforms that include the Internet of Things, allowing for real-time decision-making with respect to the drone service's surroundings with the integration of edge and cloud computing. Drones can then respond to changing traffic, weather, and emergency conditions in real time using drone software as a service (SaaS). Blockchain-enabled cryptographic traceability can authenticate flight history, safety records, maintenance data, and operational logs. In smart city mobility systems, it can also help protect and verify passenger-linked data, ensuring greater transparency, accountability, and trust.

While transformative in its potential to create a radically new type of mobility at the level of a city, the full realization of urban air mobility will involve deploying passenger drones as part of an intelligent urban transport system or a combined mode of mobility.

Table 1.2 Leading DRONE industry players.

Manufacturer	*Notable models*	*Market sectors*
DJI	Mavic, Matrice	Consumer, commercial
Parrot	Anafi	Mapping, agriculture
Skydio	Skydio X2	AI navigation, inspection
senseFly	eBee	Geospatial, survey
Autel Robotics	EVO	Public safety, logistics

1.13 LEADING INDUSTRY PLAYERS

The leading drone industry players are shown in Table 1.2.

REFERENCES

1. Grand View Research. "Drone Market Size, Share & Trends Analysis Report, 2024–2030." *Grand View Research, San Francisco*, CA, 2024, Report ID: GVR-4-68040-263-2, 150.
2. EASA. "EASA Drone Rules." *In: Easy Access Rules for Unmanned Aircraft Systems* (Regulations (EU) 2019/947 and 2019/945) (pp. 1–512), Cologne: EASA, 2025.
3. International Civil Aviation Organization (ICAO). *Manual on Remotely Piloted Aircraft Systems (RPAS)*, Montreal: ICAO, 2021, Doc 10019.
4. Kaliappan, VK, Kuppusamy, MS, Min, D, eds. Urban Air Mobility: Intelligent, Safe and Sustainable Systems for Future Transportation. 1st ed. CRC Press, 2024.
5. Austin R (2010). *Unmanned Aircraft Systems: UAVs Design, Development and Deployment.* John Wiley & Sons, Chichester, UK.
6. Valavanis KP, Vachtsevanos GJ (Eds.) (2015). *Handbook of Unmanned Aerial Vehicles.* Springer, Dordrecht. https://doi.org/10.1007/978-90-481-9707-1
7. International Civil Aviation Organization (2015). *Manual on Remotely Piloted Aircraft Systems (RPAS)* (ICAO Doc 10019). 1st ed. ICAO, Montreal, Canada.
8. European Union Aviation Safety Agency (EASA). *Easy Access Rules for Unmanned Aircraft Systems (Regulations (EU) 2019/947 and 2019/945).* Latest edition. EASA, Cologne, Germany.
9. Federal Aviation Administration (2021). *Remote Identification of Unmanned Aircraft.* Final Rule. FAA, Washington, DC, USA.
10. Outay F, Mengash HA, Adnane M, et al (2022). Applications of unmanned aerial vehicle (UAV) in wireless networks: A survey. *Peer-to-Peer Networking and Applications*, 15.
11. Mohsan MA, Othman NQH, Li Y, et al (2023). Unmanned aerial vehicles (UAVs): Practical aspects, applications, open challenges, security issues, and future trends. *Intelligent Service Robotics*, 16(1), 109–137. https://doi.org/10.1007/s11370-022-00452-4

12. Idrissi M, Salami M, Annaz F (2022). A review of quadrotor unmanned aerial vehicles: Applications, architectural design and control algorithms. *Journal of Intelligent & Robotic Systems*, 104. https://doi.org/10.1007/s10846-021-01527-7
13. Ahmed F, Mohanta JC, Keshari A, Yadav PS (2022). Recent advances in unmanned aerial vehicles: A review. *Arabian Journal for Science and Engineering*, 47(7), 7963–7984. https://doi.org/10.1007/s13369-022-06738-0
14. Liakos A, Busato P, Moshou D, Pearson S, Bochtis D (2022). Machine learning in agriculture: A review. *Information Processing in Agriculture*, 9(4), 529–545. https://doi.org/10.1016/j.inpa.2021.12.005
15. Nex F, Remondino, F (2014). "UAV for 3D mapping applications: A review," *Applied Geomatics*, 6(1), pp. 1–15, doi: 10.1007/s12518-013-0120-x
16. Sankaran S, Khot A, Espinoza CZ, et al (2015). "Low-altitude, high-resolution aerial imaging systems for row and field crop phenotyping: A review," *European Journal of Agronomy*, 70, 112–123, doi: 10.1016/j.eja.2015.07.004
17. Tsouros A, Bibi S, Sarigiannidis PG (2019). "A review on UAV-based applications for precision agriculture," *Information*, 10 (11), Art. no. 349, doi: 10.3390/info10110349
18. Guo Y, Zhang Z, et al (2022). "Deep learning for visual perception in autonomous UAVs: A survey," *Neurocomputing (Complete bibliographic details and DOI should be added from the final published article.)*
19. Zhang C, Kovacs, JM (2012). "The application of small unmanned aerial systems for precision agriculture: A review," *Precision Agriculture*, 13(6), 693–712. doi: 10.1007/s11119-012-9274-5
20. Primicerio JA et al (2012). "A flexible unmanned aerial vehicle for precision agriculture," *Precision Agriculture*, 13(4), 517–523. doi: 10.1007/s11119-012-9257-6
21. Pajares R (2015). "Overview and current status of remote sensing applications based on unmanned aerial vehicles (UAVs)," *Photogrammetric Engineering & Remote Sensing*, 81(4), 281–330. doi: 10.14358/PERS.81.4.281
22. Colomina, A, Molina P (2014). "Unmanned aerial systems for photogrammetry and remote sensing: A review," *ISPRS Journal of Photogrammetry and Remote Sensing*, 92, 79–97. doi: 10.1016/j.isprsjprs.2014.02.013
23. Zeng Y, Zhang R, Lim TJ (2016). "Wireless communications with unmanned aerial vehicles: Opportunities and challenges," *IEEE Communications Magazine*, 54(5), 36–42. doi: 10.1109/MCOM.2016.7470933
24. Zeng Y, Wu Q, Zhang R (2019). "Accessing from the sky: A tutorial on UAV communications for 5G and beyond," *Proceedings of the IEEE*, 107(12), 2327–2375. doi: 10.1109/JPROC.2019.2952892
25. Zeng Y, Lyu J, Zhang R (2019). "Cellular-connected UAV: Potential, challenges, and promising technologies," *IEEE Wireless Communications*, 26(1), 120–127. doi: 10.1109/MWC.2018.1800023
26. Hayat S, Yanmaz E, Muzaffar R (2016). "Survey on unmanned aerial vehicle networks for civil applications: A communications viewpoint," *IEEE Communications Surveys & Tutorials*, 18(4), 2624–2661. doi: 10.1109/COMST.2016.2560343
27. Bekmezci I, Sahingoz OK, Temel Ş (2013). "Flying ad hoc networks (FANETs): A survey," *Ad Hoc Networks*, 11(3), 1254–1270. doi: 10.1016/j.adhoc.2012.12.004

28. Fotouhi A, Ding M, Hassan M et al (2019). "Survey on UAV cellular communications: Practical aspects, standardization advancements, regulation, and security challenges," *IEEE Communications Surveys & Tutorials*, 21(4), 3417–3442. doi: 10.1109/COMST.2019.2906228
29. Shakhatreh H, Sawalmeh AH et al (2019). "Unmanned aerial vehicles (UAVs): A survey on civil applications and key research challenges," *IEEE Access*, 7, 48572–48634. doi: 10.1109/ACCESS.2019.2909530
30. Quan L, Han L, Zhou B et al (2020). "Survey of UAV motion planning," *IET Cyber-Systems and Robotics*, 2(1), 14–21. doi: 10.1049/iet-csr.2020.0004
31. Popović M, Vidal-Calleja T et al (2020). "An informative path planning framework for UAV-based terrain monitoring," *Autonomous Robots*, 44, 889–911. doi: 10.1007/s10514-020-09903-2
32. Giordan D et al (2020). "The use of unmanned aerial vehicles (UAVs) for engineering geology applications," *Bulletin of Engineering Geology and the Environment*, 79, 3437–3481. doi: 10.1007/s10064-020-01766-2
33. Outay F, Mengash HA, Adnan M (2020). "Applications of unmanned aerial vehicle (UAV) in road safety, traffic and highway infrastructure management: Recent advances and challenges," *Transportation Research Part A: Policy and Practice*, 141, 116–129. doi: 10.1016/j.tra.2020.09.018.
34. Kang H, Joung J et al (2020). "Protect your sky: A survey of counter unmanned aerial vehicle systems," *IEEE Access*, 8, 168671–168710. doi: 10.1109/ACCESS.2019.2909530
35. Sreenath S, Malik H et al (2020). "Assessment and use of unmanned aerial vehicle for civil structural health monitoring," *Procedia Computer Science*, 170, 656–663. doi: 10.1016/j.procs.2020.03.174
36. Goerzen N, Kong Z, Mettler B (2010). "A survey of motion planning algorithms from the perspective of autonomous UAV guidance," *Journal of Intelligent & Robotic Systems*, 57, 65–100. doi: 10.1007/s10846-009-9383-1
37. Boursianis AD et al (2022). "Internet of Things (IoT) and agricultural unmanned aerial vehicles (UAVs) in smart farming: A comprehensive review," *Internet of Things*,18, Art. no. 100187. doi: 10.1016/j.iot.2020.100187
38. Radoglou-Grammatikis R et al (2023). "Secure and reliable UAV communications for 5G/B5G wireless networks: A survey," *IEEE Communications Surveys & Tutorials*, 2023.
39. Shakhatreh H et al (2019). "Unmanned aerial vehicles (UAVs): A survey on civil applications and key research challenges," *IEEE Access*, 7, 48572–48634, 2019. doi: 10.1109/ACCESS.2019.2909530
40. Maddikunta PKR et al (2021). "Unmanned aerial vehicles in smart agriculture: Applications, requirements and challenges," *IEEE Sensors Journal*, 21(16), 17608–17619.
41. Tsouros S, Bibi A, Sarigiannidis PG (2019). "A review on UAV-based applications for precision agriculture," *Information*, 10(11), Art. no. 349. doi: 10.3390/info10110349
42. Pajares R (2015). "Overview and current status of remote sensing applications based on unmanned aerial vehicles (UAVs)," *Photogrammetric Engineering & Remote Sensing*, 81(4), 281–330. doi: 10.14358/PERS.81.4.281
43. Federal Aviation Administration (2021). *Remote Identification of Unmanned Aircraft: Final Rule*. Washington, DC, USA: Federal Aviation Administration.

44. European Union Aviation Safety Agency, *Easy Access Rules for Unmanned Aircraft Systems (Regulations (EU) 2019/947 and 2019/945)*. Cologne, Germany: European Union Aviation Safety Agency, latest consolidated ed.
45. International Civil Aviation Organization, *Manual on Remotely Piloted Aircraft Systems (RPAS)*, ICAO Doc 10019. Montreal, QC, Canada: International Civil Aviation Organization.
46. Directorate General of Civil Aviation, *Drone Rules, 2021* and *Digital Sky Platform Guidance Documents*. New Delhi, India: Directorate General of Civil Aviation, Government of India.
47. Idrissi M, Salami M, Annaz F (2022). A review of quadrotor unmanned aerial vehicles: Applications, architectural design and control algorithms. *Journal of Intelligent & Robotic Systems*, 104. https://doi.org/10.1007/s10846-021-01527-7
48. Shakhatreh H et al (2019). "Unmanned aerial vehicles (UAVs): A survey on civil applications and key research challenges," *IEEE Access*, 7,48572–48634. doi: 10.1109/ACCESS.2019.2909530
49. Colomina A, Molina P (2014). "Unmanned aerial systems for photogrammetry and remote sensing: A review," *ISPRS Journal of Photogrammetry and Remote Sensing*, 92, 79–97. doi: 10.1016/j.isprsjprs.2014.02.013
50. Federal Aviation Administration (2021). *Remote Identification of Unmanned Aircraft*. Final Rule. FAA, Washington, DC, USA.

Chapter 2

Seeing beyond

Sensor technologies in drone operations

Prerna Ajmani, Ayush Arya, Ashish Khanna, Pankaj Kumar Varshney, and Bal Virdee

2.1 INTRODUCTION

The operation of unmanned aerial vehicles (UAVs) is far beyond the predetermined path of flight, and it is a matter of knowing the position of the drone, being aware of the surrounding environment and reaching certain objectives of a mission. A drone with no sensors is like trying to attempt to make a way entirely blind, which seriously restricts its capabilities. The sensor systems equip drones with the knowledge of what is happening around them, and their presence is an important component of contemporary drone design.

The present-day drones are essentially flying platforms loaded with various types of sensors. They must examine the world; in certain instances, be able to sense what surrounds them; and monitor the things that are nearby. The reason behind working with numerous sensors is a good one. There is no single sensor that will perform everything well. There are sensors that have excelled during the day but fail during the night, and there are also those that are effective in fog or rain and cannot differentiate colors. Therefore, all these sensors should be integrated together, so that drones can be used in a variety of environments [1,2].

Capabilities of drones broadly belong to three significant categories: navigation (determining where the drone is as it travels), situational awareness (knowing about the environment around it and what is in it), and mission autonomy (making intelligent decisions without human intervention). These properties are mutually dependent on each other; that is, poor navigation implies poor awareness, and hence, it requires a human operator's intervention [3,4].

UAV sensor systems are typically classified into four categories:

1. Perception sensors, which engage in monitoring and analyzing the environment.
2. Navigation sensors, which provide information on where the drone is and its motion.

DOI: 10.1201/9781003598350-3

3. Environmental sensors, which detect the weather and air conditions such as temperature, air pressure, or gases.
4. Payload sensors, which are specifically chosen in a mission, for example, thermal cameras to detect people or multispectral sensors to scan fields.

Flying a drone has its challenges that do not occur in the ground. Rapid movements may blur up images. Changing lighting, especially at dawn or dusk, can confuse cameras. While flying high above, lots of different visual details may clutter the view, needing smart software to sort. Weather conditions like fog, rain, and dust may curtail the function of sensors. Sensors have to be small and lightweight; otherwise, drones carrying them can't fly well [5].

This chapter systematically examines these challenges and presents the sensor technologies enabling their mitigation. The discussion encompasses foundational systems (cameras and global navigation satellite system (GNSS)) through advanced technologies (light detection and ranging (LiDAR), hyperspectral imaging) that enable contemporary UAV operations.

2.2 SENSOR TAXONOMY IN DRONE SYSTEMS

2.2.1 Classification based on function

Let's start by looking at sensors based on what they actually do for the drone. This functional classification is probably the most intuitive way to understand the sensor ecosystem.

Navigation sensors are the backbone of any UAV system. The inertial measurement unit, or IMU, is probably the most critical navigation sensor. An IMU typically combines three-axis accelerometers, three-axis gyroscopes, and sometimes a three-axis magnetometer all in one package. The accelerometers measure linear acceleration along each axis, the gyroscopes measure rotational rates, and the magnetometer acts like a compass. Together, these measurements allow the drone to determine its attitude, basically, which way is up and at which direction it's facing. However, IMUs have a problem: they drift over time. The errors accumulate, which is why they need to be combined with other sensors.

GPS or GNSS is another navigation puzzle. GPS provides position and velocity information by receiving signals from satellites. When you have signals from at least four satellites, the GPS receiver can calculate its three-dimensional position pretty accurately. However, GPS has its own limitations: it doesn't work indoors; it can be jammed or spoofed; and in urban environments with tall buildings, the signals can bounce around, thus causing errors. Modern drones actually use multiple GNSS constellations—not just the American GPS, but also Russian GLONASS, European Galileo, and Chinese BeiDou—to improve reliability and accuracy.

The barometer is a simple but essential sensor that measures air pressure. Since air pressure decreases as altitude increases, you can use a barometer to estimate how high the drone is flying. This is particularly useful during takeoff and landing where you need precise altitude control. Barometers work even when GPS doesn't work, making them a crucial backup system.

The magnetometer, often integrated into the IMU, measures the Earth's magnetic field to determine heading direction. However, magnetometers can be easily disturbed by nearby metal objects or electrical currents, which is a common problem in indoor environments [1,6].

Perception sensors allow drones to see and understand their surroundings. Cameras capture visual information that can be used for obstacle avoidance, object recognition, or just recording video for a mission. Cameras come in many varieties: regular RGB (Red, Green, Blue) cameras, infrared cameras, stereo cameras for depth perception, and so on.

LiDAR is like radar, but it uses laser light instead of radio waves. It works by sending out laser pulses and measuring how long they take to bounce back, which tells the distance between objects. LiDAR can create incredibly detailed 3D maps of the environment, making it invaluable for navigation in complex spaces. The technology has been advancing rapidly, with modern LiDAR sensors becoming lighter and more affordable, though they're still more expensive than cameras.

Radar, particularly mmWave radar, operates in the millimeter wave frequency range and can detect objects even in fog, rain, or dust, conditions that would completely blind a camera. Radar uses the Doppler effect to measure not just the distance to an object but also how fast it's moving relative to the drone.

Ultrasonic sensors are the simplest of the bunch. They emit sound waves and measure the time it takes for the echo to return. They're cheap and work well for close-range obstacle detection, typically used for things like automated landing or hovering at a precise height indoors [5,1,7].

Environmental sensors measure conditions in the drone's surroundings rather than helping it navigate. Temperature and humidity sensors are straightforward; they tell you what the weather is like. Gas sensors are rather intriguing as they have the ability to smell the presence of certain chemicals in the surrounding air, which can be useful in such activities as determining the level of air pollution or locating gas seeps. Radiation sensor could appear in disaster cases when things got out of hand. Particulate sensors detect things such as the amount of smoke or dust that is present in the air [6].

Payload sensors are entirely different now. They are the instruments of specialty that you choose depending on the task that you are attempting. Just in case you have to deal with farming, you could get multispectral or hyperspectral cameras which are capable of detecting light wavelengths that humans cannot sense, and this allows determining whether crops are healthy or not. Thermal cameras are essential in search and rescue operations since they have the ability to detect the body heat. In case you need surveying or mapping, you would want high-resolution RGB cameras or even synthetic aperture radar. What this is all about is that you will select payload sensors in accordance with what actually needs to be accomplished by your mission [5,2,8].

2.2.2 Classification based on physical principle

One more possible approach to classify sensors is to consider the physical principles they apply to collect information. This way of classifying them cuts across those functional categories we just talked about, but it gives us a better understanding of what different sensor types can and can't do.

Optical sensors rely on electromagnetic radiation in or near the visible spectrum. Regular cameras are optical sensors, and thermal infrared cameras are also considered optical sensors, even though they detect infrared wavelengths that are invisible to the human eye. LiDAR also falls into this category since it uses laser light. Optical sensors generally provide high-resolution data with lots of detail, but they're sensitive to lighting conditions, weather, and line-of-sight obstructions [6].

Acoustic/ultrasonic sensors use sound waves instead of light. Ultrasonic sensors operate at frequencies above human hearing, typically around 40kHz. They're simple and cheap but have a limited range and can be affected by wind and temperature. Interestingly, there's also research into using acoustic sensors to detect drones by listening for the characteristic sounds their propellers make, though that's more about detecting other drones than helping your own drone navigate [6].

Radio frequency (RF)-based sensors use radio frequency electromagnetic waves. This includes radar, GPS/GNSS receivers, and communication systems. RF sensors have the advantage of being able to penetrate through many obstacles and work in poor visibility conditions. GPS is obviously RF-based, receiving signals from satellites. mmWave radar operates at higher frequencies (24–300GHz), which allows for better resolution while still maintaining the weather-penetration advantages of radio waves [6].

Inertial/mechanical sensors include accelerometers and gyroscopes in the IMU, as well as barometers. These sensors measure physical forces or pressures, rather than electromagnetic radiation or sound. The barometer measures air pressure through a mechanical or electronic pressure-sensing element. Accelerometers might use tiny mechanical structures that flex when experiencing acceleration, with the flexing measured electronically. These sensors are fundamental to UAV operation because they work regardless of external conditions, and you don't need satellites, visibility, or anything else. They just measure the physical motion and forces on the drone itself [6].

2.3 VISION-BASED SENSING

2.3.1 RGB cameras

When most people think about drone cameras, they're thinking about RGB cameras, which captures regular color images like you'd get from your smartphone. However, there's actually a lot more to these cameras.

The structure of an RGB camera is fairly straightforward: you've got a lens to focus light, an image sensor (usually a Complementary Metal-Oxide-Semiconductor (CMOS) sensor these days) to convert light into electrical signals, and processing electronics to turn those signals into a digital image. However, when you're putting a camera on a drone, several factors become really important that you might not think about otherwise.

Frame rate is one of them. The increased frame rate essentially translates to the fact that your video looks more lifelike, and this is very crucial when the drone is flying at a high speed. The majority of drone cameras nowadays operate at a minimum speed of 30 frames per second, and most of them have the ability to record 60 frames per second and beyond. This does not only takes into account the quality of your video, but also the image-processing algorithms that may be attempting to track objects or identify obstacles in real time.

Another thing that is of great importance is the field of view or FOV. The wide-angle lens provides you with a significantly larger picture of what is going on, and it is excellent when it comes to seeing the world around you and capturing more of it in just a single shot. Yet here is one thing to consider: the wide-angle lenses spoil the picture, particularly on the fringes. A small FOV will provide very fine details within a smaller field, but you can severely overlook any important activity going on in the external world. The vast majority of drones work with something in the middle, such as a medium-wide angle. The effect of this is, it attempts to trade coverage against the appearance of the image.

RGB cameras shine at daylight operation. With good lighting, modern cameras will be able to take extremely fine high-resolution images with colors that appear to be quite accurate like real. You can clearly observe textures, read text, and recognize specific objects, capabilities that make cameras highly useful for drone applications.

However, operating in low lights is an entirely different scenario or situation that can be imagined. As the light becomes dimmer, the image quality begins to deteriorate at a very rapid rate. You will begin to notice noise (that speckled-graininess appearance), colors will lose their precision, appearance will look strange, and it will be more difficult to discern detail. Some cameras have larger sensors or better low-light performance than others, but fundamentally, RGB cameras need light to work properly. This is why other sensing technologies become important when operating at night or in dark environments.

The role of RGB cameras in UAV operations goes way beyond just capturing pretty pictures or videos. In mapping and surveying, cameras are used to create orthomosaics, essentially, detailed maps stitched together from many overlapping photos. In SLAM (simultaneous localization and mapping), cameras help the drone figure out where it is by tracking visual features in the environment. For obstacle avoidance, cameras can detect objects in the drone's path, though this requires sophisticated image processing to work reliably [7].

One interesting thing about cameras is that they're incredibly information-rich but also computationally demanding. Every image contains millions of pixels of data, and processing all that information in real time requires significant computing power. This is why many modern drones have dedicated image-processing chips or even AI accelerators [1].

2.3.2 Infrared (thermal) cameras

If RGB cameras are like human eyes, thermal cameras have a completely different sense. They don't see reflected light; instead, they detect the infrared radiation that objects emit because of their temperature. This gives thermal cameras some unique capabilities that regular cameras just can't match.

The working principle is based on the fact that everything with a temperature above absolute zero emits infrared radiation. Hotter objects emit more infrared energy than cooler objects. A thermal camera has a special sensor (usually made of materials like vanadium oxide or amorphous silicon) that's sensitive to infrared wavelengths. The camera measures the intensity of infrared radiation coming from different parts of the scene and converts that into an image where different temperatures are shown as different colors or shades.

Thermal cameras operate in different spectral ranges depending on their design. LWIR (long-wave infrared) cameras operate in the 8–14 μm range, while MWIR (mid-wave infrared) cameras work in the 3–5 μm range. LWIR is more common for civilian drones because the sensors can operate without cooling (or with simpler thermoelectric cooling), making them lighter and cheaper. MWIR cameras generally need more sophisticated cooling but can provide better image quality in certain situations.

The applications for thermal cameras on drones are really impressive. In surveillance and security, thermal cameras can detect people or vehicles at night when RGB cameras would be useless. You can see someone hiding in bushes or spot a vehicle that's been recently driven (because the engine will still be warm).

Search and rescue is probably one of the most important applications. When someone is lost in the wilderness, they're usually warmer than their surroundings. A drone with a thermal camera can quickly scan a large area and detect that heat signature, potentially saving lives. This works in day or night, and even works to some extent through light vegetation.

Thermal inspection is another major use case. You can use a thermal camera to find heat leaks in buildings, detect hot spots in electrical equipment that might indicate a problem, or check solar panels for defective cells that run hotter than they should. Industrial facilities use thermal-equipped drones to inspect equipment like power lines or pipelines.

One limitation of thermal cameras compared to RGB cameras is resolution. While RGB cameras may achieve 20 megapixels or more, thermal cameras typically have much lower resolution, often just 640 × 512 pixels or even less. This is partly because thermal sensors are more expensive to manufacture, and also because for many thermal imaging applications, you don't need super high resolution. You're looking for temperature differences, not reading fine text [1].

2.3.3 Stereo and depth cameras

While a single camera can capture a beautiful 2D image of an object, it can't directly tell you how far it is located. That's a problem if you're trying to avoid obstacles or land precisely. This is where stereo and depth cameras play in. They add the third dimension to what the drone can see.

The stereo vision principle is actually based on how human eyes work. You have two eyes separated by a few inches, and each eye sees a slightly different view of the world. Your brain combines these two views and uses the differences between them to perceive depth. Stereo cameras do the same thing, using two camera sensors separated by a known distance (called the baseline).

The process of extracting depth from stereo images is called triangulation, and here's how it works: when you look at an object with two cameras, the object appears at a slightly different position in each camera's image. The closer the object, the bigger this difference (called disparity). By measuring the disparity and knowing the baseline distance between the cameras and their focal lengths, you can calculate the distance to the object using geometry.

However, triangulation isn't the only way to get the depth information. Structured light sensors project a known pattern (often an infrared pattern invisible to the human eye) onto the scene. The sensor then looks at how that pattern is distorted by the objects in the scene: closer objects make the pattern appear differently than distant objects. By analyzing these distortions, the sensor can calculate the depth. The Microsoft Kinect was a famous early example of this technology, though it was designed for gaming rather than drones.

Time-of-flight (ToF) sensors work differently again. They send out a pulse of light (usually infrared) and measure how long it takes for the light to bounce back from objects in the scene. Since light travels at a known speed, the time delay tells you the distance. Some ToF sensors can do this for every pixel in the image simultaneously, creating a full depth map very quickly. ToF sensors tend to work better in bright light than structured light sensors, but they can be more power-consuming.

Depth fusion refers to combining the depth information from multiple sources to get a more accurate and complete picture. For example, you might combine stereo depth with ToF depth, or fuse visual depth estimates with the data from a LiDAR sensor. Each sensing method has its strengths and weaknesses, so combining them can give you better results than relying on any single sensor.

There are definitely limitations to these depth-sensing approaches. Stereo vision struggles with textureless surfaces. If a wall is just plain white, the two cameras see essentially identical images and there's no disparity to measure. It also requires good lighting and sufficient computational power to match features between the two images. Structured light can be overwhelmed by bright sunlight outdoors. ToF sensors have limited range and can be confused by reflective surfaces. Understanding these limitations is important for choosing the right sensor for a particular application [7].

2.4 RANGE AND PROXIMITY SENSORS

2.4.1 LiDAR (light detection and ranging)

LiDAR has become one of the most important sensor technologies used for drones, especially as the technology has gotten lighter and more affordable. It's basically like radar, but instead of using radio waves, it uses laser light, hence the name, Light Detection and Ranging.

The working principle of LiDAR is pretty straightforward: you shoot out a pulse of laser light, it bounces off an object, and you measure how long it took to come back. Since light travels at a known, constant speed (about 300,000 km per second in air), the time delay indicates precisely how far away the object is. The basic equation is simple: distance equals half the speed of light times the time of flight. The value is divided by two because the light has to travel to the object and back.

There's actually two main approaches to measuring this time of flight. The direct time-of-flight (dToF) method literally measures the time between sending the pulse and receiving it. It's conceptually simple but requires very precise timing electronics because the time intervals are incredibly short, i.e., light travels a meter in about 3.3 nanoseconds.

The phase shift method is a bit different. Instead of sending short pulses, you send out a continuous wave of light whose intensity is modulated. Basically, the brightness goes up and down in a regular pattern. When this light bounces back, the pattern is shifted slightly compared to what you're sending out. By measuring this phase shift, you can calculate the distance. This method can be more accurate at short ranges but is more limited in the maximum range.

Now, there's a huge difference between different types of LiDAR systems when it comes to how they scan the environment. Rotating LiDAR, which you've probably seen on top of self-driving cars, uses a spinning mechanism to sweep the laser beam around 360 degrees. The laser is firing constantly, taking measurements in all directions as it rotates. This gives you a complete view of the surroundings but involves mechanical parts that can wear out.

Solid-state LiDAR is a newer technology that doesn't have any large moving parts. There's several ways to do solid-state scanning. MEMS (Micro-Electro-Mechanical Systems) LiDAR uses tiny mirrors that can be tilted electronically to steer the laser beam. These mirrors are microscopic, so while they move, they're much smaller and more durable than mechanical rotating systems. OPA (optical phased-array) LiDAR is even more advanced, and it steers the beam electronically by controlling the phase of light from multiple tiny emitters, like how phased-array radar works. There's no moving parts at all.

Flash LiDAR is completely different: instead of scanning a laser beam across the scene, it illuminates the entire scene at once with a wide laser pulse and captures the return with an array of sensors, like how a camera works. This allows very fast capture of the whole scene but is typically limited to shorter ranges.

The output from a LiDAR sensor is usually something called a point cloud—basically, a big collection of 3D points representing where the laser detected surfaces. Each point has X, Y, and Z coordinates indicating its position in 3D space, and sometimes it also has intensity information telling you how strongly the laser reflected from that point. Point clouds are incredibly useful for creating 3D maps, detecting obstacles, or performing SLAM [9].

Integrating LiDAR data into SLAM algorithms is one of the most powerful applications. SLAM is the problem of building a map of an unknown environment while simultaneously figuring out where you are in that environment. LiDAR is particularly good for SLAM because it directly provides accurate 3D measurements of the environment. The drone can match the current LiDAR scan against previous scans to figure out how it has moved, and gradually build up a consistent 3D map [10,9].

However, LiDAR isn't perfect, and there's trade-offs you have to consider. Range is one factor: more powerful lasers can detect objects farther away, but they also consume more power and might have safety concerns. Weight is another big issue for drones; high-end LiDAR systems can be heavy, which reduces flight time or payload capacity. Moreover, power consumption matters a lot for battery-powered drones. Generally, longer-range LiDAR with higher resolution and faster scan rates will be heavier and use more power. You have to choose the right LiDAR for your application [1].

2.4.2 Radar and mmWave sensors

Radar might seem like old technology—after all, it's been around since World War II—but modern radar, particularly millimeter-wave (mmWave) radar, is actually cutting-edge technology that's increasingly important for drones.

FMCW stands for frequency-modulated continuous wave, and it's the most common type of radar used in modern applications. Unlike older pulse radar systems that sent out short bursts of radio energy, FMCW radar transmits a continuous wave but varies (modulates) the frequency over time. Typically, the frequency ramps up linearly, creating what's called a chirp. When this transmitted signal bounces off an object and returns to the radar, the frequency has changed slightly because of the time delay. By mixing the transmitted and received signals together, you get a beat frequency that's proportional to the distance to the object [11].

What's really cool about FMCW radar is that it can also measure velocity using the Doppler effect. When the object is approaching or moving away from the radar, the frequency of the returning signal is shifted. Checking several chirps in sequential order, the radar can isolate the Doppler shift and determine the distance of the object and their velocity. This is the most helpful technology. Not only you can sense that there is something present in the way, but you can also feel whether it is in motion and how many miles per hour it is moving [11].

The main benefit of radar, particularly when compared to optical sensors such as cameras and LiDAR, is its good capability to penetrate things. Radio waves can go through fog, rain, dust, and even light vegetation, conditions that would render cameras useless and degrade LiDAR performance significantly. This makes radar particularly valuable for drones operating in challenging weather or dusty industrial environments [12].

Collision avoidance is an obvious application for drone radar. The radar can detect obstacles even in poor visibility, provide warning, and automatically trigger avoidance maneuvers. Since the radar also measures velocity, it can distinguish between static obstacles (like a building) and moving ones (like another drone), which helps in deciding how to respond [11].

Altitude estimation is another use, particularly over different surfaces. While a barometer tells you altitude based on air pressure, a radar altimeter can directly measure the height above ground or water by bouncing signals down and measuring the return. This works regardless of weather conditions and can provide very accurate height information for landing.

mmWave radar operates in the millimeter wave frequency band, typically 24–300 GHz, though for UAV applications, 60–80 GHz is becoming popular. The advantage of higher frequencies is better angular resolution, as you can more precisely determine the direction an object is in. The downside is slightly reduced range and higher power consumption. Still, modern mmWave sensors have gotten pretty efficient [1].

2.4.3 Ultrasonic sensors

Ultrasonic sensors are probably the simplest range sensors you can put on a drone; however, don't let their simplicity fool you, because they're incredibly useful for certain applications.

The ToF principle for ultrasonic sensors is exactly the same as for LiDAR or radar: you send out a pulse of sound (at frequencies above human hearing, usually around 40 kHz) and measure how long it takes for the echo to come back. Sound travels much slower than light—only about 343 m per second in air at room temperature—so the timing electronics don't need to be nearly as precise as for LiDAR.

The most common use of ultrasonic sensors on drones is indoor altimetry. When flying indoors, GPS doesn't work, and even barometers can be affected by building ventilation systems. An ultrasonic sensor pointing straight down can tell you exactly how far above the floor you are, which is perfect for maintaining a steady hover height or for autonomous landing.

Close-range navigation is another application. Many small drones use multiple ultrasonic sensors arranged around their body to detect nearby obstacles. The sensors are cheap enough that you can have four or more of them to give coverage in different directions. They're particularly good for the last meter or so when approaching an obstacle.

However, ultrasonic sensors have some significant limitations. Environmental factors like wind can affect their performance, and a strong gust can blow the sound pulse off course or make the return signal harder to detect. Temperature also matters because the speed of sound changes with temperature, which means the sensor's distance calculation can be off if you don't compensate for temperature effects.

Calibration is important too. The sensor needs to know the exact speed of sound in the current conditions to convert time to distance accurately. Some advanced ultrasonic sensors include a temperature sensor and automatically adjust their calculations, but cheaper ones might not.

The range of ultrasonic sensors is typically limited to a few meters, maybe 5–8 m maximum for the better ones. Beyond that, the sound wave has spread out too much and reflected off too many surfaces to get a clear return signal. This limited range is fine for indoor flying at low heights but wouldn't work for a drone flying outdoors at any significant altitude [1].

2.5 NAVIGATION AND POSITIONING SENSORS

2.5.1 Inertial measurement unit (IMU)

If there's one sensor that absolutely every drone needs, it's the IMU. You simply cannot fly without it because the IMU shows the flight controller which way the drone is oriented and how it's moving.

An IMU typically combines three types of sensors: accelerometers, gyroscopes, and often a magnetometer. The accelerometers measure acceleration along three axes (X, Y, and Z, or forward/back, left/right, and up/down if you prefer). The gyroscopes measure rotational rates around these same three axes (roll, pitch, and yaw rates). The magnetometer, when present, measures the Earth's magnetic field in all three axes, acting like a compass [13].

Sensor fusion is absolutely critical for getting useful information from an IMU. Raw accelerometer data tell you acceleration, but acceleration includes both movement and gravity. When the drone is sitting still, the accelerometers read about 9.8 m/s^2 pointing downward due to gravity. When the drone is moving or tilted, you're measuring a combination of gravitational acceleration and actual movement acceleration. By combining accelerometer and gyroscope data, you can separate these components and figure out the drone's actual orientation and movement.

The classic algorithm for fusing IMU data is called a complementary filter or a Kalman filter. We'll discuss more about Kalman filters later. The basic idea is that gyroscopes are good for measuring quick movements but drift over time (meaning their errors accumulate), while accelerometers don't drift but are noisy and affected by the drone's own movements. By combining them, you get the best of both: the quick response of the gyroscope with the long-term stability of the accelerometer.

Bias drift is one of the main challenges with IMUs, especially low-cost MEMS-based IMUs. Gyroscopes, in particular, suffer from this problem. Even when a gyroscope isn't rotating at all, it might read a small non-zero rotation rate, which is the bias. Worse, this bias changes slowly over time due to temperature changes or just random drift. If you try to calculate orientation by integrating gyroscope measurements, these small biases accumulate into large errors pretty quickly.

Calibration methods are used to deal with bias and other IMU errors. The simplest calibration measures the bias when the IMU is sitting perfectly still, whatever the sensors read should be zero (except for the accelerometer measuring gravity), so any non-zero reading is bias that can be subtracted out. More sophisticated calibrations also measure scale factor errors (where the sensor's output doesn't exactly match the input) and misalignment errors (where the sensor axes aren't perfectly perpendicular) [14].

The role of the IMU in attitude estimation and flight stabilization cannot be overstated. The flight controller is constantly reading IMU data, typically at rates of 100–1,000 Hz, and using it to determine the drone's current orientation. If the drone tilts unexpectedly due to a wind gust, the IMU detects that tilt immediately, and the flight controller can adjust the motor speeds to counteract it. This feedback loop happens so fast that the drone can maintain stable flight even in turbulent conditions.

For navigation, the IMU provides information about how the drone is moving. By integrating acceleration (after removing gravity), you can calculate velocity, and by integrating velocity, you can calculate position.

However, because of the drift problems we mentioned, IMU-only navigation works only for short periods, maybe a few seconds. For longer flights, you need to combine the IMU with other sensors like GPS [1,3,14].

2.5.2 GNSS (GPS, GLONASS, Galileo, BeiDou)

Most people just say "GPS," but technically GPS is just the American GNSS. There's actually four major GNSS constellations operating today: GPS (USA), GLONASS (Russia), Galileo (European Union), and BeiDou (China). Modern drone receivers can typically use satellites from multiple constellations, which improves reliability and accuracy.

The basic principle of GNSS positioning is triangulation using satellites. Each satellite broadcasts a signal that includes a timestamp showing exactly when the signal was sent (using extremely accurate atomic clocks on the satellites) and information about the satellite's orbit. Your receiver records when it received each signal. The time difference between transmission and reception tells you how far you are from that satellite (since the signal travels at the speed of light). If you know your distance from three satellites, you can calculate your position through geometric triangulation. In practice, you need at least four satellites because your receiver's clock isn't perfect, and the fourth satellite lets you solve for both your position and the timing error.

Positioning accuracy for basic GNSS is typically around 5–10 m horizontally under good conditions. That sounds pretty accurate, but for many drone operations, it's not nearly good enough. If you're trying to land precisely or fly close to structures, being off by 5 m could be disastrous.

Update rate is another important factor. Basic GPS receivers typically update at 1 Hz, once per second. For a fast-moving drone, a lot can happen in one second. Many modern receivers designed for drones can update at 5–10 Hz, providing much more responsive position information.

DGPS (differential GPS) is one technique for improving accuracy. The basic idea is to have a reference station at a known location that also receives GPS signals. Since the reference station knows exactly where it is, it can calculate the errors in the GPS signals it's receiving (due to atmospheric effects, satellite clock errors, etc.). It then broadcasts corrections that nearby receivers can use to improve their own position calculations. DGPS can improve accuracy to around 1–2 m.

RTK (real-time kinematic) positioning takes this concept further and can achieve centimeter-level accuracy. RTK uses the carrier phase of the GPS signal, not just the timing code. This requires more sophisticated processing and a nearby reference station (or access to a network of reference stations), but the results are impressive, as accuracies of 2–3 cm are typical. This is accurate enough for precision agriculture applications where you need to return exactly to the same location year after year [15].

GNSS-denied navigation is a huge research area because there are many situations where GNSS simply doesn't work, e.g., indoors, obviously, but also in urban canyons between tall buildings, under dense tree cover, or in areas where signals are intentionally jammed. When GNSS fails, drones need alternative navigation methods. This is where visual odometry (using cameras to track movement), LiDAR SLAM, or other techniques come in. Some systems use UWB (ultra-wideband) radio positioning, which is like GPS but using ground-based transmitters instead of satellites [3,16].

2.5.3 Barometric altimeter

The humble barometer might not seem as exciting as cameras or LiDAR, but it's a critical sensor for any drone that flies outdoors. Barometers are simple, cheap, and reliable and provide information that's hard to get any other way.

The pressure-to-altitude conversion is based on a simple physical principle: air pressure decreases as altitude increases. This is because there's less air above you pressing down as you go higher. The relationship isn't perfectly linear, pressure decreases exponentially with altitude, but there are well-established equations (like the International Standard Atmosphere model) that convert pressure to altitude.

In practice, barometers measure absolute pressure, typically in units like hectopascals (hPa) or millibars (mbar). At sea level, standard atmospheric pressure is about 1,013 hPa. At 1,000 m elevation, it's dropped to about 900 hPa. The challenge is that pressure varies with weather, and a passing low-pressure system can change the pressure at ground level significantly, which would make your altitude reading wrong.

That's why barometric altitude is relative, not absolute. When the drone takes off, it measures the pressure at ground level and uses that as a reference point. All subsequent altitude measurements are relative to that starting pressure. This works great for knowing how high above the takeoff point you are, but it doesn't tell you your absolute elevation above sea level (unless you know the pressure at sea level and can correct for it).

The main use in takeoff and landing stabilization is maintaining a consistent height. When a drone is hovering, it's using the barometer to detect any altitude changes and adjust motor power to compensate. During landing, the barometer helps ensure a controlled descent rate. The barometer is particularly important because it works continuously and isn't affected by visual conditions.

Sensor fusion with ultrasonic and LiDAR altitude data gives you the best of all worlds. Near the ground, ultrasonic or LiDAR sensors can provide very accurate height-above-ground information. Higher up, the barometer takes over. By fusing these sources, you get reliable altitude information throughout the entire flight envelope [13].

One thing to watch out for is that barometers can be affected by wind and the drone's own propeller downwash. When air rushes past the barometer sensor, it can create local pressure variations that don't reflect the true ambient pressure. Good sensor design puts the barometer in a location shielded from prop wash and uses filtering to remove short-term fluctuations [3].

2.6 ENVIRONMENTAL AND MISSION-SPECIFIC SENSORS

Environmental sensors greatly increase what drones can accomplish beyond just flying and avoiding obstacles. Such sensors scan numerous parameters in the region that the drone flies. They do not directly assist with flying, but they enable many useful applications.

The temperature sensors will be able to discern the temperature of the surrounding air, as whether it is hot or cold, and be useful in weather surveillance or wildfire identification. Humidity sensors detect how much moisture is present in the air, which is significant to weather monitoring and serves to identify if other sensors are influenced. There are gas sensors that detect certain gases such as carbon dioxide, methane, and even more harmful pollutants at times. Air quality detectors pay attention to such aspects as small dust particles (PM2.5 or PM10), which are significant in urban areas. Radiation sensors seek electromagnetic or nuclear radiations, and drones can fly over fragile or hazardous locations where humans cannot go. These environmental sensors enable drones to performs a wide variety of tasks, including pollution or weather surveillance, emergency response, and agriculture. In spite of these typically being special payloads, the information they carry is vital in serious environmental work [17].

Agricultural uses have literally led to much innovation in special-purpose sensors. Multispectral and hyperspectral sensors are probably the most important ones for farming. These sensors capture images in multiple narrow-wavelength bands that go way beyond just red, green, and blue. Plants reflect light differently depending on how healthy they are and what growth stage they're in, and these differences show up most clearly in the near-infrared and red-edge parts of the spectrum [18].

NDVI (normalized difference vegetation index) is one of the most com monly used measurements from multispectral imagery. It is calculated from the red and near-infrared bands and has a strong correlation with the health and biomass of plants. Healthy vegetation strongly reflects near-infrared light while soaking up red light, so a high NDVI value means healthy and strong growth, indicating that the plants are flourishing. Brown or stressed vegetation has a lower NDVI. By flying a drone over a field and calculating the NDVI, farmers can spot areas where crops are in trouble and target their interventions (like where to use fertilizer, water, pesticides) precisely where needed [18].

GNDVI (green normalized difference vegetation index) is similar but uses green and near-infrared bands instead. It is especially useful to test the amount of chlorophyll and is able to detect even small differences in the quantity of nitrogen plants. These indices are quite formidable precision agriculture tools that enable the farmers to obtain optimum yields on a minimum input basis [18].

Even the disaster management applications employ specialized sensors. Gas leak detection might use sensors tuned to methane, hydrogen, or other gases depending on what infrastructure is nearby. Smoke sensors can help assess wildfire spread and intensity. Chemical sensors might detect industrial spills or atmospheric contamination. Having drones carry these sensors means you can assess dangerous situations without putting humans at risk.

Marine and coastal monitoring presents its own sensor requirements. Salinity sensors measure the salt concentration in water, which is important for understanding ocean circulation and ecosystem health. Turbidity sensors measure water clarity, how cloudy or clear the water is, which affects light penetration and aquatic life. pH sensors measure acidity, which has become increasingly important as ocean acidification becomes a concern. While these sensors traditionally require direct water contact, some applications mount them on USVs (unmanned surface vehicles) working in coordination with UAVs, or use drones to deploy sensor packages into the water [8,19].

2.7 SENSOR FUSION AND DATA PROCESSING

Having lots of sensors is great, but you need sophisticated algorithms to combine all that data to obtain useful information. That's what sensor fusion is all about.

Kalman filters are probably the most famous sensor fusion algorithm, and they're powerful, are well understood, and work really well in many situations. A Kalman filter is essentially a way to estimate the true state of a system (like the drone's position and velocity) by combining predictions from a model with noisy measurements from sensors. The filter basically does an optimal weighted average of what you predicted would happen and what you measured actually happened, with the weights depending on how confident you are in each source [3,20].

Here's a simple example: suppose you're trying to estimate a drone's altitude. Your IMU can tell you vertical acceleration, which you can integrate to get velocity and integrate again to get altitude. However, frequent integrations accumulate errors, so after a few seconds, your altitude estimate will drift. Meanwhile, your barometer gives you an altitude measurement that doesn't drift but is noisy and slower to update. The Kalman filter combines these sources optimally, and it uses the IMU for fast, responsive tracking of altitude changes, but corrects the inevitable drift using the barometer measurements.

Extended Kalman filters (EKFs) and unscented Kalman filters are variations designed to handle nonlinear systems, as most drone dynamics exhibit nonlinear behavior. The basic Kalman filter assumes linear relationships between system states and measurements, but in reality, things like rotation are highly nonlinear (the math involving angles and rotation matrices doesn't follow linear rules). The EKF deals with this by linearizing around the current estimate. Basically, it pretends the system is linear in a small region around where you think you are. This works well only if the nonlinearities aren't too severe. The unscented Kalman filter is more advanced. It employs an ingenious sampling method to deal with nonlinearity more precisely without having to compute derivatives [20,21].

Particle filters provide a very different method for merging sensor data, particularly in cases where the number of possible states is large or where the system behavior is highly complex and nonlinear. Rather than maintain a single best guess of the drone location, particle filters maintain hundreds or even thousands of potential drone locations, known as particles, and each with a probability. On receiving new measurements, the particles that do not fit into that new measurement lose weight, and the ones that fit well receive weight. This kind of process over time brings one to the right state. These filters can perform better than extended or unscented Kalman filters, especially in situations where the drone's exact position is highly uncertain [21].

The process of sensor fusion in drones is a rather complicated matter. The lowest level is the raw data that is fed into the system. IMU measurements could be 500 times per second, GPS measurements could be 5–10 times per second, cameras could be 30–60 frames per second, but LiDAR point clouds could be at different speeds. Before combination, all these data require filtering, calibration, and even downsizing. Typically, there is a hierarchy: the low-level fusion combines such items as IMU and GPS to determine position and speed, and higher-level fusion adds camera or LiDAR data to aid in navigation and obstacle avoidance [3,20].

The problem with all this information is that it is very difficult to handle all the information in a short time to maintain control over the drone. It's pointless to know exactly where an obstacle is if the computation happens too late to steer clear. Payload limits on drones mean that only limited computing power can be onboard, so the whole system must be very efficient in crunching numbers in real time. Modern drone computers use power-efficient processors, often with dedicated hardware for specific tasks like image processing or neural network inference.

The role of AI and machine learning in sensor data interpretation is growing rapidly. Machine learning models, particularly deep neural networks, have proven incredibly effective at tasks like object detection, semantic segmentation (identifying what each pixel in an image represents), and even predicting future states. For example, a neural network can learn to detect obstacles in camera images much more reliably than traditional computer vision algorithms. However, running these models in real time on a resource-constrained drone requires careful optimization [22].

Synchronization and calibration issues are often underappreciated but absolutely critical. If your camera and LiDAR aren't precisely synchronized in time, combining their data leads to errors. When the relative position between sensors is not calibrated correctly (this is actually to say that you actually do not know where one sensor is relative to another), the combination of sensor data will provide entirely misleading data. The time synchronization and geometric calibration is a tedious task that requires a lot of time and effort in good sensor fusion systems [1,2,7].

2.8 COMMUNICATION INTERFACES AND SYNCHRONIZATION

The communication interfaces can be compared to the messengers of a drone system; that is, they receive sensor data at a location of generation and send it where it is required, e.g., in a flight controller or a main computer. Various sensors require varied types of connections depending on the amount of data they transmit, energy consumed, among other things.

I2C (Inter-Integrated Circuit) is one of the possible communication pathways. It is very basic, has two wires, and is suitable to slower sensors, such as magnetometers or pressure sensors. The good part is that a combination of devices can have the same bus, which is cleaner, but not so fast, typically only a couple of hundreds of kilobits per second [5].

SPI (serial peripheral interface) is more a step up in speed and is commonly utilized in IMUs and other sensors where the data have to be sent continuously, and not intermittently. It requires four wires and has a speed capacity of a number of megabits per second. The limitation is that it is point to point, i.e., each sensor requires its individual connection, which becomes complex [5].

Next, there is the UART (universal asynchronous receiver-transmitter), which is one of the oldest serial protocols that still exist. It is easier and reliable and is commonly applied to GPS receivers or other gadgets that send a message intermittently. Nevertheless, it is not as fast as I2C or SPI but does the job in its niche [5].

CAN (controller area network) is a strong software that is used in automotive and industrial solutions, whereby reliability is essential. CAN is commonly used in larger drones or industrial UAVs where multiple subsystems need to communicate reliably even in electrically noisy environments [5].

Ethernet is becoming more common in advanced drones, particularly for high-bandwidth sensors like cameras or LiDAR. Gigabit Ethernet can handle large amounts of data, and it's a familiar protocol with lots of existing tools and infrastructure [5,23].

Time synchronization is absolutely crucial for sensor fusion. When you're combining data from multiple sensors, you need to know exactly when each measurement was taken. Even a few milliseconds of timing error can cause problems. PTP (precision time protocol) is an Ethernet-based protocol that can synchronize clocks to sub-microsecond accuracy over a network. GPS time is another option, since GPS satellites have extremely accurate atomic clocks. The GPS signal includes precise timing information that can be used to synchronize all sensors to a common timescale.

MAVLink timestamps are used in many open-source drone systems. MAVLink is a lightweight messaging protocol designed for drones, and it includes timestamp fields in its messages. Sensors and other subsystems can use these timestamps to ensure their data are properly synchronized.

Data compression and transmission become important when you're sending data off the drone, particularly high-bandwidth data like video. Wi-Fi is common for shorter-range operation, while 5G cellular is enabling longer-range beyond-visual-line-of-sight (BVLOS) operations in some countries. Unmanned Aerial Vehicle Controller Area Network (UAVCAN) is a specialized CAN-based protocol designed specifically for drones and other unmanned vehicles, optimized for the kinds of data and communication patterns common in these systems [24,23].

2.9 CHALLENGES AND FUTURE DIRECTIONS

Despite all the impressive sensor technology we've discussed, there's still plenty of challenges to overcome and exciting developments on the horizon.

Miniaturization versus accuracy trade-offs are a fundamental challenge. Generally, bigger sensors can be more accurate, and a larger LiDAR can use a more powerful laser and more sensitive detector, giving you longer range and better accuracy. The larger sensor on the camera has the ability to freeze more light, and it also lowers noise. However, drones do not carry much weight, and therefore, you are always balancing sensor performance and its weight and size. The evolving technologies in MEMS, micro-optics, and sensor designs are slowly driving these trade-offs to a superior direction [25].

One particular area that is attempting to cope with these trade-offs is the low-SWaP sensor design (size, weight, and power). The goal is to achieve professional-grade sensor performance in a package small and light enough for small drones. This requires innovations in sensor materials, integrated circuit design, and packaging. For example, modern LiDAR modules are orders of magnitude smaller and lighter than systems from a decade ago while offering comparable performance [25].

Integration with AI edge processors is an increasingly important trend. Rather than just streaming raw sensor data to a central computer, future sensors will likely have embedded AI accelerators that can do sophisticated processing right at the sensor. A camera might not just send images, but it might send semantic information about what objects are in the scene.

This reduces communication bandwidth and processing load on the main computer.

Emerging quantum sensing technologies are on the far horizon but potentially revolutionary. Quantum sensors exploit quantum mechanical effects to achieve sensitivities impossible with classical sensors. Quantum accelerometers and gyroscopes could provide navigation-grade performance in chip-scale packages. Quantum magnetometers could detect magnetic fields with unprecedented precision. These technologies are still mostly in the lab, but they could eventually transform what's possible with small UAVs [26].

Neuromorphic sensing is another exotic possibility, sensors that work more like biological sensory systems than traditional cameras or microphones. Event cameras that only report changes rather than full frames are one example that's already making it into practical use. These can operate at much higher effective frame rates with lower data rates and power consumption.

Secure and authenticated sensor data are becoming critical as drones take on more important missions. How do you know the GPS position your drone is reporting is real and not spoofed by an adversary? How do you prevent someone from injecting false sensor data? Solutions might involve blockchain technologies for creating verifiable sensor data chains, or post-quantum cryptography (PQC) for secure communications that can't be broken even by future quantum computers [2,27].

2.10 CONCLUSION

This chapter has characterized sensor technologies as the perceptual foundation of unmanned aerial systems, establishing functional equivalence to biological sensory modalities. UAV sensor systems show a wide range of technologies, from acoustic sensors that use sound waves for measuring distances (like ultrasonic detection) to active optical systems like LiDAR that make 3D maps, and even multispectral sensors that capture light beyond what humans can see. Each of these various sensors contributes to the drone sensing setup in a different manner.

One of the concepts is that no sensor can be perfect in every case. Normal cameras perform well when lots of light is present but fail to work well in the dark. Satellite navigation (GNSS) is precise when outside, but in most cases not when indoors or in urban high-density locations. LiDAR is capable of offering precise 3D data, but it is heavier and requires increased power. Developing a powerful drone sensing system therefore is a matter of choosing the optimal combination of sensors that is suited to the mission and cleverly integrating the data they provide with fusion algorithms.

Drone sensors are experiencing rapid development. Sensors are getting smaller, lighter, and more energy efficient every day at the same time maintaining or even increasing the performance of the sensor. Others have even gotten the artificial intelligence in-board, allowing the sensors to process information and make decisions independently. With more and more tasks being performed by drones, such as package delivery, emergency assistance, or infrastructure inspections, the presence of sensor systems that are effective and reliable becomes overly important. The future of research directions is entirely new sensing techniques with quantum technology to make highly accurate measurements, brain-inspired sensors that respond to changes faster and more efficiently, and improved security protocols such as blockchain or post-quantum cryptography to ensure sensor data cannot be interfered with. All these advancements are bringing drones one step further to complete autonomy, ruggedness, and ability to take on the most difficult tasks.

REFERENCES

1. Wilson, A. N., Kumar, A., Jha, A., & Cenkeramaddi, L. R. (2021). Embedded sensors, communication technologies, computing platforms and machine learning for UAVs: A review. *IEEE Sensors Journal*, 22(3), 1807–1826.
2. Semenyuk, V., Kurmashev, I., Lupidi, A., Alyoshin, D., Kurmasheva, L., & Cantelli-Forti, A. (2025). Advances in UAV detection: Integrating multi-sensor systems and AI for enhanced accuracy and efficiency. *International Journal of Critical Infrastructure Protection*, *49*, 100744.
3. Ye, X., Song, F., Zhang, Z., & Zeng, Q. (2023). A review of small UAV navigation system based on multisource sensor fusion. *IEEE Sensors Journal*, 23(17), 18926–18948.
4. Elmokadem, T., & Savkin, A. V. (2021). Towards fully autonomous UAVs: A survey. *Sensors*, 21(18), 6223.
5. Hashim, H. A. (2025). Advances in UAV avionics systems architecture, classification and integration: A comprehensive review and future perspectives. *Results in Engineering*, *25*, 103786.
6. Balestrieri, E., Daponte, P., De Vito, L., & Lamonaca, F. (2021). Sensors and measurements for unmanned systems: An overview. *Sensors*, 21(4), 1518.
7. Xiao, J., Zhang, R., Zhang, Y., & Feroskhan, M. (2025). Vision-based learning for drones: A survey. *IEEE Transactions on Neural Networks and Learning Systems*, *36*(9), 6, 10, 14, 30.
8. Osmani, K., & Schulz, D. (2024). Comprehensive investigation of unmanned aerial vehicles (UAVs): An in-depth analysis of avionics systems. *Sensors*, 24(10), 3064.
9. Sobczak, Ł., Filus, K., Domański, A., & Domańska, J. (2021). Lidar point cloud generation for slam algorithm evaluation. *Sensors*, 21(10), 3313.
10. Seidaliyeva, U., Ilipbayeva, L., Utebayeva, D., Smailov, N., Matson, E. T., Tashtay, Y., ... Sabibolda, A. (2025). LiDAR technology for UAV detection: From fundamentals and operational principles to advanced detection and classification techniques. *Sensors*, 25(9), 2757.

11. Soumya, A., Krishna Mohan, C., & Cenkeramaddi, L. R. (2023). Recent advances in mmWave-radar-based sensing, its applications, and machine learning techniques: A review. *Sensors*, 23(21), 8901.
12. Huang, X., Xu, Z., Wu, H., Wang, J., Xia, Q., Xia, Y., ... Wang, C. (2025). L4dr: Lidar-4dradar fusion for weather-robust 3d object detection. In *Proceedings of the AAAI Conference on Artificial Intelligence* (Vol. 39, No. 4, pp. 3806–3814).
13. Li, Y., Zahran, S., Zhuang, Y., Gao, Z., Luo, Y., He, Z., ... El-Sheimy, N. (2019). IMU/magnetometer/barometer/mass-flow sensor integrated indoor quadrotor UAV localization with robust velocity updates. *Remote Sensing*, 11(7), 838.
14. Wu, H., Pei, X., Li, J., Gao, H., & Bai, Y. (2020). An improved magnetometer calibration and compensation method based on Levenberg–Marquardt algorithm for multi-rotor unmanned aerial vehicle. *Measurement and Control*, 53(3–4), 276–286.
15. Ekaso, D., Nex, F., & Kerle, N. (2020). Accuracy assessment of real-time kinematics (RTK) measurements on unmanned aerial vehicles (UAV) for direct geo-referencing. *Geo-spatial Information Science*, 23(2), 165–181.
16. Balamurugan, G., Valarmathi, J., & Naidu, V. P. S. (2016). Survey on UAV navigation in GPS denied environments. In *2016 International Conference on Signal Processing, Communication, Power and Embedded System (SCOPES)* (pp. 198–204). Paralakhemundi: IEEE.
17. Madokoro, H., Kiguchi, O., Nagayoshi, T., Chiba, T., Inoue, M., Chiyonobu, S., ... Sato, K. (2021). Development of drone-mounted multiple sensing system with advanced mobility for in situ atmospheric measurement: A case study focusing on PM2. 5 local distribution. *Sensors*, *21*(14), 4881.
18. Zheng, H., Cheng, T., Li, D., Yao, X., Tian, Y., Cao, W., & Zhu, Y. (2018). Combining unmanned aerial vehicle (UAV)-based multispectral imagery and ground-based hyperspectral data for plant nitrogen concentration estimation in rice. *Frontiers in Plant Science*, 9, 936.
19. Klemas, V. V. (2015). Coastal and environmental remote sensing from unmanned aerial vehicles: An overview. *Journal of Coastal Research*, 31(5), 1260–1267.
20. Marković, L., Kovač, M., Milijas, R., Car, M., & Bogdan, S. (2022). Error state extended kalman filter multi-sensor fusion for unmanned aerial vehicle localization in gps and magnetometer denied indoor environments. In *2022 International Conference on Unmanned Aircraft Systems (ICUAS)* (pp. 184–190). Dubrovnik: IEEE.
21. De Marina, H. G., Espinosa, F., & Santos, C. (2012). Adaptive UAV attitude estimation employing unscented Kalman filter, FOAM and low-cost MEMS sensors. *Sensors*, 12(7), 9566–9585.
22. Spasev, V., Dimitrovski, I., Chorbev, I., & Kitanovski, I. (2024). Semantic segmentation of unmanned aerial vehicle remote sensing images using SegFormer. In *International Conference on Intelligent Systems and Pattern Recognition* (pp. 108–122). Cham: Springer Nature Switzerland.

23. Ajmani, P., Singh, N., & Verma, P. (2022). Internet of vehicles taxonomy and evaluation: Architectures, protocols, and issues. In *2022 5th International Conference on Contemporary Computing and Informatics (IC3I)* (pp. 1163–1170). Uttar Pradesh: IEEE.
24. Singh, R., Ballal, K. D., Berger, M. S., & Dittmann, L. (2022). Overview of drone communication requirements in 5G. *Global IoT Summit, 13533*, 3–16.
25. Jones, K. H., & Gross, J. N. (2014). Reducing size, weight, and power (SWaP) of perception systems in small autonomous aerial systems. In *14th AIAA Aviation Technology, Integration, and Operations Conference* (p. 2705). Atlanta, GA.
26. Kumar, A., de Jesus Pacheco, D. A., Kaushik, K., & Rodrigues, J. J. (2022). Futuristic view of the internet of quantum drones: Review, challenges and research agenda. *Vehicular Communications*, 36, 100487.
27. Hafeez, S., Shawky, M. A., Al-Quraan, M., Mohjazi, L., Imran, M. A., & Sun, Y. (2024). Beta-UAV: Blockchain-based efficient authentication for secure UAV communication. *arXiv preprint arXiv:2402.15817.*

Chapter 3

Navigating the skies

UAVs and the emerging global frameworks

Prasad Gayathri, Anju Mohan, and Murugan Ramu

3.1 INTRODUCTION

Unmanned aerial vehicles (UAVs), better known as drones, have clearly reshaped diverse sectors ranging from cultivation and commerce to surveillance and defence, with their wide range of operations. Due to the overwhelming rate of spread of UAVs, the current legal systems, regulatory frameworks, and legislators are in dilemma with regard to the operations, privacy deals, and safety concerns. There is a clear-cut need for uniform laws especially when events like mid-air crashes, privacy breaches, and unauthorised airspace invasions happen due to the emergence and unmonitored use of UAVs. It is envisaged that a string of laws must follow the use of UAV, keeping in mind the advancement of technologies. While considering international aspects, there must be strategies concerning basic regulations, certificate standards, and operations, along with apt guideline, so as to incorporate the smooth amalgamation of domestic Unmanned Surface Vehicle (USV) and international airspace. Bringing in an equilibrium between scientific innovations and security and safety can be pointed out as one of the triggers in UAV regulation. Given the significant economic potential of UAV-based services, such as drone deliveries, governments and industries are investing heavily in the development of supporting technologies and regulatory frameworks

The global scope of UAV activities paves way for regulatory challenges including the cross-border connections, air traffic control issues, distinction, enforcement, and responsibility and liability issues. An unstable legal environment is supposed to be resulted from the absence of a worldwide norm that classifies UAV airspace usage, certifications, liability, etc., which makes international cooperation on drone-related and commercial UAV issues a rough patch. Another major complication in UAV regulation is cyber security, as the drones are slightly more vulnerable to signal jamming, hacking, breaches of data, and cloud computing, thereby chances of causing serious public threats are higher in an unregulated UAV situation. A regulatory setup with clauses pertaining to data security, secure communication units, and counter-UAV technologies would lessen the new risks.

 DOI: 10.1201/9781003598350-4

Moreover, the rise of UAVs, along with intervention of AI in decision making, creates several moral conundrums that require necessary legal attention. The legislative focus is at most necessary on, for example, liability in the event of UAV failure, strict adherence to international humanitarian law in matters of military applications of UAVs, moral and non-moral uses of UAVs for surveillance and law enforcement, and inclusion of ethical use of drones as a strict basis in the regulatory arena.

The author of this chapter examines how the global regulatory domain is changing in relation to UAVs, pointing out important policy issues, jurisdictional complications, and regulatory gaps. The chapter offers an intensive apprehension of the legal aspects of UAV governance and the actions that are necessary for efficient regulation in a society increasingly dominated by UAVs. For promoting regulatory and legal clarity and for the protection of cross-border operational consistency, the need for standardised international policies is also discussed in this chapter; moreover, the shaping of UAV regulations via amalgamation of regional regulatory organisations, national sovereignty, and international treaties will also be looked into. The chapter looks into the contribution of advanced technologies like blockchain and AI in enhancing UAV regulations and airspace governance, as well as addressing the future concerns, aiming to offer legal experts, policy makers, industrial activists, and stakeholders a solid foundation in order to develop a coherent and adaptable international regulatory framework for the operation and use of UAVs.

3.2 HISTORICAL DEVELOPMENT OF UAV REGULATIONS

UAV regulation's history is closely linked to its military inception (Springer 2013). At first, the main purposes of UAV deployment were combat and reconnaissance, which prompted the creation of defence-focused regulations. When early unmanned aircraft were tested for aerial surveillance during World War I, UAVs saw their first widespread application. The United States and the Soviet Union made large investments in drones for tactical and espionage purposes during the Cold War, which is when UAV technology saw major breakthroughs. Since UAVs were developed for military applications, countries have enforced stringent airspace regulations and classified protocols to control drone operations. Concerns regarding international humanitarian law, targeted killings, and sovereignty were raised by the US military's Predator and Reaper drones, which came to symbolise modern warfare. There were regulatory loopholes because civilian UAVs lacked a standardised international legal framework, whereas military UAVs operated under strict national security frameworks.

Regulatory bodies started developing thorough plans to incorporate UAVs into civilian airspace as the technology advanced beyond military uses. With the goal of developing internationally standardised laws, the International Civil Aviation Organisation became a significant force in forming UAV governance. To allow for the use of drones for both commercial and recreational purposes, the European Aviation Safety Agency, the Federal Aviation Administration (Abeyratne 2021) in the United States, and other national aviation authorities implemented laws. The expanding use of commercial UAVs in areas such as digital marketing, farming, and infrastructure monitoring has gained significant momentum in recent years. China, as the global leader in manufacturing of UAVs and drones, has implemented certification and registration protocols. Since drones with high-resolution cameras are largely available, several cases with privacy concerns emerged which compelled the policy makers to deal with the matter related to data protection and surveillance at large.

The general shift to risk-based frameworks represented a turning point in UAV law making as norms for the classification of drone activities on the basis of risk were formed by the International Civil Aviation Organisation's (ICAO) Remotely Piloted Aircraft Systems Panel, ranging from recreational drones to mega-complex UAVs. Now, this approach has greatly influenced global regulatory and normative frameworks to the extent that it caused the introduction of operational restrictions, licensing requirements, and remote identification methods.

Despite these advances, certain challenges persisted such as jurisdictional conflicts due to the traverse of UAVs across international airspace, creating legal ambiguities in enforcement and liability (Abeyratne 2021). Moreover, the growing prevalence of AI-driven UAVs and autonomous drones calls for a review of conventional aviation regulations and establishment of flexible structures of their use.

The UAV technological evolution is growing faster than ever, and as such, regulatory growth will have to keep pace with the growth-related issues of the UAV technology. Advances in the UAV technology, such as the development of drone swarms, urban air mobility, and cybersecurity, will be among the first challenges that UAV regulatory agencies will need to meet, and new challenges will emerge in the future as UAVs continue to develop. Balancing innovative advancement with safety and security is important. There has been a long history of the evolutionary development of UAV regulations necessitating a governance structure that will continue to evolve and develop the regulatory framework for addressing evolving legal, liability, and public policy issues. With UAV technology's upcoming advancements in the areas of 5G networks, AI, and a modern air traffic control system, this technology can be used to improve a variety of sectors, including emergency response, smart city, and logistics by using a proactive regulatory approach that addresses both the risks associated with the technology and provides for sustainable technological and economic growth.

3.3 INTERNATIONAL LEGAL FRAMEWORKS GOVERNING UAVs

Owing to the expansion of the application of UAVs in the global and international airspace, the demand for an accepted and recognised framework of norms and regulations has become pivotal, as UAV activities represent a clear legal obstacle compared to the old aircraft jurisdiction, including overlap problems, liability issues, and airspace integration challenges. As technologies improves further, it becomes crucial for the national agencies, regional governments, state parties, and international organisations to collaborate and develop necessary legal and regulatory standards to ensure the safety, security, and compliance of use and operation of UAVs. This chapter deeply looks into the existing international legal frameworks governing drones, treaties, tracts and agreements, case laws, and regulatory organisations shaping UAV policies while exploring the role of key bodies like the European Aviation Safety Agency and the ICAO in creation of international standards. It also delves into the latest initiatives to unify UAV laws universally and addresses important issues like air traffic control, cybersecurity risks, and moral challenges around self-piloted drone and its operations. This chapter will also explore how the international regulatory agreement is impacted by varying national security priorities. The authors also consider the challenges in applying UAV laws in areas of conflict, counterterrorism activities, and border control operations and monitoring. This chapter aims to offer a detailed insight of the legal complexities around UAVs and the pathway to internationally coordinated regulatory solutions by tackling these complex issues.

3.3.1 The role of ICAO

The UN ICAO is responsible for providing international standards for aircraft and UAVs. Although the Chicago Convention, adopted in 1944, is the primary legal document for civil aviation globally, it was not designed for new technologies such as UAVs. The introduction of the ICAO Remotely Piloted Aircraft Systems (RPAS) Standards has therefore attempted to fill these gaps and provide a framework to allow for UAV operations, safety, and certification in international airspace. However, countries are adopting the RPAS Regulatory Framework differently, leading to inconsistencies and challenges in enforcement and regulation of UAV operations internationally.

Moreover, when UAVs traverse international borders, sovereignty concerns surface. The ICAO has played a pivotal role in standardising aviation regulations. The ICAO's Convention on International Civil Aviation primarily governs manned aviation, and its recent amendments address the integration of UAVs into controlled airspace. The ICAO's Remotely Piloted Aircraft Systems Panel continues to shape international UAV regulations. Over the past decade, UAVs have advanced significantly, which has increased their application in crisis management, agriculture, logistics, surveillance, and infrastructure inspection. However, their fast spread raises questions about cybersecurity, privacy, safety, and airspace congestion (Schmitt 2017). The dangers of UAV operations might jeopardise manned aircraft safety, interfere with airport operations, and lead to security flaws in the absence of a strong regulatory framework. As a result, the ICAO's participation is essential to creating uniform regulations and operating standards for UAVs. In order to supervise the creation of international aviation standards and regulations, the ICAO was founded in 1944 under the Chicago Convention on International Civil Aviation. UAVs were not initially covered by the ICAO's framework, but as their use grew, an organised method to their regulation became necessary. The ICAO regulates UAVs in the following ways:

1. Creating recommended practices and international standards (Standards and Recommended Practices (SARPs))
2. Giving member states guidance
3. Encouraging international harmonisation and coordination
4. Ensuring security and safety procedures
5. Encouragement of technological development and research

Establishing the guidelines for safe UAV operations through the creation of standards and recommended practices is one of the ICAO's main responsibilities. Drafting regulations to allow UAVs to coexist with manned aircraft in non-segregated airspace is the responsibility of the ICAO-established RPAS panel. The ICAO's SARPs cover a number of important topics, including:

- Before being used in civilian airspace, UAVs must be certified as airworthy, which verifies that they adhere to safety regulations.
- Setting competency requirements for UAV operators through remote pilot licensing and training.
- Needs for communication, navigation, and surveillance (CNS): establishing procedures for controlling UAVs and corresponding to ATC.
- Collision avoidance systems: suggested technologies that enable UAVs to identify and steer clear of other aircraft.
- Including current air traffic management systems: outlining guidelines for UAVs.

The 193 member nations of the ICAO receive thorough assistance from the organisation to help them design national UAV laws that adhere to global best practices. This is essential to achieve uniformity across borders, facilitate UAV operations in international airspace, and avoid regulatory disparities that can impede cross-border UAV operations (Boon 2022). The Manual on Remotely Piloted Aircraft Systems (Doc 10019) and Circular 328 (Unmanned Aircraft Systems, or UAS) of the ICAO are important resources for governments creating their UAV regulatory frameworks. Furthermore, the ICAO works with national aviation agencies including the Civil Aviation Administration of China. (China), EASA (Europe), and Federal Aviation Administration (FAA) (United States) to advance uniform UAV regulations. Different nations have taken differing stances on regulation as UAV technology develops, creating disparities that make international UAV operations difficult. The ICAO is essential to the harmonisation of these laws through programmes including:

- The Global UTM (Unmanned Traffic Management) Framework enabling safe low-altitude UAV operations.
- The ICAO Drone Enable Symposia to examine developments and obstacles in UAV regulation.
- The RPAS Panel Working Groups which are specialised task groups that focus on issues including Command & Control (C2) connection standards and detect-and-avoid capabilities related to UAV integration.

The ICAO's strategies are chiefly centred on protection and mitigation of risks since mid-air accidents, loss of control, cyber threats, digital attacks, and terrorism are some of the issues and hazards linked to UAVs. The ICAO addresses these threats by:

- Endorsing risk assessment and evaluation models and frameworks to categorise UAV operations according to weight, operating area, and proximity.
- Implementing the use and deployment of geofencing technology or virtual boundary technology to limit UAVs and drones out of restricted and no-fly areas, such military zones, airports, and other prohibited areas.
- Coordinating with security organisations and law enforcement bodies and agencies to develop measures and strategies that can identify and neutralise threats from rogue UAVs.

The ICAO is actively partnering with researchers, subject-matter experts, industry leaders, and stakeholders to explore new developments and technological advancements such as artificial intelligence in UAV processes, urban air flexibility, and 5G and UAV connections. The goal of these initiatives is to improve automation, efficiency in decision making, and the practicality of air taxis and passenger drones in urban environment. The ICAO plays a major role in deploying regulation on UAV operations globally by ensuring that UAVs are integrated into airspace in a proper, safe, and uniform way by creating SARPs, advising member states, and coordinating international UAV standards, with a strong emphasis on safety, standard, and security. Despite persistent and frequent challenges, the ICAO is found to be the chief and key enabler of the future of UAV aircraft due to its ongoing continuous research and initiatives for technology developmental activities. The regulatory role of the ICAO is to be examined as extremely crucial as the UAV technology develops, as it works to guarantee a comprehensive and definitive approach that provides for innovation along with maintenance of airspace security and safety.

3.3.2 The European Union's approach

The European Union has applied a thorough, coordinated, extensive, and unified strategy that secures a balance between public interest, safety, and technological innovations in order to govern and regulate drones. The main aspect of this scheme by the EU is the conception of a unified and harmonised regulatory regime piloted by the European Aviation Safety Agency. The EU's UAV regulations aim to promote the betterment of the drone industry, which has a great potential in a number of industries, like emergency services, logistics, and farming, while ensuring that drones may be deployed with safety across Europe.

The main two rules that deal with the above-discussed frameworks are as follows:

- Regulation (EU) 2019/945, which looks into the technical specifications and market of UAV, in addition to classifying drones according to their weight, capabilities, and intended purpose, including the design, building, and maintenance of UAVs.
- Regulation (EU) 2019/947, which establishes the operating requirements for UAV operators and comprehends pilot qualifications, risk assessments, flight operation regulations, and operational authorisations.

According to EU laws, UAV operations can fall into different categories, mainly certified, specific, and open. The certified category is associated with high risk, the specific category is associated with medium risk, and the open category is low risk such as recreational use or commercial activities. The open category possesses the least risk due to the use of lightweight drones, visual line of sight navigations, compliance with no-fly and no-entry areas, and limitations in altitude. Operators who are engaging in medium-risk tasks are supposed to conduct a risk assessment that is detailed in nature and obtain approval from the respective aviation authority. For systems like drone deliveries and inspections, standard scenarios can simplify the process of approval. Moreover, matters related to large-scale deliver, transit, and cargo services, and fights over crowds and similar environments that are sensitive, fall under the certified category, which are again subject to severe regulatory oversight including mandatory authorisation for operations, licensing, advanced training, and type certification for UAVs.

The EU places a chief priority on safety after constant assessment and monitoring; hence, a standardised risk assessment procedure is supposed to be applied and followed especially by drone operators who engage in high-risk activities. A major strength of the EU approach is the conception of a unified drone market that clearly eliminates the need for drone operators adhering to different strata norms, but instead make them adhere to a unified single set of regulations. The EU also tries to attain cooperation between state parties in order to achieve the application of unified drone laws, and thereby, it promotes collaboration among national parties and encourages the advancement of drone technologies. The EU also tries to create a specific strategy to manage UAV traffic that enables safe drone operations and shares airspace with other manned aircrafts.

The EU has established an extensive drone registration system to ensure greater security and liability, which requires the operators to be certified depending upon the kind of operation the drones are engaged in, and additionally drones weighing 250 g or more ought to be registered with the concerned authority. Moreover, pilot training is made mandatory according to their operating categories wherein the certified category demands advanced training and efficiency, and the open category requires basic operational knowledge. The EU also looks into the environmental implications of UAVs, as certain specific zones and activities are either prohibited or restricted in order to protect wildlife and forest resources and to prevent and minimise pollutions, thereby striking a balance between environmental protection and advancement of the UAV technology.

Overall, the EU has a significant normative policy that is safety-oriented and environmental- and development-friendly. It guarantees safety precautions by classifying drone operations according to their level of risk so as to analyse the proportion and potential of possible hazards. However, this framework encourages a range of operations and the development of a seamless drone environment that can blend in perfectly with Europe's economy and airspace.

3.3.3 The US framework

The United States has instituted an extensive system to look into UAVs within a national airspace with oversight being set on the FAA. The FAA's regulatory strategies reflect an effective amalgamation of UAVs into the larger and integral aviation system, balancing the risk management and economic opportunities. The US Air Force regulates UAVs under Part 107 rules, which is applicable to both recreational and commercial UAV operations, of Modernisation and Reform Act of 2012. This was followed by the Small UAS Rule of 2016 that governs operations of drones under 55 pounds of weights with altitude restriction and other operational limits.

For recreational drone usage, the FAA requires drones weighing more than 0.55 pounds to be registered and to adhere to safety regulations, which include flying within line of sight, not exceeding 400 ft, and avoiding airspace near airports and regulated airspace without a permit.

Drone operators are generally allowed waivers by the FAA in zones including regulated airspace, night operations, flying over persons, and beyond visual line of sight activities. Bigger drones and operations transporting persons or goods are subject to Part 135 laws, which regulate commercial activities such as air taxis and drone-based package delivery. Similar regulations that govern manned aircraft also apply to these drones, such as those pertaining to type certification, pilot certificates, airworthiness requirements, and operating clearances. To guarantee that manned aircraft can coexist safely, the FAA is integrating airspace. UAS traffic management systems created in collaboration with commercial businesses and drone remote identification regulations are important endeavours. Remote ID enables tracking of drones in real time, thereby giving authorities and safety officials clear and pivotal information for managing the airspace. However, to qualify and operate for profit, a remote pilot certificate is necessary along with the conditions of attaining the age of at least 16 years, passing a background check, and passing an FAA-approved test.

For matters relating to search and rescue, disaster response, firefighting, and surveillance, the FAA has established detailed rules and regulations governing the use of UAVs by public safety bodies, including police, fire departments, and emergency response teams excluding missions that demand controlled and regulated airspace or planes beyond visual line of sight. To enhance liability and accountability, drones weighing more than 0.55 pounds are needed to be registered in a national database, whereas in case of violation, this registration allows the identification of operators and guarantees traceability. Additionally, the FAA conducts tests and trains pilots, especially for drones that operate in challenging and high-risk areas, which enhances advanced flight skills training for specialised missions, remote pilot certification training, and partnerships with business and academic institutions. Hence, by means of test sites and research programmes, which includes partnerships with companies on autonomous flying systems, urban air mobility pilot trials, and environmental impact and effect analysis, the FAA greatly promotes technology and innovation in the UAV field.

The Federal Trade Commission addresses privacy-related and ethical dimensions of UAV application, while the FAA oversights drone airspace, safety, and liability. In addition to several states, local governments have also brought in additional regulations pertaining to data collecting, monitoring, and privacy protection. It is supposed that the use and safety of UAVs are found in equilibrium by laws like Part 107, pilot certification programmes, and integration into the National Airspace System. Thereby in near future, UAVs will play a big role in aviation because of the framework's consistent evolution to make better the automation, traffic control, and airspace integration.

3.3.4 The role of the United Nations and other multilateral efforts

As UAVs have become more common and available and their applications extent to certain fields like disaster relief, peacekeeping, surveillance, and humanitarian aid, the United Nations and other multilateral organisations play a crucial and essential role in the regulation, management, overseeing, and promotion of UAVs as international norms are generally set by the UN for the safe, proper, and responsible use of UAVs.

The ICAO is the chief and primary body in charge of the creation of international aviation standards, and has been actively developing regulations ever since for safe integration of UAVs and manned aircrafts. Efforts of the ICAO include the creation of universal standards and recommended practices for UAVs, including automated operations, safety requirements, remote identification, and UAV airspace integration. Since the use of drones has drastically extended to several fields like disaster relief, peacekeeping, surveillance, and humanitarian assistance, the United Nations and other multilateral organisations are essential bodies to create and apply apt regulations, management, and governance of UAVs.

The global coalition of governments, business leaders, and international organisations acts majorly in advancing safe and liable use of UAVs world wide, and such initiatives focus on establishing a global setup for managing UAV traffic, implementing international mechanism for UAV production and operation, and providing security measures to stop UAVs from being abused for harmful purposes. Another influential actor in shaping global discussions on UAVs is the World Economic Forum, which pinpoints the amalgamation of drones into urban mobility networks and disaster relief works. Initiatives from the World Economic Forum comprises the International Telecommunication Union collaboration, regulatory development, and global drone collaboration. Many governmental bodies and international entities also aim to channel the drone technology for humanitarian and security mission with the idea of minimum risk of abuse. Organisations like World Food Programme, United Nations' Development Programme, and Red Cross employ UAVs for a better range of humanitarian ideas, like supply chain management, disaster response, peacekeeping, and security (Heyns et al. 2016), and they join forces to provide best practices to ensure moral and secure deployment of UAVs.

The ethical and moral application of UAVs has turned out to be more and more of a global concern, especially when it comes to matters relating to privacy, decision-making autonomy, and the use of UAVs for surveillance (Zirkel 2016). Global regulatory frameworks, morals, drone ethics, and conception of rules for autonomous drone application to avoid mishaps or abuse in war or peacekeeping scenarios are found to be the major plot for discussion within multilateral organisations (Crootof 2017). It has turned out to be evident that the existence of global organisations and multi-level cooperation is necessary to ensure the safe, ethical, and effective deployment of UAVs cross-border, as the UAV technology develops steadfast and finds use in a variety of fields, including humanitarian assistance, environmental monitoring, defence, and commercial logistics. Establishing globally accurate and accepted standards for UAV operations turns out to be a major priority of the United Nations through its specialised agencies such as the ICAO, UNOOSA, and International Telecommunication Union (ITU) as they make sure that UAVs may be securely incorporated into national airspaces and international environments.

For the achievement of free and safe UAV operations internationally, there should be a unified internationally harmonised legislation on that behalf, which stands essential in avoiding disputes and misunderstandings that may result from divergent state laws. This harmonisation helps prevent threats to national-level security and safety in airspace, along with advancing humanitarian cooperation, cross-border commercial trade, and the enhancement of global economy. Ensuring the ethical use of UAVs is another greatest aspect of the UN's role, clearly in delicate situations like military surveillance, conflict zones, and data privacy. Since the scenario of growing use of drones is prevalent for military and even civilian purposes, international organisations need to work together to address and resolve any issues pointing to autonomy, surveillance, and accountability, as these dialogues are crucial in bringing in peacekeeping, international peace, and cooperation, upholding human rights in the era of deployment of UAVs.

Furthermore, fostering innovation demands a global collaboration through initiatives like the Global Drone Coalition, the World Economic Forum, and other business alliances. To formulate best practices for the commercial application of UAVs and its integration into urban air mobility systems, like drone deliveries or urban air taxis, these platforms play a major role by bringing together governments, private stakeholders, and academic institutions. Since these programmes prioritise the safety, sustainability, and technological advancement of UAVs in addition to legal structure development, the UN can significantly bring up the reach and effectiveness of activities at times of necessity by collaborating the application of UAVs with these international NGOs, local governments, and other initiatives (Hodgkinson & Johnston 2018). As UAVs becoming substantially embedded into day-to-day affairs, the multilateral cooperation will be crucial to ensure the responsible, safe, and equitable deployment globally. These organisations' rigorous efforts will ultimately influence the development of UAVs, and such advantages will be realised in a way that advances world peace, security, and progress.

3.4 NATIONAL REGULATIONS AND DIVERGING APPROACHES

Even though international efforts aim for regulatory cohesion, national UAV norms continue to be highly varied across the globe. Countries such as China, India, and Australia have established distinct legal frameworks shaped by unique security, commercial interests, and operational context as such. Liability, privacy, and ethical considerations being the hidden force in the global drone ecosystem, India has held on to the promise of UAVs for many industries, including infrastructure, agriculture, disaster relief, logistics, scrutiny, and surveillance. In order to ensure UAVs' safe and efficient incorporation into the airspace, the Indian government has acknowledged the significance of regulating them while simultaneously looking into the legal matters like privacy, security, and economic development issues. Due to the presence of a complicated socio-political context, steadfast evolution of technology landscape, and concerns relating to public safety, privacy protection, and national security, India's regulatory stance to UAV regulation has found to be changed over time. Since India's regulatory environment stands different from other regions that prioritise uniformity, our national interests need to strike a balance between safety and security concerns and innovation.

The Directorate General of Civil Aviation, functioning under the national aviation regulatory authority of the Ministry of Civil Aviation, plays a central role in formulating and overseeing India's legal approach to UAVs. The Directorate General of Civil Aviation (DGCA) is in charge of making and applying rules to guarantee the safe use of UAVs and support the expansion of drone and aviation zones. Initially, India's approach to UAV regulation was unstructured and minimal, with few laws or regulations governing the use of drones for business and personal uses. The DGCA later on introduced more formal, systematic, and structured rules in the middle of the late 2010s because of security and safety concerns, and the Civil Aviation Requirements for Remotely Piloted Aircraft was the first major step. Since commercial drone usage was the main focus of these earlier laws, the operators were supposed to get authorisation for aircraft that were seen to be used outside the purview of leisure perspectives. Moreover, drones could not be flown in certain controlled airspace or close to critical areas according to the earlier guidelines and regulations. Later on, the DGCA, due to the increase of UAV activities and operations, started creating more rigid regulations to handle the cases as drone usage spread quickly across most of the industries including the agricultural sector, cinematography, construction assessment, surveillance, and logistics. The year 2021 saw the introduction of the Drone Rules which are new, more comprehensive drone laws that were adopted by the DGCA to optimise drone operations in India with a defined and structured framework while taking into account best practices and technical advancements around the globe.

According to the 2021 rule, drones are categorised into operational categories according to their weight, flight danger, and intended use:

- Drones weighing less than 250 g
- Microdrones weighing between 250 and 2 kg
- Drones that are small (2–25 kg)
- Drones weighing between 25 and 150 kg
- Drones weighing more than 150 kg

However, there are certain specific operational criteria, such as limitations on access of airspace and flying authorisation, applicable to each category.

- Certification and Registration: All drones and their operators must be registered with the DGCA in accordance with the Drone Rules 2021. Additionally, operators of drones weighing more than a specific amount must have a Remote Pilot Licence.
- Remote Pilot Licensing: The DGCA also mandated certification and training for remote pilots. A Remote Pilot Licence is necessary for operators of tiny drones weighing more than 2 kg. The DGCA licences training organisations, and various drone pilot categories have distinct training requirements.
- No Permission, No Take-off: According to the No Permission, No Takeoff regulation, drones weighing more than 250 g are required to have a permission-to-fly software installed. This software enables the operator to get real-time airspace clearance prior to flying in certain areas. The DGCA's Digital Sky, India's official drone registration platform, is accessible through the software interface.
- Operational Restrictions: Drones are only allowed to operate in certain locations and under particular circumstances, according the new regulations. Only drones that have remote identification and geofencing installed are permitted to fly in the regulated airspace. Drones must also be flown within the optical line of sight, unless the operator has been given special permission to operate outside of VLOS.
- Prohibited Areas: Some locations, such as airports, high-security zones, government facilities, and military zones, are still off-limits to drones. The DGCA establishes no-fly zones, which are incorporated into the Digital Sky platform.

- Drone Delivery: Laws governing drone delivery providers are still being drafted. Nonetheless, India has been investigating the employment of drones for e-commerce and medical supply delivery, and test flights have been conducted to determine if such operations are feasible.

While regulating UAVs, the Indian government has to handle two important concerns: privacy and national security. The quick spread of drone technology has sparked worries about data privacy and monitoring. Specifically, drones that include cameras and sensors may be used for eavesdropping or gathering personal information.

Privacy Concerns: In India, the Ministry of Electronics and Information Technology's regulatory guidelines and the country's current data protection legislation combine to regulate the privacy problem around drones. The information technology data protection guidelines must be followed by drones used for surveillance or data collecting.

Additionally, the application of UAVs by law enforcement bodies is subject to restrictions under Indian Telegraph Act of 1885 and the Indian Penal Code (IPC), which regulate wiretapping and monitoring. Concerns have also been raised regarding the implications of UAVs for national security, as they may be used as tools for terrorist activities, espionage, and unauthorized information gathering. Hence, India has historically been wary of their deployment close to sensitive areas, especially along its borders with China and Pakistan. To remove these risks, to guarantee that drones cannot be used for espionage, or to jeopardise national security, the DGCA and other security authorities enforce stringent laws and regulation of airspace that should be strictly maintained in areas of sensitive and key infrastructure, military zones, and border regions. Along with these measures, the Indian government has also made investments in counter-UAV systems to guard against illegal drone activities, particularly in the sensitive zones, vicinity of military installations, and other vital infrastructure, so that these devices identify, follow, and destroy rogue drones that might endanger public safety.

The UAV sector of India has grown significantly across many industries, including disaster management, logistics, infrastructure inspection, and agriculture. The Indian government has shown notable interest in using drones to further economic growth and welfare of the general public:

- Precision agriculture is employing drones more and more to map fields, apply pesticides, and monitor crops. These UAVs support farmers in managing large-scale agricultural operations more effectively, increasing crop yields, and reducing expenses.
- In areas hit by floods, earthquakes, and forest fires, UAVs have shown to be invaluable and quite helpful in disaster relief operations. They enable rescue efforts by offering quick situational awareness, aiding in damage assessment, and giving real-time aerial images.

- Bridges, electricity lines, solar farms, and telecommunication towers are just a few examples of the vital infrastructure that UAVs are frequently employed to check. Because of their high-resolution cameras and sensors, these drones enable operators to evaluate the state of infrastructure without endangering people.
- In metropolitan areas, India has started experimenting with drone delivery services. To investigate the possibility of employing drones for package delivery, companies such as DHL, Blue Dart, and Zomato have carried out test flights. Drones are being utilised for quick transit.

India has made great strides in UAV regulation, but there are still many obstacles and weaknesses in the country's strategy (Mehta 2021). The primary difficulties consist of:

- Fragmented Enforcement: Across the nation, local differences exist in how UAV laws are enforced. Rural areas and smaller towns frequently lack adequate monitoring and enforcement methods, even if big cities like Delhi and Bangalore may have greater adherence to the law.
- Regulatory Ambiguities: Some parts of the Drone Rules 2021, such as the precise rules controlling drone delivery services and Beyond Visual Line of Sight (BVLOS) operations, are still in the planning stage or are not yet clear.
- Data Privacy Issues: The use of drones with cameras and sensors to collect personal data is a serious problem since India lacks a comprehensive data protection law. A framework at the national level to address drone surveillance and personal data protection is still lacking.

India presents the drone sector with substantial potential prospects, as well as obstacles. There is a huge market for UAV applications because of the nation's big population, extensive geographic area, and expanding infrastructure and surveillance needs. India occupies a unique position to utilising the UAV technology in fields including disaster relief, delivery services, and agriculture, and India's UAV regulations clearly strike a balance between protecting public safety, encouraging technical advancement, and settling privacy and national security matters. Even though India has made efficient strides towards creating a coherent set of legislation, challenges persist in enforcement and privacy, and the need for more explanation in particular areas, such as drone deliveries and BVLOS operations, still exists (Bhatia 2014). The chief objective is to provide a framework that supports innovation, permits safe drone actions, and provides the safety of its residents as India's regulatory scenario undergoes necessary changes. To make sure that India is placed at the forefront of drone technology and protect its national interests and public safety and security, further legislative movements and advancements and a more efficient and robust enforcement system would be essential to the growth and acceptance of UAVs in the country.

3.4.1 The UAVs and liability issues

In India, UAVs, or drones, have witnessed a rapid style of adoption in a variety of industries, from personal to military and commercial uses. The popularity of UAVs has, however, raised a number of legal and liability concerns, including as trespassing, property damage, privacy violations, cybersecurity risks, and infractions of aviation laws. The existing legal environment built under criminal law, torts, aviation rules, and constitutional law provides guidance and solution in handling these challenges, even if India is still in conception stage with regard to particular case laws pertaining to UAV liability in which the invasion of people's privacy is one of the biggest legal issues with drones in India. In the seminal Puttaswamy case (*Justice K. S. Puttaswamy (Retd.) v. Union of India* n.d.), the Supreme Court affirmed the right to privacy guaranteed by Article 21 of the Indian Constitution. Because of this lawsuit, people now have the basic right to privacy, which includes protection against unapproved surveillance.

Drones with infrared imaging and high-resolution cameras may easily enter private areas and take pictures and videos without permission. Section 66E of the Information Technology Act 2000 makes it illegal to take, publish, or send pictures of private spaces without permission. Under this law, people, the media, or even businesses may be held accountable if they deploy drones for illegal monitoring. The judiciary has acknowledged the seriousness of privacy violations through cases like Shreya Singhal (*Shreya Singhal v. Union of India* n.d.), in which the Supreme Court invalidated Section 66A but maintained other privacy protections under the IT Act, even though there are no direct case laws pertaining to drone-related privacy breaches in India. Due to their ability to access private property without permission, drones raise trespassing-related legal issues. Criminal trespass is defined in Section 441 of the IPC 1860 as the unauthorised intrusion into another person's property with the purpose of committing an offence or causing annoyance. Even though traditional trespassing laws were made to prevent physical intrusion, it is still ambiguous how they should be interpreted in relation to UAVs. In the matter of Mustafabad Oil Mills (*Ramlal v. Mustafabad Oil Mills* 1968), the court ruled that it is considered as trespassing if an offender enters another person's property without their personal presence. Applying this logic, unpermitted UAV flights over private property may result in criminal and civil penalties. Moreover, drones that generate disruptions like noise pollution or blocking public places can be considered a public nuisance under Section 268 of the IPC.

In events of failure, operator carelessness, or airspace breaches, drones become a serious threat to both property and personals. Individuals or organisations causing injury due to reckless use of UAVs are clearly subject to responsibility under IPC Sections 304A, which deals with death by negligence, and 336–338, which deals with acts endangering life or inflicting severe hurt. Drone-related accidents can be dealt by using the negligence rules that were developed in landmark cases such as *K. Krishnan v. State of Tamil Nadu* (n.d.). In this instance, the court determined that severe legal culpability results from negligence that causes tragic accidents. Under the civil tort law for damages and the criminal law for reckless endangerment, liability may emerge if a drone crashes into a person or piece of property as a result of the operator's incompetence, contempt for aviation regulations, or equipment malfunction.

Furthermore, insurance for drone activities is still a relatively new topic in India (A et al. 2021). Although liability coverage for road accidents is provided under the Motor Vehicles Act, victims of drone-related accidents may find it difficult to receive compensation due to the lack of a comparable comprehensive liability framework for UAVs. Because they may be used for illegal data collecting, hacking, and even espionage, drones—especially those with data transmission and storage capabilities—pose a cybersecurity risk. Drones used to gather private information without permission may be subject to the criminal penalties for unauthorised access to personal information under Section 72 of the Information Technology Act 2000. The Supreme Court emphasised the need for privacy safeguards in the digital age in the 2015 case of *Shreya Singhal v. Union of India*. Under the IT Act, using a drone for illegal and unauthorised monitoring of business or government areas, data theft, or hacking may result in legal actions and penalties. Additionally, drones that collect personal data without consent and authorisation may face further legal scrutiny as per India's Personal Data Protection Bill.

Drones are generally found to be governed by aviation laws as they operate in regulated airspace. UAV operations in India are governed by the Directorate General of Civil Aviation in tune with the Aircraft Act 1934 and the most recent Drone Rules 2021. In order to stop illegal access to restricted airspace, these regulations specifically include drone registration, no-fly zones, remote pilot certification, and geofencing. The Indian Apex Court discussed the significance of rigorous aviation safety compliance in the *Air India Ltd. v. Cochin International Airport Ltd.* (n.d.) case, which can be further applied to airspace infractions involving UAVs, and it could be concluded that unauthorised drone operations close to military installations, airports, or sensitive government locations may be prosecuted and subject to fines and imprisonment.

In addition, no-fly zones have been established by the DGCA around sensitive areas including the vicinity of airports, military bases, nuclear plants, and national security installations. The Aircraft Rules 1937 and the IPC Sections 188 and 287 for violation of legitimate authority and negligent conduct with regard to machinery might be used to punish violations of these restrictions. In an effort to streamline drone operations while maintaining safety and compliance, India has implemented the Drone Rules 2021 in response to the growing usage of drones in e-commerce, agriculture, law enforcement, and disaster management. Future case laws will probably provide more precise legal precedents, as the liability framework for UAVs is still developing. Drone makers and operators are now subject to liability under a mixture of criminal, civil, and constitutional laws. In contrast to the FAA regulations in the United States, which explicitly define roles and consequences for UAV-related infractions, India could require specialised drone liability legislation. Ethical and legal challenges will continue to emerge as drone technology develops, which necessitates new interpretations of current laws by Indian courts. It is essential for the judiciary to strike a balance between innovation and safety in order to guarantee that UAVs may be utilised for advantageous purposes without violating both public safety and individual rights. Although there aren't any case laws in India specifically addressing accountability and liability for UAVs, the country's current cybersecurity, privacy, trespassing, and aviation rules offer a guideline for handling drone-related conflicts and concerns.

Landmark judgements in the cases of Shreya Singhal and Justice K.S. Puttaswamy reinforce cybersecurity and privacy safeguards, which may be expanded to include the abuse of UAVs eventually, and also people impacted by drone-related infractions have legal remedies under the IPC's trespassing, carelessness, and public nuisance provisions. Attaining proper legal clarity on liability problems will be necessary for companies, law enforcement, and people operating UAVs, given the fast advancement of drone technology. Furthermore, India will be able to achieve a balance between encouraging innovation and maintaining public safety by maintaining regulatory frameworks and enacting particular UAV liability laws.

3.5 THE LEGAL PRECEDENTS AND JUDICIAL INTERPRETATIONS OF UAV REGULATIONS: A NATIONAL AND INTERNATIONAL PERSPECTIVE

UAVs are becoming increasingly prevalent throughout the world, thereby raising complex legal issues in relation to liability, responsibility, privacy, security, and airspace management. Although some countries have enacted comprehensive legislation controlling UAV activities, the judicial interpretations are still crucial in establishing precedents. The judiciary have looked at issues relating to national security concerns, trespassing of the airspace, illegal spying and surveillance, and privacy infringement. Reviewing national and international case laws often provides necessary and important insights into how various states parties handle UAV-related legal challenges.

In India, presently there are no official judicial decisions that address the problems with UAVs, but many landmark judgements have looked into several general legal concerns that may be relevant to UAV activities, such as privacy, liability, scrutiny, surveillance, and airspace rights. The apex Court of India has specifically identified the right to privacy as a basic fundamental right under Article 21 of the Constitution in one of the most important cases (*Justice K.S. Puttaswamy v. Union of India*). Moreover, this decision is applicable to drone surveillance since according to the judgement, each and every invasion of privacy must be proportionate, necessary, and legally justified. Drone surveillance may be contested on the grounds of breaking this law if it is carried out by the government or commercial organisations without authorisation and proper consent.

Yet another relevant case in this aspect is *People's Union for Civil Liberties v. Union of India* (n.d.), which is often referred to as a telephone-tapping case. In this case, the Supreme Court emphasised that in order to prevent arbitrary invasions of privacy, even the governmental monitoring must be subject to checks and balances. The above-mentioned case is mostly concerned with wiretapping, but in matters pertaining to UAV monitoring, this case can be a benefit from its lessons. The court underlined the crushing need for judicial supervision and procedural protections in surveillance techniques, and arguments against the uncontrolled use of drones for wide span scrutiny and monitoring can draw on this history. The Supreme Court made a judgement in *R. Rajagopal v. State of Tamil Nadu* (n.d.) about the boundaries and potential restrictions of a person's right to privacy against media interference. According to the decision, unless it is in the public interest, private information cannot be collected or gathered without consent or authorisation. This principle can be applied in journalism with drones, as unauthorised aerial imagery may be viewed as a violation of privacy and result in legal actions. While India's regulatory system has provided guidelines that apply to UAV operations, there is still a lacuna of case law specifically addressing drone-related conflicts and issues. Judicial interpretations of privacy, trespass, and security statutes will clearly determine future litigation on UAV-related issues.

The legal challenges posed by UAVs have been addressed by several litigations globally, providing proper insight into how different countries tackle with matters regarding drone operations, invasions of privacy, trespassing into airspace, and damages responsibility. *Boggs v. Merideth* (n.d.) was a landmark case involving UAV trespassing rights in the United States, wherein, David Boggs, the plaintiff, sued William Meredith, his neighbour, alleging that the latter had violated his private property by shooting down his drone. Meredith further argued that he had the right to protect his property as the drone was intruding on his lawn. Even though the case brought up chief issues about privacy rights and airspace ownership and the court emphasised the necessity for clear regulations, the court rejected the case owing to jurisdictional restrictions.

The US Court of Appeals in *Taylor v. Federal Aviation Administration* (n.d.) made another noteworthy decision for the District of Columbia Circuit against the mandatory UAV registration requirements for recreational drone operators. In this matter, it was argued that the FAA had approached its jurisdiction to demand registration for small drones used for private use. This ruling further strengthened the discussion on striking a balance between drone operators' rights and national security considerations, which resulted in policy and strategy changes in the United States.

In Europe, privacy issues have been discussed with regard to the General Data Protection Regulation framework in the matter of UAV monitoring. Historically, the European Court of Human Rights has given decisions in matters pertaining to surveillance that have an indirect impact on UAV operations. European Court of Human Rights (ECHR) further decided that the installation of surveillance and security cameras in a university lecture hall without permission has clearly violated the privacy right under Article 8 of the European Convention on Human Rights, in the 2017 case of *Antović and Mirković v. Montenegro* (n.d.). This ruling stressed that all kinds of monitoring, including UAV surveillance, must stick to the legislational requirements of necessity and proportionality, even if drones were not used in this case.

The United Kingdom has also tackled the legal and regulatory issues pertaining to UAVs all the way long. A drone operator, the defendant, was charged in *R v. Rogers* (2020) for breaking national aviation regulations by operating a UAV close to Gatwick Airport which is a sensitive zone. The UK court ruled that the strict adherence of UAV regulations and restrictions should be practised to preserve public safety, and even though the defendant strongly claimed that the drone presented no serious security threats, it was advised to abide by the existing regulations as such. This case projects the importance of keeping no-fly and no-entry zones in and around airports and other sensitive zones so as to mitigate potential threats and hazards.

Australia has addressed laws in relation to consumer protection and aviation safety to settle concerns involving UAVs, and as seen in The SkyView case (*The Civil Aviation Safety Authority v. Marsh* 2014) case, the court penalised an Australian drone service operator for conducting aerial photography without the required permissions and consents, thereby reaffirming that UAV operators must adhere to license regulations and certifications.

The Chinese government has adopted a stringent strategy against UAV violations, with courts severely penalising unapproved drone operations. In 2020, a defendant was given an imprisonment as sentence by a Chinese court for deploying UAV close to a military facility and thus violating air traffic control regulations, illustrating the nation's zero-tolerance approach towards acts threatening or endangering national security in China.

An extensive and comparative analysis of case laws across the globe shows the divergent legal issues surrounding drone operations. While some nations prioritise privacy issues, others might focus on mitigating the dangers to aviation safety and security. Due to the lack of a 'one-size-fits-all' global regulation, UAV laws have been interpreted differently throughout the globe, which causes heavy inconsistency of laws and regulations. In India, there is a solid basis for handling UAV-related problems due to the presence of judicial precedents pertaining to privacy and surveillance, yet the absence of direct judicial review is still a problem. In matters of airspace ownership in the United States, the imminent need to look into the property owners' rights regarding drone trespass is strongly highlighted. Ensuring that drone monitoring aligns with data protection regulations, the European courts have reinforced privacy protections (Villasenor 2015).

Globally, the trend of growing recognition of the necessity of specific UAV rules is very significant in the present context. Internationally, judiciary have underlined the importance of maintaining an equilibrium or a balance between technological improvements, privacy rights, and security concerns, through attaining clarity in statutes. Areas having regions with strong privacy laws like the EU have placed an emphasis on individual rights, whereas countries with stricter UAV laws, like China and the United Kingdom, have shown a strong propensity towards protecting national security, showcasing that the increase in regulatory conundrums around UAV/drone technology is reflected in the development of precedents pertaining to UAVs. Although India is yet to witness a judicial review for straightforwardly resolving and handling privacy breaches, trespassing allegations, and security issues on UAV disputes, the existing doctrines around UAV legislation provide a corner stone for that aspect. Courts across the world have shaped UAV laws by interpreting property rights, surveillance legislation, and aviation safety standards. As the era of drone technology rises, the legal systems throughout the globe will need to adapt by creating apt legal and regulatory environment for handling new issues while striking a balance between privacy, security, and innovation (Rakhecha 2021). For the future UAV developments, a combination of technical standards, judicial activism, and legislative amendments will probably influence the reform.

3.6 THE FUTURE OF UAV REGULATION

As UAV capabilities continue to expand, legal and regulatory frameworks must evolve to accommodate and address the emerging technologies such as urban air mobility, drone delivery services, and AI-driven autonomous operations. In order to manage the UAV traffic in the future, the necessity of establishing an internationally recognised UAV traffic management system is a crucial milestone in regulatory growth, and similarly, a comprehensive and extensive legal framework is strongly advised since the usage of UAVs evolves in the military, commercial, and recreational sectors. Ever since the imminent growth of the drone technology is noticed, many nations' current legal frameworks, which raises questions concerning liability, privacy, security, and airspace management, need to be amended accordingly so as to fit into the situation. Furthermore, governments, state parties, and international organisations are attempting to find a proportion in encouraging innovation with making sure that UAV operations don't violate people's rights, jeopardise public safety, or endanger national security. Developing a standardised international strategy will be a crucial element of future UAV regulation (Scott 2022). The current laws that regulate UAVs greatly vary among the countries across the globe, which clearly causes irregularities while conducting cross-border actions especially in the matter of enforcement of the laws. To address these irregularities found globally, entities like the ICAO are working to form a uniform set of regulations that fit all. It can be further observed that the plight of UAV governance will probably be greatly dependent on the implementation of flying lanes, universal drone-identifying systems, and remote pilot certifications along with data protection and privacy being another key area that requires priority. Issues relating to the emergence of the idea of mass surveillance and unauthorised data gathering are escalating as drones are clothed with more sophisticated cameras, face recognition software, and artificial intelligence.

In India, the Digital Personal Data Protection Act should coordinate the UAV-specific concerns of privacy to stop the exploitation of data obtained by UAVs. The future goals include stringent transparency measures, requiring drone operators to report their monitoring operations and obtain prior approval when necessary. Strengthening national security through improving real-time scrutiny, geofencing, and enforcement of restricted zones will stand as a key legal goal as UAVs get closer to both public and private operations. Regulators may respond by enforcing no-fly zone enforcement procedures, improved real-time tracking systems, and more stringent geofencing technologies.

Law enforcement organisations will also probably be given more power to stop rogue drones using anti-drone technology including cyber-defence tools, drone interception systems, and electromagnetic jammers. Liability and insurance systems will also need to be addressed in future UAV regulations. Determining who is responsible for accidents or damages has become harder as drone deliveries, aerial inspections, and autonomous UAV activities has become more widespread. Clear responsibility structures, drone insurance requirements for business operators, and compensation procedures for impacted parties are all expected to be introduced by legal frameworks. Regulations must continue to be flexible and progressive as the UAV technology develops. Swarm drones, artificial intelligence, and urban air transportation will all present new difficulties that call for the ongoing legal analysis (Sehrawat 2021). Nations that take the initiative to create thorough and adaptable UAV regulations will be in a better position to capitalise on drone technology's advantages while reducing its hazards. In order to create a UAV ecosystem that is safe, effective, and compliant with the law, governments, IT companies, and international regulators must work closely together going ahead.

3.7 CONCLUSION

It can be observed that the rapid proliferation of UAVs has changed several sectors and industries by intensifying efficiency in agriculture, logistics, surveillance, and disaster management. However, this technological advancement brings in chief and major legal, ethical, and security challenges that require immediate legal attention. Moreover, privacy concerns caused by high-resolution cameras, facial recognition, and AI-powered tracking systems pose serious substantial threats to individual rights. Without the presence of adequate legal safeguards, the risk of UAVs leading to widespread surveillance, unauthorised data collection, and breaches of privacy is a high priority matter of concern. While jurisdictions like the United States and the EU have introduced robust privacy regulations, India is still sustaining a regulatory gap in this aspect.

To mitigate the above-discussed risks, it is necessary to establish a robust and well-defined legislative framework that ensures mandatory consent, data retention limits, and constant judicial oversight. The presence of equally pressing security challenges, with events of UAV deployment without prior consent which lead to disruption of airport, cyberwarfare, smuggling, etc., demands a well-structured normative framework on this behalf. In order to strengthen the counter-drone activities, certain technologies such as geofencing, real-time UAV tracking, and anti-drone defence systems are essential to ensure crucial infrastructure and national security. Simultaneously, these upcoming regulations should guarantee that only authorised operators deploy UAV activities in sensitive zones to prevent misuse while allowing legal applications to flourish. Another critical concern in this arena is airspace management, as many drones are supposed to share airspace with manned aircraft and hence, seamless integration into the existing air traffic management systems is critical to prevent collisions, unauthorised flights, and air congestion. Automated drone identification, streamlined authorisation protocols, and UTM systems will play major roles in creating safe co-existence between UAVs and traditional manned aviation. In addition, liability frameworks must evolve to define liability and accountability in cases of accidents, injuries, or damages caused by UAVs and establish clear responsibility among drone manufacturers, operators, and service providers, alongside compulsory insurance requirements, which will provide fair compensations for affected parties.

It can be concluded that the future of UAV governance requires dynamic, robust, and adaptable regulatory mechanisms capable of keeping pace with technological advancements in the area of UAV and similar technologies. The regulators must proactively develop frameworks that find a proportion between technological progress and ethical outcome as emerging technologies such as autonomous swarm drones, blockchain-based UAV tracking, and AI-driven decision making gain importance. The advent of urban air mobility, including passenger drones and drone taxis, further clearly highlights the urgency for futuristic legal frameworks that encourage safety while fostering technological growth.

These multifaceted futuristic challenges show that the international harmonisation of UAV regulations and the ICAO and other regulatory bodies must take a lead in putting efforts to establish globally consistent standards, ensuring safety in the operation of UAVs, ethical and legal compliance, and non-problematic cross-border UAV operations. In conclusion, the incorporation of UAVs into global airspace presents vast potential but face complex regulatory, ethical, and safety challenges. A globally coordinated and legally driven approach towards innovation and technological advancements is essential to unlock the maximum benefits of UAV operations, while along with mitigating potential risks associated with it, striking a balance between privacy, security, and innovation will pave the way for a safer, more efficient, and ethically responsible UAV ecosystem in future years.

ACKNOWLEDGEMENTS

We hereby express our sincere gratitude to Dr. Asha Sundaram, Principal of Saveetha School of Law, SIMATS, Chennai, for her immense and valuable moral support and insightful feedbacks and comments, which have greatly enhanced the quality of this manuscript. We extend our heartfelt thanks to our research supervisor for their valuable guidance and expertise throughout the research process and course of the study. Our heartfelt gratitude goes to the learned research fellows for their thoughtful contributions and insights. Finally, we are deeply grateful to Saveetha School of Law and Saveetha Institute of Medical and Technical Science, the alma mater, for their unwavering support and dedication during this endeavour making this research possible.

REFERENCES

Abeyratne, R. *Legal and Regulatory Issues of UAVs in International Airspace*. Springer, 2021.

Air India Ltd. v. Cochin International Airport Ltd. AIR 2000 SC 801 (n.d.).

Antović and Mirković v. Montenegro. Application no. 70838/13 (n.d.).

Bhatia, G. "State Surveillance and the Right to Privacy in India: A Constitutional Biography." *National Law School of India Review* Vol. 26, No. 2, pp. 127–160, 2014.

Boggs v. Merideth. No. 3:2016cv00006 (n.d.).

Boon, K. *Drones and International Law: A Critical Analysis*. Oxford University Press, 2022.

Chatterjee, Vasudha & Ambika Vishwanath. "Fast-Tracking the Flight of India's Drone Industry." Observer Research Foundation (ORF) Issue Brief, 2023. Available at: https://www.orfonline.org/research/fast-tracking-the-flight-of-india-s-drone-industry

Civil Aviation Safety Authority v Marsh (2014) FCA 1253.

Crootof, R. "War Torts: Accountability for Autonomous Weapon Systems in International Law." *Cornell Law Review*, Vol. 102, pp. 1347–1403, 2017.

Heyns, C., Akande, D., Hill-Cawthorne, L. & Chengeta, T. "The International Law Framework Regulating the Use of Armed Drones." *International & Comparative Law Quarterly*, Vol. 65, No. 4, pp. 791–827, 2016.

Hodgkinson, D. & Johnston, R. *Aviation Law and Drones: Unmanned Aircraft and the Future of Aviation*. Routledge 2018.

Justice K. S. Puttaswamy (Retd.) v. Union of India. (2017) 10 SCC 1. (n.d.).

Krishnan v. State of Tamil Nadu. AIR 2006 SC 3037 (n.d.).

People's Union for Civil Liberties v. Union of India. AIR 1997 SC 568 (n.d.).

R v. Rogers. NSWCCA 305 (2020).

R. Rajagopal v. State of Tamil Nadu. (1994) 6 SCC 632. (n.d.).

Rajagopalan, Rajeswari Pillai & Rahul Krishna. "Drones: Guidelines, Regulations, and Policy Gaps in India." Observer Research Foundation (ORF) Occasional Paper, March 5, 2018. Available at: https://www.orfonline.org/research/drones-guidelines-regulations-and-policy-gaps-in-india

Rakhecha, V. "Privacy, the Shield that Protects or the Sword that Strikes Freedom of Speech?" *Indian Law Review*, Vol. 5, No. 1, pp. 62–84, 2021.

Ramlal v. Mustafabad Oil Mills. AIR 178 (1968).

Schmitt, M. N. *Allinn Manual 2.0 on the International Law Applicable to Cyber Operations*. Cambridge University Press, 2017.

Scott, BI. *The Law of Unmanned Aircraft Systems* (2nd edn). Kluwer Law International, 2022.

Sehrawat, V. *Drones and the Law: International Responses to Rapid Drone Advancement*. Routledge, 2021.

Shreya Singhal v. Union of India. AIR 2015 SC 1523 (n.d.).

Springer, P. J. *Military Robots and Drones: A Reference Handbook*. ABC-CLIO, 2013.

Taylor v. Federal Aviation Administration. 351 F. Supp. 3d 97 (D.D.C. 2018) (n.d.).

Villasenor, J. "Observations from Above: Unmanned Aircraft Systems and Privacy." *Harvard Journal of Law & Public Policy*, Vol. 36, No. 2, pp. 457–517, 2015.

Zirkel, F. J. "Unmanned Aerial Vehicles and Privacy: A Legal Perspective." *Harvard Journal of Law & Technology*, Vol. 29, No. 1, pp. 1–42, 2016.

Part 2

Unmanned aerial vehicles simulation and modeling

Chapter 4

Simulators for UAVs

Naresh Tripathi, Kalpana Johari, Neelima Gupta, and Sandhya Aneja

4.1 INTRODUCTION

Simulation plays a very crucial role in the field of UAV research and its progress. This importance arises from the ability to test UAVs in simulated real-world environments before actual deployment. Simulations allow for repeatable tests across different physical and environmental conditions. Due to their flexibility and repeatability, simulations are also cost effective.

These simulators replicate various subsystems of UAVs such as control systems, aerodynamics, and sensors, as well as the surrounding environment. They provide valuable tools for:

Algorithm development
Pilot training
Mission rehearsal
Hardware and software integration testing
Artificial Intelligence and Machine Learning testing

4.2 TYPES OF UAV SIMULATORS

There are different types of simulators used in UAV development and testing. Figure 4.1 illustrates the principal simulator types: SITL, HITL, and visual/physics-based simulators.

4.2.1 Software-in-the-loop (SITL) simulators [1–3]

Software-in-the-Loop (SITL) is a simulation technique in which the UAV firmware runs on the host computer as it would on actual flight hardware. However, instead of receiving real sensor and environment data, it receives fully simulated inputs.

Unlike traditional setups where firmware is uploaded to a physical autopilot board, the firmware is compiled and executed entirely on the host machine in SITL. This allows for flexible and safe development and testing

DOI: 10.1201/9781003598350-6

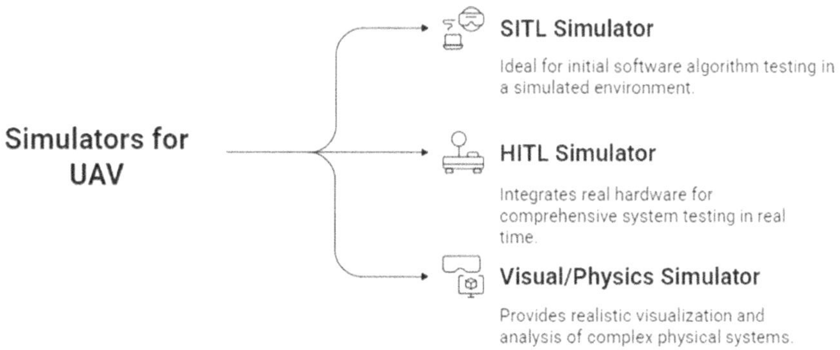

Figure 4.1 Types of UAV simulators.

without the need for real hardware. SITL is preferred for its safe testing environment as it does not involve any risk of crashing the real drone, developers can make changes to the flight control code without flashing the hardware, it also saves money as we do not need the real hardware for testing the software created for the hardware as we can just run it over the software and check whether it works or not, and we can repeat the simulation as many times as we want for comparing the algorithm for the same environment conditions or for debugging.

Even after being so effective, SITL still comes with its limitations.

It is significantly challenging to reproduce hardware problems, such as electrical noise, communication delay, or actuator failure, accurately in a simulation environment. To account for any of these factors, the simulator must be linked to a high-fidelity physics engine; otherwise, reliable modeling of realistic dynamics becomes difficult. Moreover, SITL simulators often have difficulty simulating sophisticated hardware components, such as LiDARs or onboard cameras. These components require advanced 3D simulators that are capable of generating reliable sensor data and visual environments to test perception and navigation algorithms adequately.

Some commonly used SITL environments consist of:

JMAVSim: Lightweight simulations with PX4 that can be used to quickly test UAV flight dynamics and control algorithms only.

Gazebo: A more advanced 3D robotics simulation that allows sensor emulation, physics modeling, and realistic complex environmental interactions, which is suitable for UAV applications that rely on specific full-range vision and LiDAR capabilities.

4.2.2 Hardware-in-the-loop (HITL) simulators [4–7]

Hardware-in-the-Loop (HITL) is a powerful method in UAV development, where actual flight hardware is used in conjunction with a simulated environment to evaluate system performance.

Unlike SITL, which operates entirely on the host system, HITL involves a real autopilot or remote controller connected to the simulator. The simulator mimics the UAV's surroundings and provides sensor feedback, allowing developers to test hardware responses in a virtual, controlled, and repeatable environment. HITL effectively bridges the gap between simulation and real-world flight testing.

It is ideal for checking how the flight controller responds to flight conditions, input from the sensors, and control commands in real time. This helps in catching any bug related to hardware, which is not possible using SITL.

In HITL, the simulator would run a virtual drone which in turn produces synthetic data from the virtual drone, which is fed to the autopilot hardware of a remote control. The hardware would react in the same way as it would have reacted when it would have received the data in real time. When the hardware receives the data, it would react by running control loops, updating state, and sending motor commands. This setup also includes a ground control system (GCS), which monitors telemetry, sends commands, and logs data like in a real flight. The benefit of using HITL is that we get to know the response of the hardware including the timing, delays, quirks related to the hardware, sensor fidelity, and integrity of interface before flying. It is also safe for testing as it lets us evaluate critical maneuvers, failsafe, and custom logic without worrying about the crash. It also helps us in checking the performance bottlenecks, which we are not able to check in SITL.

Even after being so effective, HITL still comes with its limitations.

It has a very complex setup which requires correct firmware configuration and reliable simulator integration. The synchronization between the simulator and the hardware must be perfect; otherwise, the quality of simulation will suffer. As this also includes hardware, it becomes resource intensive as it needs hardware power supply, the cooling setup for hardware, and space to store it. It is hard to run multi-HITL setup simultaneously compared to an SITL.

Some of the environments of HITL are:

Gazebo (when configured for HITL testing)
X-Plane
RealFlight
FlightGear
Visual or physics-based simulators [8–11]

These kind of simulators are designed not only to mimic the environment around a UAV but also the physical force applied on it. This helps in making the testing, development, or research over a UAV more effective and realistic. It combines two main aspects:

Physics Engines: These are used to mimic the real-world forces such as gravity, drag, thrust, and lift over a UAV in a simulated environment. This helps in validating the aerodynamics and control systems.

Visual Rendering: It gives the ability to create a realistic environment which could be used for training, flying, and testing of UAVs. This is the most effective in achieving the computer vision–based goals such as object detection, SLAM, and autonomous navigation.

This requires high-performance computing power with a dedicated graphics processing unit (GPU), including dependencies Robot Operating System (ROS) or PX4, Unreal Engine, and Python and C++ libraries. Some simulators may need a good-insight robotics middleware and command-line tools, and they all have simulator-specific configurations. The benefit of using the visual or physics-based simulators is that they give a realistic physics modelling and visually accurate environments, making them suitable for both flight testing and vision-based features of UAVs. It also provides a risk-free environment where we can test UAVs for crashes and bug without worrying about the real hardware, and hence, it is also cost effective. This is also ideally very excellent for developing camera vision–based projects like object tracking, SLAM, and obstacle avoidance using cameras or LiDAR. This is possible as they allow extensive customization to the environment and drone models, sensors, whether, and terrain. They can also mimic a wide range of sensors to test data fusion and navigation algorithms.

Even after being so effective, the visual or physics-based simulators still come with limitations.

Simulation may not be perfectly matched to the real-world hardware limitations. Requirements of the high-end system are very cost extensive as even a little lack in the hardware presence may cause lag in the simulation and we may not be able to achieve desired output. We cannot prepare the simulation for the uncertainties which we may face in the real world like the electromagnetic interference, GPS jamming, wind turbulence, or sensor calibration errors. They are only good for the initial testing of the system, and hence, it cannot replace the final field testing which includes both software and hardware. Setting up the simulators for customized and detailed environments is a time-consuming task and may require expertise in simulation framework.

Some of simulators are as follows:

Microsoft AirSim is an open-source simulator that provides many options to researchers for studying computer vision, deep learning, and reinforcement learning while developing autonomous systems.

Gazebo is one of the most popular robotics simulators on the market today, delivering highly accurate 3D environments and sensor emulation for testing UAV navigation.

X-Plane is a professional-grade flight simulator that utilizes high-end graphics to deliver excellent add-on opportunities for visualizing UAV missions.

RealFlight is a commercial UAV/remote control simulator designed specifically for training pilots, offering realistic simulations of flight physics and issues that occur in a realistic environment.

Velocidrone is a first person view(FPV)-focused racing drone simulator that is highly regarded for providing realistic simulations of the look and feel of flying a drone and providing extremely fast-paced training scenarios.

DroneSim Pro is another simulator that is designed specifically for training purposes, offering highly detailed, photo-realistic environments to safely practice the operation of a UAV.

CopterSim is a multirotor-specific simulator that uses 3D imagery and allows users to practice their FPV skills in the safest manner possible.

DroneSim Pro merges visuals that present a realistic appearance with a strong emphasis on flight physics to train drone pilots in an experience that simulates real flight conditions.

CopterSim delivers the physics of actual flying simulations of quadcopters and other multirotor drones, making it a great tool for operator to train and study flight dynamics.

4.3 SIMULATORS PRESENT IN THE MARKETPLACE

While discussing about different kinds of simulators above, we saw that many environments overlap in the types of simulators. This proves that one environment can be considered in more than one category. For instance, Gazebo can be categorized as SITL, HITL, and visual-based simulators. Thus, at this point, we will refer back to all the environments of simulators that were mentioned earlier. Figure 4.2 summarizes the UAV simulator variants discussed above and their principal applications.

4.3.1 Gazebo+PX4

Gazebo is an open-source 2D/3D robotics simulator developed by Open Source Robotics Foundation. Its development started in 2002, and its development was split into two directions in 2017. The first one was "Gazebo", the original one, and the other is "Ignition", which had moved to become the collection of loosely coupled libraries. In 2022, Open Source Robotics Foundation got a trademark issue with the name "Ignition", and so the foundation took the opportunity to switch the names of the version with the original version as "Gazebo Classic", and the new modern version was simply named as "Gazebo". It is used in the field of robotics and UAV for researching as it provides high-accuracy simulated environment for testing algorithm, control system, and even multi-robot and multi-drone coordination, all without risking real hardware. "Gazebo Classic" was integrated with Open Dynamics Engine (ODE), OpenGL rendering, and support code

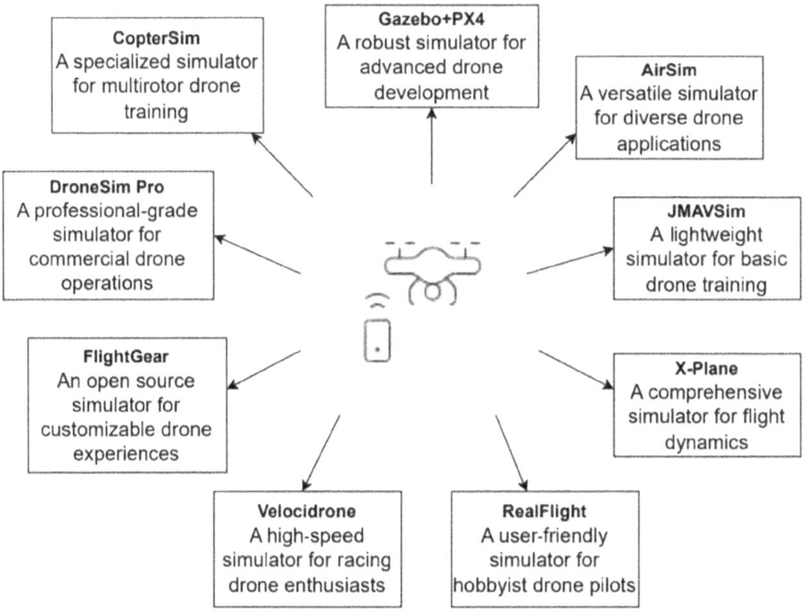

Figure 4.2 UAV simulator variants.

for sensor simulation and actuator control. It was discontinued in 2025 and replaced with the modern version "Gazebo". Gazebo can use multiple physics engine such as ODE physics engine and Bullet. This helps it simulate real-world physics such as lighting and environmental conditions, which makes it suitable for both indoor and outdoor simulations.

4.3.1.1 Key features

- Gazebo uses physics-based simulation. For this kind of simulation, it uses powerful physics engines such as ODE, Bullet, or Dynamic Animation and Robotics Toolkit (DART). This simulates the motion and interaction of drones with the environment.
- In Gazebo, drones can be equipped through simulation with realistic sensors such as GPS, inertial measurement unit, barometer, magnetometer, RGB and depth cameras, and LiDAR. These sensors also mimic the noise and drift found in them in the real world.
- Gezebo's integration with flight stacks such as PX4 and ArduPilot is very smooth, which allows UAVs to run in simulation. This enables SITL testing, where the real firmware is tested in the fully simulated environment.
- Users can use custom C++ or Python plugins for defining the behavior of a UAV, custom sensors, controllers, or environmental interactions.

This also includes extensively customized 3D environments—indoor, outdoor, urban, or rural. We can also import these environments from open-source libraries.

4.3.1.2 Workflow of Gazebo for UAV simulation

1. Model the UAV using SDF (simulation description format) or URDF (unified robot description format).
2. Load the environment by choosing or creating your own custom world.
3. Connect PX4 for flight control and use ROS for sensor data, mission planning, and real-time control.
4. Launch Gazebo and observe the UAV behavior.
5. Collect synthetic data from simulated sensors, flight logs, and ground truth for using in further analysis and machine learning model training.

4.3.1.3 Gazebo+PX4+ROS integration [9,12–15]

1. PX4 supplies the firmware capable of being executed in the SITL mode. PX4 serves as the autopilot firmware, controlling the behavior of UAV during flight. In a simulation, PX4 runs in SITL mode. In SITL, it behaves as if it was running real hardware, but it interacts with the simulated environment. This means a developer is able to execute their flight control logic through simulation without having to operate a real drone.
2. ROS is used for communication between simulated sensors and drone algorithms. The ROS runs as the middleware between the UAV's simulated sensors (e.g., GPS, IMU, LiDAR, or camera sensors) and higher-level algorithms. It connects components such as mission planning, perception, and navigation by mobilizing data between components in real time.
3. Architecture: The setup consists of the following components working together:
   - Gazebo runs the UAV in a realistic 3D simulation environment that applies physics and sensor models.
   - PX4 runs at a lower level and is responsible for low-level flight control tasks around stabilization, motor control, and executing flight commands.
   - ROS nodes run high-level tasks that include communication among sensors, processing of data, making decisions, and executing missions.

The architecture creates a complete simulation loop in the sense that Gazebo provides the environment for the UAV, PX4 keeps the UAV stable during flight, and ROS is responsible for coordinating intelligent decision making.

4.3.2 AirSim [8,16–19]

AirSim (Aerial Informatics and Robotics Simulation) is an open-source simulator created by Microsoft Research. It is a cross-platform simulator including drones and ground vehicles such as cars. It is built on Epic Games privately owned by Unreal Engine 4 with the aim to be used for AI research. Its main focus is on creating realistic environment for AI, computer vision, and deep reinforcement learning applications for autonomous vehicles. AirSim provides about 12 km of road and about 20 city blocks with APIs to get the data and control UAVs in a platform-independent way. These APIs can be accessed using different programming languages such as C#, C++, Python, and Java. This environment comes under the category of HITL, and it also supports middleware such as PX4 and ROS.

4.3.2.1 Key features

- Using Unreal Engine's high graphic capabilities, AirSim can create a highly detailed environment, which includes shadows, lighting, and texture details, accurate enough to train neural networks for object detection, segmentation, and navigation.
- Since it is primarily built for research in the field of AI, it supports autonomous navigation, object tracking, and reinforcement learning.
- It also simulates various sensors of UAVs such as RGB camera, depth camera, segmentation and infrared cameras, IMU, barometer, GPS, magnetometer, and LiDAR.
- We can control drones, collect data, inject noise, or set mission using simple scripts and also access the APIs provided using C++, C#, and Python. REST- and RPC-based communication allows integration with Machine Learning (ML) pipelines and robotics framework.
- This also allows for multi-drone simulation in the shared environment.

4.3.2.2 Workflow of AirSim for UAV simulation

1. Setup an environment in the Unreal Engine importing custom-designed or prebuilt environments form the libraries.
2. It supports multirotor and fixed-wing UAVs, and the parameters of the drone can be configured using the settings.json.
3. We can use the keyboard or remote control for flight control for the purpose of pilot training, and we can also use APIs where we can use programming languages such as C#, C++, and Python for writing the script for autonomous behavior of the drone such as:
    - Takeoff and landing maneuvers.
    - Setting the GPS location.
    - Velocity and altitude.
    - Automatically capturing images or collecting LiDAR data.

4.3.3 JMAVSim—lightweight UAV simulation for PX4 [20–22]

JMAVSim is a lightweight tool developed in Java with a focus on working with the PX4 flight stack. It is included in PX4 and is the default SITL simulator, so it is easy to set up and can be run quickly for testing turnaround. JMAVSim is focused primarily on multirotors or quadrotors, allowing developers to quickly and easily simulate a basic flight control functionality. Developers can simulate basic behaviors, like taking off, hovering, flying maneuvers, and landing, and can learn how a drone might respond to failure or control conditions. JMAVSim does not come close to the advanced graphics or sensor simulation of a 3D simulator like Gazebo; however, it does not require advanced development or computing resources, and so, its simplicity and integration into PX4 are excellent for developing, debugging, and validating control logic quickly.

4.3.3.1 Key features

- This comes tightly bounded with PX4 and is launched automatically when we are running the PX4 in the SITL mode. This lets users test PX4 in the virtual drone without needing an original hardware.
- It is not designed for graphically intense and very deep realistic environment simulation, but enough for visualization for the 3D movement of the drone in the simulation.
- This is very lightweight, and so, it can be operated on a system with a very modest hardware.
- It is helpful in mimicking key sensors such as IMU, GPS, magnetometer, and barometer, and after collecting data from these sensors, it sends it back to PX4 as if it was being driven from an analog hardware.

4.3.3.2 Workflow of JMAVSim for UAV simulation

1. We start by initializing PX4 in the SITL mode on the host computer, and as the JMAVSim comes as the default bundle with PX4, it is launched internally by PX4 as it is the default simulation backend.
2. JMAVSim comes with a basic multirotor and quadrotor physics model, which can be configured using PX4 parameter settings. No 3D editor is required as it is code based.
3. It runs the simulation loop that simulates basic physics such as gravity, thrust, drag, motor response, and simple environmental forces. PX4 receives data about them from sensors as if it received it from real sensors.
4. Flight stack calculates actuator outputs and sends them back to JMAVSim, and then the simulated drone responds according to the input received from PX4. We can monitor the drone via the following:
   - QGround Control (Ground Control System)
   - Command-Line Interface
   - MAVSDK/MAVROS for scripting or external control

5. A simple 3D environment shows the drone flying, which is not a very detailed environment, but this simple 3D simulation satisfies our needs of debugging and verifying flight control behaviour.
6. PX4 stores the sensor data which can be used later for understanding the system performance.

4.3.4 X-Plane [23–26]

X-Plane is a highly accurate flight simulator, which was launched by Laminar Research in 1995. It is so accurate that it is also approved by aviation authorities like Federal Aviation Administration(FAA) for pilot training. It was primarily designed for fixed-wing aircraft, but it also supports UAV simulation through plugin systems and custom aircraft models. It has a commercial desktop version that is sold for macOS, Windows, and Linux. Since 2009, a mobile version is also available for Android, iOS, and WebOS.

4.3.4.1 Key features

- X-Plane simulates the dynamics depending upon the physics of each surface of aircraft rather than precomputed data.
- It comes with preset support for UAV and is given through plugins such as MAVLink, XPlaneConnect, or custom-developed software which integrate autopilot software like PX4 or Autopilot.
- There are real-time changes in weather, day-and-night cycle, and high-resolution environment, helping in suitable testing of aircraft in the most realistic conditions.
- It is also compatible with the hardware of aircraft, which makes it ideal for mimicking ground station or pilot in loop testing.

4.3.4.2 Workflow of the X-plane for UAV

1. Install the X-Plane on the host computer system, and choose a suitable aircraft for which you want to do the testing. For accurate UAV simulation, modify the specification of the aircraft model using the Plane maker, which is a built-in aircraft design tool.
2. Decide over the autopilot software you want to use, ArduPilot or PX4, and configure X-Plane to communicate with the autopilot firmware through the user datagram protocol (UDP) network interface.
3. X-Plane sends the flight state data through the MAVLink or custom UDP channels to the firmware, which in response processes these data and communicates with motor commands, and these commands are received by the X-Plane and applied over the aircraft in real time.
4. Connect the Mission Planner (for ArduPilot) and QGroundControl (for PX4) if needed.

5. Run autonomous mission and test manual flight using physical remote controller or game controller.
6. X-Plane and Autopilot both log the records of flight data, which can be analyzed later for tuning the parameters, checking mission success, or evaluating control performance.

4.3.5 RealFlight [27,28]

RealFlight is a commercial-grade flight simulator that focuses mainly on the radio-controlled aircraft. It also includes multirotor and fixed-wing drones. It was developed by Knife Edge Software started in 1996. This was developed by aiming at the hobbyists, flight instructors, and commercial drone pilots preparing for real-world flight.

4.3.5.1 Key features

- This comes with a ready-to-fly model that behaves as if it is flying in the real world.
- This also comes with various training scenarios such as obstacle course, windy conditions, and camera-based navigation.
- It also provides crashes and damages which are physics based, and this helps to improve flight discipline.
- It also supports the tracking and FPV mode.

4.3.5.2 Workflow of the RealFlight for UAV

1. Install the RealFlight version which supports simulation through MAVLink.
2. Select or customize an aircraft in RealFlight to mimic your UAV.
3. Bridge the RealFlight with a firmware like ArduPilot using plugins like Flight Axis Interface or Serial Bridge. RealFlight relay sensor data to a firmware that computes actuator outputs and sends the control signal back to the aircraft.
4. Run test flights, plan a mission, and tune parameters depending upon the data collected.
5. Analyze the flight behavior using the logs collected by ArduPilot for debugging and improvements.

4.3.6 Velocidrone [29,30]

It is a simulator built for First-Person View drone racing. It is highly used by professional racers of drone community to sharpen the reflexes and improve control over drones while flying in high-speed scenarios. It was developed by Ashley Davis and the team at Bat Cave Games in 2016. It is known for its high-accuracy physics.

4.3.6.1 Key features

- It has a very accurate physics and is very responsive like a real racing drone.
- We can design custom race course in it to mimic the real-world environment.
- We can use it in multi-user scenarios by connecting it to other racers online and competing with them.
- It gives us freedom to fully customize our drone according to our needs.

4.3.6.2 Workflow of the Velocidrone for UAV

1. Install the simulator over the host computer, and choose from the prebuilt model or configure according to your own.
2. Use a USB compatible remote control transmitter or gamepad for controlling the drone, and calibrate controls using the simulator.
3. Choose from various race tracks or freestyle environment, or customize one for your needs and also pick a flight mode like Acro, Angle, or Horizon.
4. Practice manual control, racing, or freestyle flying.
5. Review your performance and repeat to improve.

4.3.7 FlightGear [31–33]

FlightGear is a robust open-source simulator used for both aerospace research and industry. While its focus is mainly on the commercial and general aviation, it also supports UAV simulation through custom aircraft model and scripting. David Murr started this project on the 8th of April 1996, and its first release was in 1997, with continued development.

4.3.7.1 Key features

- The user can make their own drone by defining configurations in XML or JSBSim (flight dynamics engine).
- It supports Telnet and socket-based controls which make the UAV autopilot simulation externally possible.
- It comes with real-world topography, terrain, and navigation data, and it is customizable via scripting.
- It includes basic features that can be included using custom scripting.

4.3.7.2 Workflow of the FlightGear for UAV

1. Install the simulator over the computer.
2. Select the preexisting aircraft, and model it according to your UAV requirements.
3. Configure FGFS (Flight Gear Property Server) to receive input via UDP or transmission control protocol (TCP), because we need to connect it with a firmware like ArduPilot or PX4 for real-time data exchange.

4. Collect flight data using FlightGear, and send them to the firmware which will compute actuator outputs and send them back to the flight gear to apply them over the UAV.
5. Perform the test and evaluate the responses, and log the data for later analysis and tuning in the parameters.

4.3.8 DroneSim Pro [34]

It is a paid simulator used for real-world pilot training, epically in areas like photography, inspections, and search and rescue. It mainly focuses on training as if the drone is in the real environment.

4.3.8.1 Key features

- It comes with different training scenarios that include advanced mission like photography, target search, and survey of paths.
- It does the accurate modeling of real drone flight physics.
- It does not need high-end system specifications for operating over a computer system.
- It works with a USB flight controller.

4.3.8.2 Workflow of the DroneSim Pro for UAV

1. Install the simulator over a PC.
2. Select the training environment from the existing ones, and also choose the existing drone type as it comes with limited preset choices.
3. Use a supported remote controller or a USB controller for realistic control.
4. Perform flight training by doing tasks such as lying through obstacles, hovering, taking photos, or simulating rescue operations.
5. Use the feedback system of the simulator to improve control and precision.

4.3.9 CopterSim [35,36]

This is a basic quadcopter simulator made available via the web and intended primarily for educational and demonstration purposes. The simulator has a straightforward and accessible interface that enables users to examine the basic principles of the dynamics of flight for drones and practice basic flight controls. Due to its simplicity, it is especially beneficial to beginners, students, and hobbyists who want to learn how quadcopters respond to commands such as throttle, pitch, roll, and yaw. It does not simulate the full complexity of sophisticated UAV simulators, but its ease-of-use is an effective tool for demonstrating the core principles of drone operation without the need for proprietary development tools or special hardware or software installation.

4.3.9.1 Key features

- The biggest benefit is that we do not need to install visual interface locally as it can run over any modern web browsers. However, we need to install the core simulation engine for the portion which handles physics, control systems, and sensor modelling.
- It offers a very simple interface to mimic and simulate pitch, roll, yaw, and throttle.
- It is a very good simulator for teaching flight control of drones to students who are new to drone use.

4.3.9.2 Workflow of the CopterSim for UAV

1. Install the simulators with dependencies which are often used with MATLAB/Simulink or ROS.
2. Define a UAV model with needful parameters.
3. Integrate simulator with flight control algorithms often developed in Simulink or ROS. We use this setup to simulate sensor data and UAV dynamics.
4. Run the simulation for testing control algorithms, navigation logic, or estimation under variable conditions.
5. Analyze UAV behaviors in the 3D space, and tune the controller and UAV model accordingly.

We have put the comparison of all the simulators depending on the following factors (Table 4.1):

- Platform
- Type

Table 4.1 Comparison of UAV simulators

Simulator	*Platform*	*Type*	*Key use*	*Open source*	*Supports PX4/ArduPilot*
Gazebo+PX4	Linux	3D Physics	R&D	Yes	Yes
AirSim	Win/Linux	Visual/AI	Vision/ML	Yes	No (custom integration)
JMAVSim	Cross-platform	Lightweight	SITL	Yes	Yes
X-Plane	Win/Linux/Mac	High-fidelity	Testing	No	Yes (via plugins)
RealFlight	Windows	Training	Practice	No	No
Velocidrone	Windows	Racing	FPV	No	No
FlightGear	Cross-platform	Simulation	Research	Yes	Yes (custom)
DroneSim Pro	Windows/Mac	Training	Basic Flight	No	No
CopterSim	Browser	Basic	Learning	Yes	No

- Key Use
- Open Source
- Supports PX4 or ArduPilot [24,27,31,34]

4.4 CONCLUSION

UAV simulators are important tools for today's drone development and operations. They provide a safe, inexpensive, and flexible method to test, validate, and train in a scenario without putting real hardware at risk. Whether your choice is AirSim for advanced AI-based flight testing, or Gazebo and X-Plane for near-realistic flight dynamics and behavior, there are a multitude of simulators designed specific to use cases. The simulator that is chosen depends heavily on the mission. Researchers may want to focus on testing algorithms and integration with sensors, developers may need control system validation or mission planning, students will take advantage of training simulators and so they can practice flying in a safe environment, and drone racers can take high-speed FPV training and race practice. By selecting the appropriate simulator, users can accelerate innovation, improve safety, and gain confidence prior to flying a UAV into the real world.

REFERENCES

1. ArduPilot Developers: SITL Simulator (Software in the Loop)—Dev documentation. https://ardupilot.org/dev/docs/sitl-simulator-software-in-the-loop.html. Accessed: 2025-08-10 (2024).
2. Dojo for Drones: SITL Drones—A Drone Development MUST. https://dojofordrones.com/sitl-drone/?utm source=chatgpt.com. Accessed: 2025-08-10 (2022).
3. ArduPilot Developers: Simulation—Dev documentation - ArduPilot. https://ardupilot.org/dev/docs/simulation-2.html?utm source=chatgpt.com. Accessed: 2025-08-10 (2024).
4. Wikipedia Contributors: Hardware-in-the-loop simulation. https://en.wikipedia.org/wiki/Hardware-in-the-loop simulation. Accessed: 2025-08-10 (2025).
5. PX4 Developers: Hardware in the Loop Simulation (HITL)—PX4 User Guide. https://docs.px4.io/main/en/simulation/hitl.html. Accessed: 2025-08-11 (2025).
6. Jain, S., Kamali, C., National Aerospace Laboratories, I.: Hardware-In-The Loop Simulation for Unmanned Aerial Vehicles. MATLAB EXPO 2016, National Aerospace Laboratories, CSIR-NAL. Accessed: 2025-08-11 (2016).
7. UAV Navigation-Grupo Oes´ıa, Unmanned Systems Technology: Hardware-in-the-Loop Simulation Test for Drones Unmanned Systems. https://www.unmannedsystemstechnology.com/expo/hardware-in-the-loop-hil/. Accessed: 2025-08-11 (2025).
8. Shah, S., Dey, D., Lovett, C., Kapoor, A.: Airsim: High-fidelity visual and physical simulation for autonomous vehicles. arXiv preprint arXiv:1705.05065 (2017).

9. Wikipedia Contributors: Gazebo (simulator). https://en.wikipedia.org/wiki/Gazebo_(simulator). Accessed: 2025-08-11 (2025).
10. Unknown, A.: A review of nine physics engines for reinforcement learning including gazebo. arXiv preprint 2407.08590 (2024).
11. X-Plane Developers: The world's most advanced flight simulator (X-Plane). https://www.xplane.com/. Accessed: 2025-08-11 (2025).
12. Number Analytics: Gazebo AI Robotics – Ultimate Guide. https://www.numberanalytics.com/blog/ultimate-guide-gazebo-ai-robotics.Accessed: 2025-08-11 (2025).
13. PX4 Developers: ROS with Gazebo Simulation—PX4 User Guide. https://docs.px4.io/v1.12/en/simulation/ros–interface.html. Accessed: 2025-08-11 (2025).
14. Dimmig, C.A., et al.: Survey of simulators for aerial robots: An overview and in-depth systematic comparisons. *IEEE Robotics Automation Magazine* (2024). https://doi.org/10.1109/MRA.2024.3433171
15. Intrinsic AI/Open Robotics: Advancing real-world robotics through simulation with Gazebo Ionic. https://www.intrinsic.ai/blog/posts/advancing-real-world-robotics-through-simulation-with-gazebo-ionic. Accessed: 2025-08-11 (2024).
16. Wikipedia Contributors: AirSim. https://en.wikipedia.org/wiki/AirSim. Accessed: 2025-08-11 (2025).
17. Wired UK: Microsoft's open-source sim will let you crash self-driving cars. https://www.wired.com/story/microsoft-airsim-crash-drone-car. Accessed: 2025-08-11 (2017).
18. AirSim Project Documentation: AirSim Documentation—Multiple Vehicles. https://buildmedia.readthedocs.org/media/pdf/airsim-fork/docs/airsimfork.pdf. Accessed: 2025-08-11 (2018).
19. PX4 Developers: AirSim Simulation—PX4 Guide. https://docs.px4.io/main/en/sim_airsim/index. Accessed: 2025-08-11 (2025).
20. PX4 Developers: jMAVSim with SITL. https://docs.px4.io/v1.12/en/simulation/jmavsim. Accessed: 2025-08-11 (2025).
21. PX4 Developers: jMAVSim Simulation (DevGuide). https ://bkueng.gitbooks.io/px4-devguide/en/simulation/jmavsim.html. Accessed: 2025-08-11 (2025).
22. PX4 Developers: Simulation Overview (PX4 Guide). https://docs.px4.io/main/en/simulation/. Accessed: 2025-08-11 (2025).
23. Horri, N., Pietraszko, M.: A tutorial and review on flight control co-simulation using Matlab/Simulink and flight simulators. Automation, 3, 486–510 (2022). Note: MDPI stays neutral with regard to jurisdictional claims in published (2022).
24. PX4 X-Plane Plugin Team: PX4 X-Plane Plugin—GitHub. https://github.com/alireza787b/px4xplane. Accessed: 2025-08-11 (2024).
25. Al´aez, D., Olaz, X., Prieto, M., Porcellinis, P., Villadangos, J.: Hil flight simulator for vtol-uav pilot training using x-plane. *Information*, 13(12), 585 (2022).
26. ArduPilot Developers: Using SITL with X-Plane 10/11. https://ardupilot.org/dev/docs/sitl-with-xplane.html. Accessed: 2025-08-11 (2025).
27. RealFlight Team: RealFlight RC Flight Simulator—1 RC Simulator. https://www.realflight.com/. Accessed: 2025-08-11 (2025).

28. ArduPilot Developers: Using SITL with RealFlight (FlightAxis interface). https://ardupilot.org/dev/docs/sitl-with-realflight.html. Accessed: 2025-08-11 (2025).
29. VelociDrone Developers: VelociDrone FPV Simulator—Features and Track Editor. https://www.velocidrone.com/features. Accessed: 2025-08-11 (2025).
30. Reddit user (r/fpv): Velocidrone is the official sim for the MultiGP racing league. Reddit discussion thread. Accessed: 2025-08-11 (2024).
31. Wikipedia Contributors: FlightGear (Flight Simulator). https://en.wikipedia.org/wiki/FlightGear. Accessed: 2025-08-11 (2025).
32. FlightGear Project Documentation: FlightGear JSBSim Flight Dynamics Autopilot Support. https://wiki.flightgear.org/Autopilot. Accessed: 2025-08-11 (2025).
33. PX4 Developers: FlightGear Simulation Setup in PX4 SITL. https://docs.px4.io/v1.12/en/simulation/flightgear. Accessed: 2025-08-11 (2025).
34. Reagan, J.: Cobb to launch 'real-world' drone simulator. DroneLife. Accessed: 2025-08-11 (2015).
35. RflySim Documentation: Introduction to Core Components: CopterSim (HIL Simulation Software). https://rflysim.com/doc/en/1/CoreAppIntro.html. Accessed: 2025-08-11 (2025).
36. RflySim Documentation: HIL Simulation Platform: CopterSim and 3DDisplay. https://rflysim.com/en/3_Using/HILSimulator.html. Accessed: 2025-08-11 (2025).

Chapter 5

Flight controller selection

An investigation for the development of drones for specific applications

N.C. Ajay Vishwath, Arvind R. Yadav, Roshan Kumar, and Udaykumar V. Joshi

5.1 INTRODUCTION

The flight controller is a key embedded electronic component that serves as the central unit of a drone. Its primary function is to handle essential operations such as autonomous control and navigation. Different types of flight controllers exist, each with its own set of features and characteristics [1].

The need of appropriate flight controller (FC) selection is due to the demand of cutting-edge drones which can fly in vilest environments anywhere and anytime [2]. Substantial innovations have been made in today's drones for applications such as surveillance, firefighting, medical, logistics, agriculture, and defence applications. However, the challenge lies ahead is controlling these gadgets for getting real-time responses and managing complex systems [3].

One of the reasons the challenges remain unresolved is the unavailability of standards for FC architectures and on top of that exclusive controller algorithm's that govern unmanned aerial vehicles (UAVs). Thus, codes written for an FC may not work well for another FC, and added testing and development time is needed for the task [4]. The PID (proportional, integral, and derivative) controllers are mainly used in outmoded FCs for stability but with the help of human interference [5].

Timing guarantees are essential for embedded and cyber-physical systems, as they need to ensure predictable delays between sensing, processing, and actuation. For instance, in a multirotor drone's FC, data from a gyro or inertial sensor are collected and processed to calculate the aircraft's attitude. These sensor data are then fused and used to generate control signals that adjust the rotor speeds to control the drone's flight. If the time between sensor input and actuation is not constrained, the drone may lose control and could fail to stay in flight [6].

Global positioning system (GPS) data are the main source of information for tracking drones during flight. These data are processed and filtered by the FC software, after being combined with data from other sensors, to estimate the drone's position, velocity, and orientation [7]. FCs available on the market perform the control of toys and even serve as the autopilot

DOI: 10.1201/9781003598350-7

of large aeroplanes and helicopters. However, selecting a FC is not a trivial task because the information between manufacturers is not standardised [8]. FCs are crucial for maintaining stability, performance, and control throughout a drone's flight. It is of utmost importance to prudently choose an appropriate FC for an application under consideration, but the negligence in doing so may result in putting the payload and drone in severe jeopardy [9].

FCs are the brain of UAVs, and they are responsible for the overall flight operation. The predefined route to be followed by a drone is programmed into GPS for autonomous flying. Research is ongoing to provide optimal solutions for "Automatic Trajectory Tracking", conflict detection and resolution, and navigation and swarm formation with the help of FCs [10].

5.1.1 Role of FCs in UAV systems

The user input and sensors input are processed by FCs for real-time flight manoeuvres to adjust and control the motor of UAVs. It ensures:

- Flight stability and control
- Navigation and positioning
- Communication with the pilot or ground control
- Execution of autonomous missions

5.1.2 Core components of an FC

S. No.	*Core component*	*Function*
1	Microcontroller Unit (MCU)	• Acts as the central processing unit. • Runs the flight control firmware (e.g., PX4, ArduPilot, Betaflight).
2	Inertial Measurement Unit (IMU)	• Combines gyroscopes, accelerometers, and sometimes magnetometers. • Measures orientation, acceleration, and angular velocity.
3	GPS Module (sometimes external)	• Provides location data for navigation and geofencing.
4	Barometer	• Measures altitude using atmospheric pressure.
5	Compass (Magnetometer)	• Measures heading relative to the Earth's magnetic field.
6	Electronic Speed Controller (ESC) and Motor Outputs	• Sends Pulse Width Modulation (PWM) signals to ESCs to control motor speeds.
7	Input/Output Ports	• For connecting peripherals like Radio Control (RC) receivers, telemetry modules, cameras, or LiDAR.
8	Power Management Circuitry	• Ensures that the FC receives stable voltage from the UAV's power supply.

5.1.3 Core functions of FCs

The data generated by the IMU sensor are used for stabilising the UAV and maintaining its altitude. Furthermore, the data collected from diverse sensors of the UAV are integrated/fused to estimate the velocity, position, and orientation of UAVs. The FC is responsible for navigation and path control with the help of GPS data. In case of sensor failure or signal loss, the FC allows UAV to behave as a reliable device by means of a return-to-home (RTH) function. The commands given by the pilot or generated from earth station are processed by the FC for appropriate action, and it also has data recording, analysis, and real-time data transmission facility. The FC also manages gimbals, cameras, and other on-board equipment.

5.2 CLASSIFICATION OF DIFFERENT FCS

There is a wide variety of FCs available, ranging from simple models to more advanced, expensive systems. To make it easier to understand, they can be grouped into four categories based on their target users:

- FCs for Hobbyists/DIY Builders: These controllers are easy to install and ideal for individuals who want an affordable option without a complex setup.
- Racing FCs: Designed for lightweight, precise, and highly responsive performance.
- FCs in Filming Category: The drones used for these filming purposes are in general camera drones, which are specifically designed for stability to capture high-quality video footage and to ease controls to the pilot.
- Commercial Drone FCs: This drone category represents the technological developments for advanced drone applications, generally used in transporting valuable cargo and also ensuring safe operation even in complex environments.

5.3 FCS FOR HOBBYISTS/DIY BUILDERS

These FCs come in affordable options that range from beginners-friendly to high-end tech models with a wide range of features. Depending on the budget, cheaper options may lack features such as built-in on-screen display (OSD) and integrated GPS. FCs with instinctive software interfaces, which are easier to tune and work with, are suitable for beginners.

5.3.1 K.K.2.1.5

The K.K.2.1.5 FC is the most popular option for multirotor drones, especially in the hobbyist community. It is a part of the K.K. series, designed and built by K.K. Multicopter and known for being a more straightforward,

Figure 5.1 Pins and ports of K.K.2.1.5 FC.

user-friendly controller feature for beginners and also intermediate builders. The K.K.2.1.5 model FC specifically offers various features adapted to multirotor configurations. All the pins and ports used in K.K.2.1.5 FC are shown in Figure 5.1.

5.3.1.1 Challenges

The K.K.2.1.5 FC has limited compatibility with necessary modern drone features. The K.K.2.1.5 FC does not support Betaflight, iNav, or GPS integration, which are some of the advanced features found in newer FCs. It lacks support for modern flight stabilisation features like DShot (digital ESC protocol), Smart Audio for Video Transmitter (VTX) control, and other advanced flight modes that are standard in newer FCs. The FC comes with relatively basic firmware. While it is easy to set up for basic flying, it lacks the depth of customisation and tuning options available in more modern controllers. The lack of advanced settings may limit your ability to fine-tune the drone for optimal performance, especially if you're looking to do freestyle or racing. The K.K.2.1.5 lacks an integrated OSD, meaning that you won't be able to see essential flight data (battery voltage, Received Signal Strength Indication (RSSI), flight mode, etc.) on your First Person View (FPV) feed. Thus, an OSD unit may be integrated with it if the need arises. The K.K.2.1.5 FC requires careful soldering and wiring the components correctly, which is a challenge for novice users in this domain.

Furthermore, it has a lesser number of Universal Asynchronous Receiver/Transmitter (UARTs) and I/O ports, and thus, limited peripherals can be interfaced with this FC, which creates an issue when you have to interface several sensors/modules for specific applications.

5.3.1.2 Calibration

The calibration of an accelerometer is needed for the appropriate flight operation. The calibration step requires a UAV to be kept on flat-levelled surface to ensure that the FC has clear-cut idea about the drone's orientation. Furthermore, the ESC has to be calibrated for synchronised and smooth motor operation by providing appropriate throttle range to motors. The ESC is calibrated by means of throttle stick.

While the K.K.2.1.5 FC doesn't have the sophisticated tuning options of newer controllers, you can still adjust PID values to modify how the drone behaves during flight. However, the range of adjustments is much more limited compared to modern firmware like Betaflight. Some settings may need to be manually tuned via physical adjustments or via basic control inputs. The lack of advanced software means that many fine adjustments that could be made on a computer might have to be handled directly by the user on the drone. The calibration of K.K.2.1.5 is shown in Figure 5.2.

5.3.1.3 Pros

The K.K.2.1.5 is simple to set up, which is one of its major selling points, especially for beginners. If you're looking for a simple FC with the basic configuration needed, the K.K.2.1.5 can be the best option. The K.K.2.1.5 is the most budget-friendly option among the other feature-rich FCs. For the hobbyists

Figure 5.2 Calibration of K.K.2.1.5 FC.

or drone enthusiasts on a budget, it provides a cost-effective solution without any of the complex calibration procedures that one faces with expensive FCs.

The K.K.2.1.5 is a compact and lightweight module, making it a good choice for simple drone builds, such as those designed for basic FPV flying, prototyping, or training. Its lightweight physical design helps in reducing the overall weight of the drone built, and this contributes directly to better flight times and less stress to motors. Though it comes under more basic FC, the K.K.2.1.5 offers stable flight performance for its class. It's a good choice for basic flying operations and training purposes, especially when you're starting drone flying just as a hobby. The K.K.2.1.5 can support basic flight modes, like Stabilise and Acro mode, which are suitable for basic flying and learning purposes. It's perfect for enthusiasts who want to practice and improve their flight skills without the need of advanced features like GPS or altitude hold.

5.3.1.4 Cons

As stated earlier, the K.K.2.1.5 could lack many of the advanced features which other newer FCs could support, most important advanced PID tuning, OSD integration, GPS support, and some multirotor flight modes (like altitude hold or RTH). This limits the potential of the drone in operation as you progress in the hobby. The K.K.2.1.5 does not natively support GPS or related functions like RTH or position hold, unlike the other modern FCs. This can be a major drawback if one wants to fly more advanced flights, particularly for long-range flights.

The K.K.2.1.5 also does not support advanced ESC protocols like DShot (digital ESC protocol), which is one of the popular features in the modern FCs for a more precise motor controlling. It uses only the older PWM or OneShot protocols, which will not offer the same performance or responsiveness as the DShot protocols, particularly in high-performance or racing flights. While you can adjust the PID settings in the K.K.2.1.5, it does not offer the same extensive tuning capabilities that are available in Betaflight- or iNav-enabled FCs. Advanced settings like filtering, motor layout options, or detailed flight behaviour adjustments are missing in the K.K.2.1.5.

The K.K.2.1.5 FC does not come with an on-board OSD feature, and the absence of this feature is a huge downside for FPV flying capabilities. Most of the modern FCs now include an OSD, which lets one overlay essential telemetry data (battery voltage, RSSI, flight mode) onto the FPV video feed.

5.3.2 Thunder QQ Super

Thunder QQ Super FC is one of the FCs from the Thunder series, which is dedicatedly designed for multirotor drone flights like quadcopters, hexacopters, and octocopters. This FC is intended to offer good performance for various types of builds, though being easy to set up, which makes it interesting for both beginners and intermediate drone enthusiasts.

5.3.2.1 Challenges

The Thunder QQ Super FC is rather basic in terms of features compared to other modern FCs. Moreover, it lacks advanced flight features such as GPS mode, altitude hold, position hold, and RTH, making it less suitable for long-range or autonomous flights. The inbuilt firmware (KK2.1.x) on the Thunder QQ Super is fairly simple, and it provides the basic PID tuning options but does not offer the other advanced tuning options available in modern FCs that run on Betaflight or iNav. This gives only a tuned performance to a limited degree.

The Thunder QQ Super does not support any telemetry systems, and this limits the ability to receive any in-flight data like battery voltage, GPS status, or even signal strength. Additionally, it does not support any modern DShot ESC protocols as well, which can lead to a less precise motor controlling, particularly during aggressive flights or high-speed manoeuvres. All the setup and configurations are done through an on-board LCD screen and buttons provided on the FC, which can be cumbersome and unintuitive as compared to other modern FCs that already have the user-friendly interfaces like Betaflight Configurator. This might be challenging for the beginners who are used to the PC-based setup methods.

The FC has limited input–output ports and does not easily support the integration of any additional peripherals, especially GPS, telemetry, or extra sensors. This could be a drawback for enthusiasts looking to create more complex builds with technically advanced operations.

5.3.2.2 Calibration

This FC requires calibration of the accelerometer for a proper orientation and a stable flight. This calibration is typically done by placing the drone on a flat level surface and using the on-board interface to perform the calibration effectively. Appropriate calibration is essential to ensure the flight stability, especially in modes like Stabilise and Acro. Unlike most FCs, Thunder QQ Super demands ESC calibration to ensure the operation of each ESC within the same range of throttle. Usually, Manual Calibration is performed by following a specific procedure involving throttle stick movements keeping the drone powered on.

The LCD screen and buttons can be utilised for configuration and calibration process which is a bit challenging for the users accustomed with Graphical User Interface (GUI) on a computer/mobile. The PID controller values determine the characteristics of the drone flight (determines the response to the yaw, roll, and pitch). Thus, compared to cutting-edge UAVs, limited calibration/configuration options are available with Thunder QQ Super.

5.3.2.3 Pros

It is a cost-effective, budget-friendly FC for building drones for novice users. It is comparatively easy to configure for elementary flying manoeuvres. This is appropriate for users who focus on flying the drone, rather than pondering upon complex configuration task. The Thunder QQ Super provides firm performance for flight modes such as Acro and Stabilise. The condensed size and lightweight of the FC allow it to be used in micro/mini-drone and is a great choice for drill and leisure flight for amateurs. The Thunder QQ Super provides support for educating multirotor flights and thus gives a lot of options to users to learn and build a drone for basic operations.

5.3.2.4 Cons

It doesn't have RTH, GPS, and altitude hold features, which limits the drone operation capability, and therefore, it is not appropriate for long-range/autonomous flight operations. The unavailability of integrated OSD makes it unsuitable for racing/long-distance flying operations as real-time flight data are critical for such applications. Basic PID tuning feature is possible here, but the granular tuning is not presented as accessible in advanced FCs. Fine-tuning of drone for optimal performance is difficult as sophisticated filters are not present in this FC. Last but not the least, the setup done through on-board LCD and keypad makes it a challenging task for beginners. Other advanced FCs typically offer easier configuration through software like Betaflight Configurator, which allows for quicker setup and tuning.

The Thunder QQ Super has fewer input and output ports for adding peripherals like that of GPS, telemetry, or extra sensors. This means one may not be able to expand the drone's capabilities beyond basic flying without the usage of additional hardware or other FCs. One of the major drawbacks is that the Thunder QQ Super does not support modern DShot ESC protocols, which are essential and widely used in advanced FCs for more precise motor control. It still relies on older ESC protocols like PWM or OneShot, which will not deliver the similar performance in high-speed or aggressive flights.

5.3.3 ArduPilot Mega (APM)

The APM (ArduPilot Mega) FC is one of the most popular open-source FCs in the market, particularly favoured for its versatility in supporting various aircraft and vehicle types, including fixed-wing aeroplanes, helicopters, multirotors, and even boats, cars, and rovers. It offers many technologically advanced features for autonomous flight and is widely used by both hobbyists and professionals in various fields. The APM FC is shown in Figure 5.3.

Figure 5.3 The APM (ArduPilot Mega) FC.

5.3.3.1 Challenges

The APM FC may be little difficult to configure for beginners, especially for the users unfamiliar with the intricacies of autopilot systems. FC setup requires connecting the controller to Mission Planner firmware or APM Planner on a PC, configuring parameters on it, and calibrating the required sensors. It may be a steep learning curve for amateurs compared to other simpler, plug-and-play systems. Though the APM is a capable controller, it is not as powerful as newer controllers like the Pixhawk or modern Betaflight systems. For more advanced autonomous flight operations, particularly with complex sensors or heavy unmanned vehicles, this limited processing power might lead to slower response times or limitations on the flight performance.

APM FC is somewhat older, and though it supports the basic autonomous features like the common waypoint navigation, advance RTH functions, and geofencing, it does not support more recent innovations like blackbox logging (for advanced flight data analysis), advanced ESC protocols (e.g., DShot), or modern telemetry and GPS feature out of the box without additional peripherals. The ArduPilot firmware employed in APM FCs are found to have compatibility issues with the latest software/hardware. Though ArduPilot firmware is an open source, it brings in lot of issues for the latest hardware, especially for the latest sensors/communication protocols. Most of the features, like compass calibration, OSD, advanced telemetry, or advanced GPS functions, require external modules to be added to the FC. This comes with complexity and extra costs.

5.3.3.2 Calibration

To ensure proper stable flight performance, the accelerometer and gyroscope must be calibrated through the Mission Planner software accurately. This is typically done by placing the drone on a flat surface for the accelerometer calibration and also moving it in the specific directions for getting the gyroscope calibration done. The magnetometer or the compass also needs to be calibrated, especially if GPS-enabled navigation is used. Such calibration is performed by rotating the aircraft in all different orientations to allow the software to measure the local magnetic field and adjust for the compass offsets. The ESCs also need to be calibrated to ensure that they respond properly to the motor throttle commands. This is done after powering on the drone, in the transmitter – moving the throttle stick to maximum – and then powering off to sync ESCs.

The radio control transmitter also needs to be calibrated through Mission Planner firmware. This ensures that the correct throttle, pitch, roll, and yaw commands are properly transmitted to the APM FC. One also need to set the failsafe parameters to avoid unexpected behaviour. The APM FC supports a variety of flight modes, and also each and every mode must be calibrated (such as switching between Stabilise, Loiter, Altitude Hold, Acro, and more). This is always done through the software interface, where one can map the modes to a specific switch on the radio transmitter.

5.3.3.3 Pros

The APM FC supports a wide variety of aircraft and vehicle types and is based on the open-source ArduPilot firmware. This provides good flexibility, as you can modify the code as you want, add new customised features, and integrate more additional sensors or modules that suit your needs. Furthermore, it has features like RTH and supports geofencing, waypoint navigation, and task scheduling, which is needed for applications such as long-range autonomous flight, areial surveying, and mapping. Some of the significant features offered by the APM FC are real-time drone data (such as battery voltage, position, and altitude) communication to the earth station – critical information for long-distance UAVs.

Compared to other FCs which emphasise multirotors, APM FCs widely usable in fixed-wing aeroplanes, UAVs, and Unmanned Ground Vehicles, and thus are an excellent choice for a variety of robotics and UAV domains. It's an open-source FC, and thus a lot of learning materials, libraries, programming, and a troubleshooting guideline are available for learners to solve any issue.

5.3.3.4 Cons

When compared to Pixhwak/Betaflight, the APM FC is out of date as it has a restricted processing capacity. Unavailability of features like OSD and

DShot (cutting-edge ESC protocol) makes it less attractive for applications such as racing and complex sensor integration. Setting up the APM FC is a troublesome process for novice users. Software such as Mission/APM Planner are used for sensor calibration and another parameter configuration. This can be a bottleneck when handling far more complex operations, like multiple sensor inputs, high-definition cameras, or advanced flight control algorithms.

The APM does not support modern communication protocols and standards like DShot ESCs, SBUS (Serial Bus), or Crossfire telemetry. This can limit the controller's performance for high-speed or competitive FPV applications. With the rise of more advanced FCs (e.g., Pixhawk, Betaflight, iNav), the APM is becoming less common for new drone builds. It may not be the best choice if you're looking for the latest features or cutting-edge performance.

5.3.4 Pixhawk

In the drone domain, the Pixhawk 2.4.8 FC is at present the most popular choice, predominantly for individuals looking for a highly customisable and sophisticated autopilot system. Several manual and autonomous UAV types, such as multicopters, fixed-wing aircraft, unmanned boats, and unmanned cars, are supported by Pixhawk 2.4.8 FCs. As part of the PX4 ecosystem, Pixhawk supports the ArduPilot firmware, which is portrayed in Figure 5.4 and provides the most sophisticated autonomous flight features and far more dependable performance.

5.3.4.1 Challenges

The Pixhawk 2.4.8 FC requires extensive setup and tuning, mostly for novice users in the domain of drone. For individuals who are unfamiliar with autonomous flying, the entire process can be astounding thanks to the graphical interfaces offered by the Mission Planner (for ArduPilot) and QGroundControl (for PX4). Sensors, flight modes, and tuning are examples of complex parameter settings. Even though the Pixhawk 2.4.8 works well with a varied range of sensors and accessories, certain hardware may still cause problems. To guarantee correct integration, some more recent sensors or ESCs might need firmware or software updates, but one has to ensure the compatibility of the same.

Compared to Betaflight FC used in racing drone, Pixhawk 2.4.8 is bigger and heavyweight. Pixhawk 2.4.8 is not appropriate for applications such as racing or mini-drones (where the weight and size of FCs are critical). However, this FC encompasses many useful features, but the unavailability of real-time kinematic (RTK), airspeed sensors, and GPS may force users to buy additional modules that shall not only increase the cost but will also

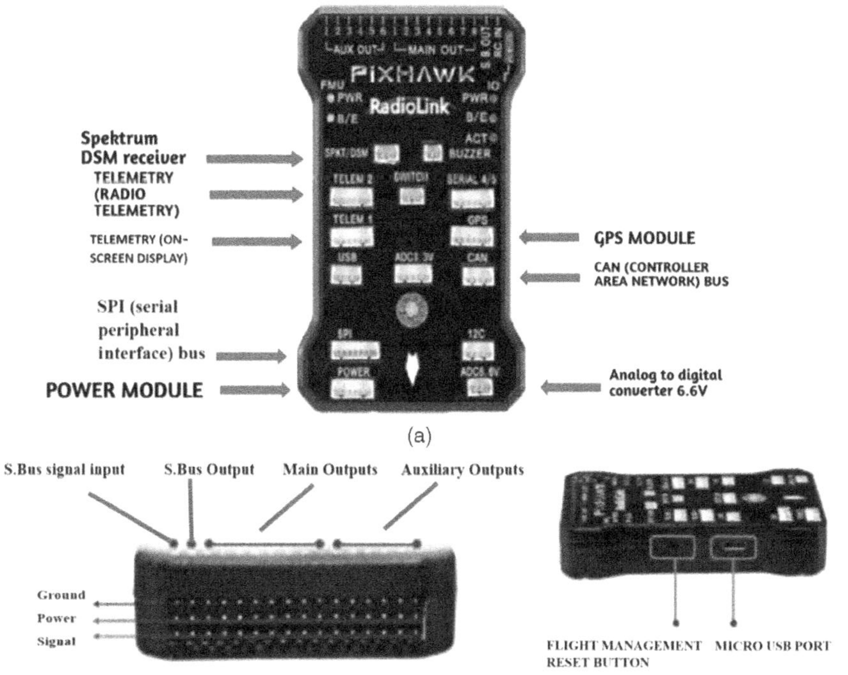

Figure 5.4 Pixhawk 2.4.8 FC.

increase the complexity of the system. This FC is a power-consuming unit because of its powerful processor, and thus it requires a power module that can source sufficient power for the functioning of the FC and associate peripherals.

5.3.4.2 Calibration

Mission Planner and QGroundControl is used to calibrate accelerometers. To ensure accurate sensor readings, the software will ask you to change the drone's orientation while it is being calibrated on a level surface. To guarantee accurate movement and stabilisation, the gyroscope must be calibrated by keeping the drone in still position on a level surface. The software itself recognises the calibration completion process. Compass calibration is performed for accurate GPS navigation. The compass sensor must be calibrated by revolving the drone in several directions as Pixhawk 2.4.8 uses magnetometer to measure the local magnetic field. The ESCs require to be calibrated to ensure that FC is properly synchronised with the motor. This task is performed by switching on the drone, then keeping the throttle to its maximum level, and allowing the Electronic Speed Controllers (ECS) to fine-tune the throttle range. Last but not the least, the radio system requires to be calibrated for ensuring proper communication by revolving the transmitter section such that it locates the minimum and maximum signals. The designated switch on the transmitter section is used to change between various flight modes, and the FC must respond appropriately to each switching action.

5.3.4.3 Pros

It has advanced features, namely, RTH, automatic pilot, geofencing, self-governing waypoint steering, and mission scheduling. The ArduPilot open-source firmware is compatible with Pixhawk 2.4.8 FC, which is widely used for surveillance and mapping applications and long-range hovering. The users can change the firmware as per their need, thus enhancing the system performance. This FC has an RTK GPS that offers the position accuracy in centimetre. Some of the applications where this FC is widely used are UAV with multirotor, fixed-wing aircraft, and submarines. It measures and displays the battery level, altitude, and GPS location. The data are logged into an Secure Digital (SD) card which can be later analysed for performance enrichment and troubleshooting. Its failsafe mechanism ensures that the drone survives in failure situations and return safely to home position in dreadful situations. This FC is compatible with most of the present-day sensors and therefore widely used to build professional drones. This FC has various community/blogs that assist in troubleshooting all the types of issues related to Pixhawk 2.4.8 from setup to problem-solving levels.

5.3.4.4 Cons

Calibration of Pixhawk 2.4.8 FC is a challenging task for novice users. The calibration requires use of software, namely, Mission Planner/QGroundControl, to fine-tune various parameters to maximise the performance of the system. It cannot be employed to build smaller drones because of its substantial weight and vast size. Furthermore, additional modules may add cost and complexity to the system. It consumes extra power compared to basic FCs while operating various sensors and modules, and thus demands a power unit that can suffice the power need. It is not suitable for long-distance hovering and racing applications because of its significant weight.

5.3.5 DJI NAZA M Lite

DJI NAZA M Lite is the most popular FC amongst the drone operators, especially for multirotor functionality, ease of use and trustworthiness, and appropriateness for stable flight operation with simple setup procedures.

5.3.5.1 Challenges

It doesn't have features such as blackbox for data logging, extensive telemetry operation, and high-end sensor integration capability. Thus, this FC is not appropriate for users who wish to interface several sensors and high-end features in the drone. As it has proprietary DJI firmware, tailor-made modification and flexibility similar to open-source firmware are difficult to achieve. Tuning procedure is straightforward in the NAZA M Lite FC, which is good for novice, but experienced users may find it difficult to optimise the performance of the system. It requires firmware update to be compatible with the advanced and sophisticated modules to be interfaced with this FC, which shorten the lifespan of the FC as it is unable to cope with advancements in the drone technology. It is apt for mini-drones but doesn't suite for bigger-size UAVs.

5.3.5.2 Calibration

The accelerometer is useful for steady flight operation, and thus, the accelerometer should be calibrated meticulously on a level surface by means of DJI-assisted technology. The drone's angular movement is precisely measured with the help of a gyroscope. The gyroscope is calibrated by the DJI Assistant software by keeping drone in steady condition. The GPS calibration is performed by positioning drone in an open ground under clear sky to verify correct GPS functionality. The transmitter section needs to be calibrated to interpret the control signal authentically. The ESCs require to be calibrated to ensure that the FC is properly synchronised with the motor. This task is performed by switching on the drone, then keeping the throttle to full, and then apply power to the system.

5.3.5.3 Pros

The DJI Assistant software is a useful tool for novice users to perform calibrations, firmware updates, and other configurations without much efforts. It is appropriate for airborne photography and cinematography due to excellent balancing capabilities of the FC. It allows undisturbed video recording under windy conditions too. The RTH capability allows improved safety as the drone returns to home location in case of low-battery condition or signal loss situation, thus improving the system's reliability. The home lock and course lock features of the FC allows to manage and control the drone's orientation simply under robust wind or complicated manoeuvres.

5.3.5.4 Cons

Compared to Pixhawk/Betaflight-restricted PID, tuning facility is available with DJI. Its use is restricted to mini-, small-, or mid-size drones and thus

Figure 5.5 DJI NAZA M Lite.

not suitable for heavy-payload drones. Having proprietary firmware system doesn't give flexibility to open-source changes. NAZA M Lite doesn't support unconventional sensors and peripherals. For professional-grade applications that call for integration with specialised equipment, this makes it less appropriate. The NAZA M Lite supports basic telemetry functions, but unlike some other FCs, it does not support advanced telemetry. Blackbox logging and comprehensive in-flight diagnostics, which are crucial for performance optimisation and troubleshooting, are absent.

Figure 5.5 illustrates the NAZA M Lite, which is generally costlier than more straightforward FCs like Betaflight-based systems. Even though it has more features, pilots who don't require DJI's entire feature set might find the price to be too high.

5.4 RACING FCS

In FPV racing drones, the FC is meant to make drones steady, transmit collected data/signal to pilots, and empower precise flight movements. This section explores FCs and criteria for the selection of FCs as presented hereunder:

- Selection of frame size and available mounting options based upon the application.
- If there is need of GPS communication and requirement of compass; furthermore, if there is support for UART/USART and I2C communication protocol; and, if it provides Bluetooth communication, LED, and Buzzer support.
- Inbuilt Barometer.
- The FC is compatible with power supply provided by the available battery.
- USB port.

The multirotor flight controllers are generally categorised into regular FC and All-in-One FCs. The All-In-One (AIO) is preferred when compact size and less weight are desired, but if trustworthiness is required, then the regular FC/ESC shall be chosen.

5.4.1 SpeedyBee F405 V4

The SpeedyBee F405 V4 is one of the best value FC stacks on the market, providing solid quality and performance. It comes with most of the essential modern features, including support for an analogue FPV system with an on-board Betaflight OSD chip, and is also "plug and play" compatible with digital FPV systems. We've been almost exclusively using this stack for our 5″ and 7″ builds in the past year because it is affordable and suitable for freestyle, racing, long-range, and cinematic flying. While there are other budget FC stacks on the market, they typically use the F722 chip, which has only half the memory of the F405 in the SpeedyBee. Less memory restricts the number of features you can use in Betaflight and makes the system less future-proof.

5.4.1.1 Challenges

The SpeedyBee F405 V4 supports Betaflight and iNav, but like many FCs, ensuring that the firmware is up to date and compatible with your peripherals (e.g., ESC, GPS, VTX) is important. Flashing a wrong version or an incompatible firmware could cause issues with motor control and flight performance.

It's difficult to configure FC if one is not acquainted with Betaflight. It is found that the ESC/Processor of SpeedyBee F405 V4 gets heated up when it is flown for more duration, and thus it requires cooling arrangements when operating at high voltage or load. The UARTs and I/O ports available in F405 V4 restrict the number of peripherals to be attached, thus denying complex build-up. The Wi-Fi connection sometimes offers connectivity issues in SpeedyBee app. The USB connectivity is not a reliable solution for tuning and configuration.

5.4.1.2 Calibration

For precise motor behaviour, throttle range, and to avert motor desynchronisation, the SpeedyBee F405 V4 requires the calibration of ESC unit. To ensure smooth drone hover, the accelerometer and gyro have to be calibrated. To avert the flight volatility, the Betaflight has to be used to calibrate the accelerometer. To augment the flight performance, PID fine-tuning is needed.

5.4.1.3 Pros

- It allows wireless configuration of FCs with SpeedyBee app.
- It has ease of parameter tuning within no time in comparison of traditional approaches.
- Due to compact form factor, it is preferred for lightweight and smaller drones.

- The use of STM32F405 processor offers higher throughput, which allows implementation of complex flight algorithms and sensors and thus confirms steady flight performance.
- The F405 V4 supports iNav and Betaflight firmware options and thus offers a lot of litheness depending on one's flying style.
- The on-board OSD of SpeedyBee F405 V4 is used for exhibiting significant data such as RSSI, battery voltage, and flight mode. Thus, SpeedyBee F405 V4 is a better choice at affordable price for drone lovers.

5.4.1.4 Cons

- The unified 12A ESCs cannot handle high-current motors/aggressive flying and thus is not suitable for larger drones.
- Available UARTs make it difficult to interface additional units to the drone, which is needed for building complex systems.
- Its compact design makes Soldering a difficult task for novice users.
- It lacks wireless firmware flashing for FCs.
- It has limited ESC Support and Obsolete Components.
- Some users have raised concerns regarding the trustworthiness of the FC.
- Thermal throttling might result from insufficient cooling and airflow, which could decline flight performance under high-power conformations.

5.4.2 SpeedyBee F405 Mini

It is an excellent choice for ultralight 3- to-5-inch drones (20×20mm stack). Costing less than 5,000 INR, the SpeedyBee F405 offers a feature-rich FC and ESC.

5.4.2.1 Challenges

- Like other FCs, apprising and configuring the firmware are complicated, particularly through Betaflight.
- Soldering and assembling connections are difficult because of its compact size.
- It cannot handle multiple ESCs, GPS, etc., as it has fewer outputs in contrast to large controllers.
- Sporadic connection problems are reported by users in wireless configuration with SpeedyBee app's F405 Mini.

5.4.2.2 Calibration

Calibration is required to be performed when first time setting up the system or when the ESC/Motor or other components are changed. The ESC is required to be calibrated for unswerving throttle response. The Betaflight can be used for calibration. The Gyro sensor needs to be calibrated first, followed by calibration of the drone level to ensure the drift avoidance during hovering. Furthermore, PID calibration has to be performed for improving flight characteristics.

5.4.2.3 *Pros*

It presents a condensed design, making it apt for racing/micro FPV drones. It has a Bluetooth support that allows users to configure drones using SpeedyBee app easily. The STM32F405 processor offers high-performance processing capacity. It is versatile as it integrates Betaflight and other FPV tools, it supports various flight modes, and it is budget friendly for novice users in this domain.

5.4.2.4 *Cons*

Less input/output ports and unavailability of barometer restrict it not to be used for complex system. Soldering and handling gentle electronics elements are difficult due to its tiny size. External barometer is required for altitude handling.

5.4.3 iFlight Beast F7 AIO 55A

It is suitable for 3.5″ to 5 ″ drone assembly. The compact and lightweight features make it appealing for racing applications. The 55A ESC rating makes it exceptionally durable as an AIO board, but not as ESCs on separate boards. This FC has all the essential connections and features as available in Betaflight FCs.

5.4.3.1 *Challenges*

In spite of the FCs backing for Betaflight, flashing the firmware necessitates attention due to FC-integrated ESCs. Faults in the firmware/ESC flashing procedure may result in boot failure/configuration snags. For greenhorns, the combination of an AIO FCs and 55A ESCs may be difficult to handle. It can be difficult to ensure that all wiring, tuning, and calibration activities are performed appropriately, especially for inexperienced users.

The components may heat up when exposed to high loads, which draws high current during destructive flying. Averting thermal throttling involves adequate cooling, which can be achieved by heat sinks/airflow. Compared to discrete ESCs and segmental FCs, upgrading ESC is a problematic procedure. The entire AIO needs to be replaced even if its single part fails. However, ESCs are integrated into the AIO, but proper calibration is still required.

5.4.3.2 Calibration

During Beast F7 AIO configuration, the ESCs must be calibrated by Betaflight Configurator to correctly respond to the throttle input. Furthermore, the accelerometer and gyro also have to be calibrated to ensure that the drone's orientation is accurate during flight. Subsequently, fine-tuning the PIDs of the setup is critical. With iFlight's app/companionable wireless configurators, the computer is not necessary for calibration.

5.4.3.3 Pros

Integration of FCs and ESCs as a single unit not only minimises weight but also reduces wiring complexity, and thus, it is a potential choice for lightweight system and ideal choice for racing and heavy payloads. Integrated Bluetooth offers wireless configuration using iFlight's app. This controller also integrates impeccably with Betaflight, allowing access to various flight modes and tuning options easily. Furthermore, telemetry support provides OSD data and radio link value response.

5.4.3.4 Cons

As everything is integrated in a single board, if any component fails, the complete unit needs to be replaced. Under heavy-load-handling situations, excessive heat can be generated which can result in component failure. Therefore, an excellent cooling mechanism is necessary for proper functioning of the system. The Beast F7 AIO can be costlier than basic FCs and discrete ESCs, making it less attractive for low-budget users.

5.4.4 Flywoo GOKU GN745 AIO V3

The GOKU GN745 is a durable substitute to the iFlight Beast and is suitable for lightweight 3″, 4″, and 5″ drone builds. The unique attributes are F7 processor, BLHeli32 ESC, bidirectional DSHOT, RPM filtering, and added cutting-edge BLHeli32 features. It has seven full UARTs, barometer, and 8 MB of on-board blackbox memory.

5.4.4.1 Challenges

- The limited number of UARTs and I/O ports makes it difficult to interface several peripherals to the unit.
- Integration of ESCs and FCs in a single unit results in overheating under heavy-load operating conditions. Therefore, appropriate cooling arrangement is required for smoother functioning of the system.
- It can be difficult to ensure that all wiring, tuning, and calibration activities are performed appropriately, especially for inexperienced users.

- The Bluetooth protocol sometimes offers connectivity issues during the initial setup.

5.4.4.2 *Calibration*

- Suitable ESC calibration is essential to confirm unswerving motor performance. The Betaflight/Flywoo app can be used for establishing the correct throttle range. Furthermore, accelerometer and gyro can be configured with Betaflight Configurator.
- To avert the flight instability, the drone should be calibrated on a level surface.
- Fine-tuning of the PID is required as per the flying style of the user.
- The Bluetooth protocol sometimes offers connectivity issues during the initial setup.

5.4.4.3 *Pros*

It is most suitable for micro/small drones because of small form factor. Furthermore, integration of FC and ESC not only makes it compact but reduces weight and wiring complexity. It's easy to tune/adjust parameters on the fly with integral Bluetooth support for the Flywoo app/compatible configuration tools. The integration of telemetry with Betaflight OSD offers monitoring of battery voltage, signal strength, and flight mode, thus improving situational mindfulness during flight operation. Availability of barometer improves drones' stability when operated in GPS mode/Hovering condition. The unification of 45A ESCs and BLHeli_32 firmware delivers accurate control in power distribution and thus enhances the drone efficiency by offering better flight times/precise control.

5.4.4.4 *Cons*

Unavailability of a more number of UART/Input–Output Ports because of small form factor interfacing multiple peripherals such as GPS/telemetry/cameras is a significant constraint. Heat dissipation is a significant concern when the drone is operated under heavy-load conditions. This necessitates proper cooling; otherwise, it leads to thermal throttling/reduced performance. It can be problematic to ensure that all wiring, tuning, and calibration activities are performed appropriately, especially for inexperienced users. Connectivity issues are reported by users when configuring Flywoo app and Bluetooth.

5.4.5 HappyModel X12 AIO

The HappyModel X12AIO is a good choice if requirement is to make lighter machines of the order like 1S to 2S or 2″ and 3″ toothpicks. This model has 400mW VTX with a built-in ExpressLRS receiver.

5.4.5.1 Challenges

The 12A ESCs are suitable only for lightweight builds and maximum up to mid-load power setups. It fails to perform with heavy-build, higher-current motors. The AIO design results in simple wiring, but tiny pads soldering of the order of 20×20 mm mounting size is a challenge and demands precise soldering skills. Similar to other controllers, compatibility condition must be satisfied with this too. Sometimes unresponsive motors and underperformance are recorded due to incorrect or incomplete firmware. With HappyModel X12, the 12A ESCs could heat up if the drone is operated aggressively. Their compact size (20×20 mm) limits the number of UARTs and I/O options, and so limits the extended peripherals such as GPS, telemetry, or external cameras.

5.4.5.2 Calibration

The 12A ESCs in the HappyModel X12 need calibration, and so all four motors run in the same way. Calibration can be done in Betaflight or HappyModel applications. Proper ESC calibration helps avoid uneven throttle and motor shaking to maintain drone stable. The accelerometer and gyro are also calibrated to keep the drone flying straight without drifting. Betaflight provides many simple options to tune the drone for better performance. Using Bluetooth, the settings can be changed in the HappyModel app even when the drone is outdoors.

5.4.5.3 Pros

The HappyModel X12 AIO is a lightweight and small-size controller. Hence, it is a good choice for micro-drones where low weight is important. It works well for tiny whoops and for 2- or 3-inch builds too. The 12A ESCs fit to low- and mid-power motors and give good efficiency in small drones and longer flight time. The board has Bluetooth that facilitates wireless alterations in settings using HappyModel app. The HappyModel X12 AIO is a low-cost option and is a good choice for people who want a strong micro-drone at little cost. It is also good for indoor flying and simple freestyle or racing with small drones.

5.4.5.4 Cons

The 12A ESCs are limited to give enough power to bigger motors and heavy machine builds. It may result in unnecessary overheating and instability. The AIO design makes wiring simple, but the small solder pads on the 20×20 mm board need extreme care and work finish. Due to the limited space (20×20 mm size), it is difficult to add extra components such as telemetry, GPS, sensors, or cameras.

5.4.6 Flywoo GOKU F405 HD 1–2S 12A AIO V2 FC

For an ultralight micro quad with digital FPV systems like Walksnail, DJI O3, or HDZero, the Flywoo GOKU F405 12A AIO V2 FC may be considered as a competitive option. It delivers powerful flight control in this small-size group. It works with both 1S and 2S, so it fits many micro builds too. It is not made for analogue FPV systems because it has no OSD chip.

5.4.6.1 *Challenges*

The 12A ESCs in the GOKU F405 V2 are made for small and low-power builds. So, when larger motors or higher power parts are used with heavy loads, the ESCs may lack performance and may lead to ESC failure. Like many AIO boards, the small size and tiny pads on the 20×20mm layout make soldering difficult. The small pads need careful and steady work, especially when connecting the performance-intensive parts and motors. Because the board is very compact, the number of UARTs and I/O ports is again limited for extra parts like GPS, telemetry, or other sensors. The 12A ESCs may face high current during hard flying. Too much heat can reduce power and even damage ESCs. The GOKU F405 is designed for 1–2S LiPo batteries that are fine for light and small drones, but it also limits the total power.

5.4.6.2 *Calibration*

The 12A ESCs in the Flywoo GOKU F405 need calibration for smooth motor operation and is done using Betaflight or in the Flywoo app. The ESC calibration is important as it allows all ESCs throttle range in the required limit and thus allows motor to perform better. Furthermore, Betaflight is employed for the calibration of gyro sensor and accelerometer. Integrated Bluetooth allows wireless configuration with the help of Flywoo app.

5.4.6.3 *Pros*

It is more suitable for many applications because of its less weight around 3–4g, which allows better control and more flight time. Integrated ESC and FC avert more wiring, and small frames are needed to build the system. Significant parameters such as voltage of the battery and mode of operation can be displayed on FPV by employing Betaflight OSD, which helps pilot in taking better decision. Wireless configuration is possible with Flywoo app for tuning purpose.

When an integrated VTX or a supported VTX is used, SmartAudio allows VTX settings like power level and channel to be changed in Betaflight. This is the advantage for racing or changing channels quickly. This board gives a good value when making tiny FPV drones or light freestyle builds.

5.4.6.4 Cons

The 12A ESCs are not fit for high-power builds due to the risk of overheating and burnout with heavy load. The 20×20mm size also gives only a small number of ports that limit the connection to more parts like GPS, telemetry, or extra sensors. The small and compact design makes heat control harder and limit flying time too, due to the risk of overheating, and even damages ESCs. The GOKU F405 AIO V2 works only with 1–2S LiPo batteries and hence limits power and flight time.

5.5 DRONE FILMING FCS

The drone for filming and shooting applications are desired to meet expectations of size, payload, and preferred flight modes. To get smooth and controlled footages, drone controllers with accurate GPS and advanced stabilisation features are in demand.

5.5.1 Holybro Pixhawk 4

The Holybro Pixhawk 4 is a popular FC. It is often used for drones. It offers precise control and stability, and includes advanced features.

5.5.1.1 Challenges

Utmost care should be taken while setting up and adjusting. This is an important aspect for new developers who are just starting to build drones or FCs. It connects with many sensors and accessories. However, some configurations might be necessary for specific setups. It is compatible with PX4 and ArduPilot firmware. The Pixhawk 4 is larger and heavier compared to other FCs. Choosing the right one and its configuration can be challenging for those unfamiliar with drone firmware customisation. This might make it difficult to use in tiny drone designs.

5.5.1.2 Calibration

To calibrate sensors such as the IMU, compass, and accelerometer, proper tools and understanding are required. Improper calibration may not set things right and may lead to unsteady flights. Technicians who don't often set up advanced drone systems often struggle with flying on their own, navigating to specific points, and handling safety measures in case if any problem encounters.

The drone calibration is done by keeping it on a flat surface. Then, the accelerometer and gyroscope are calibrated using QGroundControl or Mission Planner software. The calibration is ensured by maintaining X and Y axes at zero followed by adjusting the Z axis to zero. To calibrate the drone compass working well in magnetic fields, the drone is rotated in different directions and the magnetic interference is recorded. Properly set transmitter and receiver and well-fitted ESCs to FC help the control input to respond properly. The perfect calibration of GPS module ensures navigation and drone position accurately.

5.5.1.3 Pros

The Pixhawk 4 can be used for a wide range of drones, from small hobby models to large industrial setups. Widely employed for videography and surveillance because of self-flying capability, transmission of real-time data, and a pattern follower. It offers stable, accurate flying experience, and RTH and failsafe landing capabilities.

5.5.1.4 Disadvantages

For a novice operator, it's a complex system to understand, and it is bit heavy and thus cannot to be employed for applications having lightweight requirement. Additional modules such as GPS and power unit add to the cost of the system and increase the complexity.

5.5.2 SpeedyBee F7 V3

It is a widely used FC because of its faster signal-processing capability as it has F7 chip and has the appetite of interfacing with GPS, analogue and digital FPV, and telemetry unit. Bluetooth and Wi-Fi protocols make the configuration easy with an app. With many ports and simple plug connections, it fits many drone types like racing and freestyle.

5.5.2.1 Calibration

Many features are included, but setting everything up can be tricky for new users of FPV drones. This involves configuring all the input and output ports along with the Betaflight profiles. Sometimes, due to interference or bugs in the software, there may be loose connections to the Bluetooth and Wi-Fi system, and so, it can lose connection which may hamper smooth operation. The calibration can be carried out through the "Calibration" menu in Betaflight and by placing the FC on a flat surface before calibration. The gyroscope settings can be done in Betaflight, usually needed after the changes made in hardware or to set up the controller for the first time.

5.5.2.2 Pros

The STM32F7 CPU used is stable and fast, and it helps the drone fly smoothly even in tough conditions. It is feasible to upgrade and tune the firmware because Bluetooth and Wi-Fi are built-in. Betaflight can help adjust PID tuning, change flight modes, and modify other settings. The SpeedyBee F7 V3 is a good choice for smaller drones. It is compact and lightweight, which makes it ideal for designs that require saving space and weight. For long flights, the built-in OSD assists to monitor critical information like battery level and signal strength at FPV view. It is compatible with many ESC protocols like DShot and Multishot that make it easy to use with various motors and ESCs.

5.5.2.3 Disadvantages

The F7 V3 has many features but may be overwhelming for beginners in Betaflight and drone adjustments. Sometimes ample troubleshooting or firmware updates required to properly connect to Bluetooth or Wi-Fi. A separate GPS module is required to use GPS features like RTH even if controller device can support external GPS. The SpeedyBee F7 V3 costs more than some basic FCs due to advanced processors, Bluetooth, and Wi-Fi.

5.5.3 DJI NAZA N3

The DJI NAZA N3 is an advanced FC made by DJI. It is used in both racing and professional drones. It is known for being reliable and giving stable flight. It provides smooth control and precise navigation due to inbuilt GPS and is preferred by users for ease of functioning and safety.

5.5.3.1 Challenges

It is costlier compared to other FCs because of additional features such as GPS and novel sensors. It doesn't offer much customisation as offered by Betaflight/INAV FCs. Furthermore, users find it difficult to slightly modify setting because of limited tuning options available in DJI NAZA N3. Sometimes during the firmware updates/setup process, users find it difficult to establish connection with DJI Assistant software.

5.5.3.2 Calibration

The modules such as IMU and compass have to be necessarily calibrated, whereas the GPS unit and ESC tuning are discretionary but endorsed. It is necessary to keep away metallic and magnetic objects while calibrating compass. The IMU calibration is necessary when the drone is operated at a new location or the firmware was modernised.

5.5.3.3 Pros

The built-in IMU and GPS modules of NAZA N3 make drones flying stably and operate smoothly under windy environments. Under communication cut-off and low-battery circumstances, RTH safely brings the drone to the starting point. The smaller size of the FC allows it to be used in small-size drones. Furthermore, the DJI Assistant software makes calibration and communication easy, and this FC can log important data such as battery level, height, and signal status of GPS, which helps make crucial decisions.

5.5.3.4 Cons

It is quite costly, has limited customisation and tuning features, and is not compatible with other brands. Furthermore, some users have noticed software bugs establishing communication with DJI Assistant software.

5.5.4 DJI NAZA A3

This FC is mostly suitable for high-end and heavy-payload-capacity applications, and it can hover smoothly under windy climates with utmost stability.

5.5.4.1 Challenges

It is expensive and works well with DJI parts only. Only DJI software is required to set up and tune important units of the drone, but it is bit difficult to operate for novice users. Due to larger size of the FC, it is not preferred in smaller drones.

5.5.4.2 Calibration

The gyroscope and the accelerometer have to be calibrated to precisely know the level of the drone. The GPS calibration has to be performed in the absence of cloudy condition only. In order to run the motors in sync, the ESC calibration has to be performed. Furthermore, the safety mode calibration is necessary to bring back the drone safely to the home location under unforeseen conditions.

5.5.4.3 Pros

It offers stable flight operation in stormy conditions due to the IMU and GPS features, and it is best suitable for surveillance applications that require camera as a sensor. It has capability to produce real-time battery status, signal status of the GPS, and height at which the drone is operated. This FC is lightweight, consumes less power, and also supports autonomous flight mode.

5.5.4.4 Cons

Being costly makes it less attractive to the novice users for normal applications. This FC works well with DJI models; therefore, it is not compatible with other open-source hardware modules. Compared to Betaflight FC, fewer discussion groups are available for this FC; therefore, novice users receive thus less support in resolving problem related to the FC.

5.5.5 DJI NAZA H

The DJI NAZA H is a FC used in six motor drones. It is part of the NAZA line made by DJI. It helps the drone fly steady and not shake much in air. It is easy to use and good for people who want safe and simple flying.

5.5.5.1 Challenges

The NAZA H is more expensive than many other boards like Betaflight and iNav. So, it may not be suitable for playtime flyers looking for low-cost options. It has fewer setting changes, and so people who like deep tuning may not like it. It works mostly with DJI parts, and so using other brand parts may not be possible. Even though it is of smaller size, it is heavy for small drones.

5.5.5.2 Calibration

Before flying, the gyro and the accelerometer should be set properly. In order to ensure that the motors work in synchronisation, the ESC needs calibration.

5.5.5.3 Pros

Being a steady FC (due to the IMU and GPS features), it is apt for surveying and filming operations. The RTH feature is useful under startling conditions. It supports auto flight mode operation such as flying firmly at a height on a prespecified path.

5.5.5.4 Cons

It has limited tuning features, works well with DJI boards but not with other makes, and exclusive software and costly FC make it less attractive for novice users.

5.6 COMMERCIAL FCS

A strong and stable FC is required for heavy-payload-capacity drones. It should have cutting-edge features, ease of control, and safe flying condition for applications like agriculture, land mapping and inspection, medical logistics, etc. In this section, some of the FCs suitable for such applications are presented.

5.6.1 DJI Matrice 600

It is equipped with DJI NAZA M Controller to handle heavy payload with steady hovering, and it largely finds application in power-line inspection, preparation of map, land-use survey, and filming.

5.6.1.1 *Challenges*

It is highly expensive and unsuitable for trivial jobs; moreover, acquaintance with the DJI software is must for configuring the drone, making the process time-consuming. It works well with DJI brand parts and has limited control for flight setting as compared to Pixhawk.

5.6.1.2 *Calibration*

The DJI Assistant software should be used to configure the IMU for smooth flight operation. Other units such as GPS and compass must be calibrated effectively for RTH functionality. The motors are required to be configured such that sufficient power is made available to them to work effectively. Smart battery must be carefully inspected specially for heavy-payload condition.

5.6.1.3 *Pros*

Matrice 600 is capable of handling up to 6 kg payload easily with steady flight condition during windy situation also. With a light load, it can stay in the air for about 40 min. It also has auto return and waypoint flying for safe use. Two batteries and backup systems are advantages that give extra safety. It works very well with HD cameras for taking pictures and videos.

5.6.1.4 *Disadvantages*

The price is too high for small users. Flight time becomes short when it carries heavy load, and so extra batteries are needed. It works best inside DJI's own system, and so it is less effective with outside parts or open software.

5.6.2 Pixhawk 2.1 (Cube)

Pixhawk 2.1, also called the Cube, is a very common open-source FC used in many kinds of drones. People use it for map work, land survey, film shooting, and carrying heavy loads. It is popular because it gives more control and can work with many drone types like multirotor, fixed-wing, and hybrid models. It is available as open source and facilitates users to change and set it as per the requirements of their projects.

5.6.2.1 Challenges

Still, it is tricky to use at first. The setup process is not straightforward and so time consuming. Additional units namely radio units, GPS, supply unit, and appropriate sensors have to be interfaced correctly, and testing of the same is to be done carefully for sound functioning of the drone.

5.6.2.2 Calibration

One of the most difficult tasks is calibration of modules, namely, ESC, compass, IMU, and GPS. Calibration of GPS and compass units in the presence of metallic object or weak signals may disrupt the flying operation of the drone. Motors and sensors used with this drone require careful calibration and extra setting to function properly. It requires frequent software updates, which is difficult for new users who are not proficient with firmware.

5.6.2.3 Pros

PX4 and ArduPilot (open-source software) can be used for the setup and configuration of Pixhawk 2.1. It supports GPS and RTH, and has inbuilt 2-IMU and Barometer units, which makes the flight operation more stable. Cortex M7 processor improves processing capacity most suitable for complex operations. Furthermore, a lot of support is available online, which is a bonus for novice users for solving the problems faced during setup/update.

5.6.2.4 Disadvantages

For novice users getting expertise in setting up is a complex task. The interfacing of several sensors may be a tricky part and requires the knowledge of required updates.

5.6.3 Vector V2 by Eagle Tree systems

It is considered to be one of the most stable flying systems, is easy to configure, and offers superior performance. Easy access to Eagle Tree's telemetry tools for fetching the vital data useful for flight operations.

5.6.3.1 Challenges

It doesn't support customisation beyond a limit as compared to open-source boards and works well with the genuine parts produced by Vector V2 and not compatible with other brands' modules. It may get interfaced, but problems may be arise in the functioning of the unit. The firmware update may reset the configuration; Thus, all the important units of the drone require recalibration and are most expensive compared to other open-source controllers.

5.6.3.2 Calibration

Vector ground station software is needed for calibration. The drone must be kept on flat surface/ground, thereafter the software directs the user the next steps for calibration. The gyroscope and accelerometer calibration requires the drone to move in different directions. Furthermore, the compass calibration requires rotation of the drone in various directions as long as good signal is not received. The GPS calibration has to be done in open ground. The ESC and radio unit calibration has to be performed for precise flight operation and communication purpose.

5.6.3.3 Pros

It weighs around 30 g, and thus it is appropriate for racing/mini-drone-based applications. It has functionalities like GPS hold, RTH, and altitude hold. It gives real-time flight data such as battery power, GPS, and height, which helps in safe flying. The controller works well for people who want stable and reliable flight. It can perform jobs like filming and land survey due to the ability to remain steady even in long flights.

5.6.3.4 Disadvantages

It cannot be changed deeply like open-source boards. Users looking for full control or want to write their own flight code have very limited freedom. The price is high, and using only Eagle Tree parts makes it less flexible. Vector V2 fits better for only users who want an easy and ready system.

5.6.3.5 Conclusion

This chapter presents the viewpoint of a drone expert. The use of the drone and the level of proficiency will determine which FC is most suitable for a person. The K K 2.1.5 and Thunder QQ Super are reasonable, stable to fly, and easy to set up for novice users. While they lack modern-day features like GPS support, OSD, sophisticated ESC protocols, and autonomous modes, they are perfect for learning the fundamentals of multirotor flying. They are good for recreational flying, but builders who want long-range performance, racing, or sophisticated tuning will soon outgrow them. Waypoint navigation, autonomous missions, and a variety of vehicle types are supported by more authoritative open-source substitutes such as APM and Pixhawk 2.4.8. Although APM is still in use, it is outdated. The complexity and setup time can be more for novice users. Pixhawk 2.4.8 still offers appealing features and is better suited for large/professional platforms. The DJI NAZA M Lite offers dependable GPS hold and RTH features for filming setups and for users looking for stability with less customisation.

FCs such as SpeedyBee F405 V4, SpeedyBee F405 Mini, iFlight Beast F7 AIO, Flywoo GOKU GN745, HappyModel X12, and Flywoo GOKU F405 HD offer lightweight designs, wireless configuration, and integrated ESCs for FPV racing and micro-drones. These controllers work well in compact, high-performance designs, but they often sacrifice heat tolerance and expansion capability in favour of compactness. Additionally, heavy motors and aggressive freestyle flying are not appropriate for them due to their current restrictions.

The FCs SpeedyBee F7V3, DJI NAZA3, NAZA H, and Holybro Pixhawk 4 offer telemetry, autonomous operating mode, and GPS-based stability, which makes it more suitable for film industry and professional use. The reliability, redundancy, and customisation options available with Pixhawk 2.1 (Cube), DJI Matrice 600, and Vector V2 make it more suitable for research and industrial applications.

REFERENCES

1. Ebeid, E., Skriver, M., & Jin, J. (2017). A survey on open-source flight control platforms of unmanned aerial vehicle. In *2017 Euromicro Conference on Digital System Design (DSD)*, Vienna, Austria (pp. 396–402). IEEE.
2. Hassanalian, M., & Abdelkefi, A. (2017). Classifications, applications, and design challenges of drones: A review. *Progress in Aerospace Sciences*, 91, 99–131.
3. Kangunde, V., Jamisola Jr, R. S., & Theophilus, E. K. (2021). A review on drones controlled in real-time. *International Journal of Dynamics and Control*, 9(4), 1832–1846.
4. Ebeid, E., Skriver, M., Terkildsen, K. H., Jensen, K., & Schultz, U. P. (2018). A survey of open-source UAV flight controllers and flight simulators. *Microprocessors and Microsystems*, 61, 11–20.
5. Khan, F. S., Mohd, M. N. H., Larik, R. M., Khan, M. D., Abbasi, M. I., & Bagchi, S. (2021). A smart flight controller based on reinforcement learning for unmanned aerial vehicle (UAV). In *2021 IEEE International Conference on Signal and Image Processing Applications (ICSIPA)*, Kuala Terengganu, Malaysia (pp. 203–208). IEEE.
6. Cheng, Z., West, R., & Einstein, C. (2018). End-to-end analysis and design of a drone flight controller. *IEEE Transactions on Computer-Aided Design of Integrated Circuits and Systems*, 37(11), 2404–2415.
7. Khan, A., Ivaki, N., & Madeira, H. (2022). Are UAVs' flight controller software reliable?. In *2022 IEEE 27th Pacific Rim International Symposium on Dependable Computing (PRDC)*, Beijing, China (pp. 194–204). IEEE.
8. Pütsep, K., & Rassõlkin, A. (2021). Methodology for flight controllers for nano, micro and mini drones classification. In *2021 International Conference on Engineering and Emerging Technologies (ICEET)*, Istanbul, Turkey (pp. 1–8). IEEE.

9. Vishwath, N. A., Yadav, A. R., Mehra, V., Patel, V., Patel, M. K. B., Sondagar, V., & Singh, G. (2024). Computational analysis and designing of flight controllers to improve the drone performance for a novice pilot. *International Journal on Interactive Design and Manufacturing* (IJIDeM), 19(2), 1395–1408.
10. Nguyen, D. D., Rohacs, J., & Rohacs, D. (2021). Autonomous flight trajectory control system for drones in smart city traffic management. *ISPRS International Journal of Geo-Information*, 10(5), 338.

Chapter 6

Introduction to the Robot Operating System (ROS)

Sarla More, Prerna Ajmani, Durgesh Mishra, and Ebrahim A. Mattar

6.1 INTRODUCTION TO THE ROBOT OPERATING SYSTEM (ROS)

The Robot Operating System (ROS) is an open-source operating system that aims at making robotic software development simpler. It also offers platforms and libraries to build sophisticated and trusted robotic uses [1]. The operating system, being a fundamental service, provides hardware abstraction without preventing developers to develop interoperable software components. The ROS poses as a valuable robotics collaboration tool through the standardization measures and modularity. Willow Garage Robotics Research Facility developed the ROS and released it in 2007.

The platform was quickly adopted by research institutions and then by academic establishments to construct flexible protocols of robot development [2]. ROS 1 then became available to the community in 2010 after which its tools and libraries were available to developers. With the further development of robotic technology, the need of Group-based Social Navigation (GSON) framework arose and the development of ROS 2 was made. ROS 2 was developed to provide better functionality in comparison to ROS 1, and it has enhanced security measures with time operation capabilities and extended embedded system support.

In 2017, ROS 2 version was launched with significant modifications to ROS 1, considering the constriction experienced from the first version. The open-source ROS framework innovation saw a lot of progress in robotics software development. The software development of robotics is made easier by the fact that the ROS provides the required tools and libraries in addition to the standards that reduce the level of complexity when programming a robot. Robotic systems depend on the ROS as their foundational component to achieve reliability in both academic investigations and industrial operations. People benefit from many vital aspects that result from ROS usage.

The robotic development process requires the ROS because it separates engineers from hardware details and thus enables them to focus on creative development. The modular construction of the ROS and its expansive

 DOI: 10.1201/9781003598350-8

library support enables developers to build reusable tools that facilitate international robot collaboration. The absolute ROS was introduced to Willow Garage to understand its original development purpose. The robotics research organization Willow Garage began developing the ROS in 2007 because of its trailblazing work. Despite initial development goals for being a flexible framework, the ROS became renowned because its adaptability and user-centered design appealed to many users.

6.1.1 ROS 1 to ROS 2 evolution

ROS version 1 underwent major deployment for teaching purposes and educational research after its 2010 introduction. Robotics science progressed by evolving its fundamental requirements. This development led ROS developers to release ROS 2 in 2017 which resolved critical security real-time performance and embedded system compatibility issues, thus making the ROS suitable for the fast-changing technological landscape.

Key distinctions between ROS 1 and ROS 2 are as follows:

- Real-Time Functionalities represent one of ROS 2's major additions because the new version includes real-time middleware, but ROS 1 did not provide a real-time solution, which inhibited its usefulness to driverless car applications.
- Better Security: Authorized interaction combined with encryption made security solutions to ROS 2 and resolved the underlying effects of ROS 1 systems.
- Embedded Systems Support: The current market of robotics is dependent on embedded systems. ROS 2 may be utilized in a constrained system as embedded systems can be used.

6.1.2 Various prominent characteristics concerned with the Robot Operating System

- Modularity: The ROS has a modular structure, enabling the software development teams to create and reuse specific software modules, which are called nodes. System architecture helps to add new features and share nodes across different modules.
- Hardware Abstraction: The ROS system separates hardware interface details to developers and have a uniform access to different hardware components. The technical design to achieve this objective is described on device drivers and hardware abstraction layers.
- Message-Passing Interface: ROS nodes allow to provide a successful consistent transmission of data in a robotic system through a message-passing interface. Such an approach helps to decouple the components through strengthening the system resilience.

- Tools of simulation and Visualization: Before rolling out the real hardware, programmers can use simulators like Gazebo and RViz to provide tests and debug their software. These tools can offer the robotics designers with a safe and cost-effective mechanism of improving robotic designs and algorithms.

6.1.3 Real-world case studies

By integrating the ROS into the unmanned aircraft systems (UAS), the key sectors of the industrial advancement are influenced in aerial surveying and delivery as well as defensive services. Agricultural tasks and environmental observation have been renewed by utilizing the ROS-controlled drones that combine Light Detection and Ranging (LiDAR) systems with GPS and high-resolution cameras to perform accurate scientific surveying and land mapping.

ROS Navigation Stack will allow drones to optimize their surveying procedures by cutting the time needed to complete them by half and increasing the accuracy of the received data, which simplifies airborne inspections and makes them less costly. The ROS makes it possible to create self-driving delivery drones like those utilized in such companies like Amazon Prime Air and Zipline that are used in the logistics industry. The drone delivery systems implement swarm coordination systems with secure communication protocols and artificial intelligence-based obstacle detection in real time to decrease the delivery time by 30% in particular to those with minimal infrastructure.

ROS-enabled drones serve the surveillance and reconnaissance needs of defense and security organizations including Defense Advanced Research Projects Agency (DARPA) and the U.S. Department of Defense. ROS 2 manages live data analysis along with group collaboration techniques and sensor information combination through a decentralized approach to achieve better intelligence collection and minimize direct human exposure. Real-world cases showcase the ROS as a tool that makes UAS applications more efficient and cost-effective while delivering autonomous features between multiple domains.

6.2 ROS ARCHITECTURE

Both distributed robotic systems and flexible scalability and adaptability are possible when using the ROS architecture [3]. The design of the system is tiered such that there is good interaction between the software and the hardware components to make the component interaction successful. The following is a brief summary of ROS architectural components and primary layers.

Nodes: In the ROS, the basic unit of the robotic application is called a node, and it performs autonomous calculation. The autonomous robot processes are independent of each other in the robotic system since robotic nodes enable them to be developed in a modular manner [4]. Complex robotic systems with the modular approach improve the flexibility and scalability of systems and allow them to be easier to service:

Publish–Subscribe Messaging Paradigm: The basic ROS principle of the publish–subscribe paradigm is that nodes can share data with one another, with no communication taking place between nodes [5]. In this model, Publishers and Nodes create and disseminate messages of data regarding certain topics.

Subscribers: Nodes have to subscribe important subjects in order to access data streams. A sensor node with the publish–subscribe model transmits laser scans to a subscribed processing node to be analyzed at the node. This is because node decoupling makes asynchronous communication possible within this framework and enhances the efficiency of integrating components.

Services and Actions: The publish–subscribe paradigm [6] is an outstanding framework in terms of continuous data streams, but in some of the applications node-to-node messaging is needed.

The ROS Provides Two Alternatives to This: Network nodes can use the synchronous communication mechanism. Network nodes can use services to send requests to other nodes that are willing to wait. An ideal use of synchronous communication is to ask a robot what is its current state using services.

Actions: These asynchronous modes of communication enable nodes to initiate lengthy processes and monitor their execution activity by activity. The navigation missions include defining goals and delayed feedback to task accomplishment, which makes these actions an optimal solution.

Parameter Server: The ROS has a single configuration parameter system, called the parameter server, that has the service of storing configuration parameters that nodes are able to access through runtime access [7]. Wheel diameter or sensor offsets and environment-specific information such as mapping data and purpose positions are system parameters.

The Parameter Server of the ROS acts as a central repository of configuration parameters that both allow configuration mastery and remove the redundancy of parameter implementation in individual nodes. Adaptive behavior and enhanced debugging capability are made possible by node-based runtime parameter reading and modification.

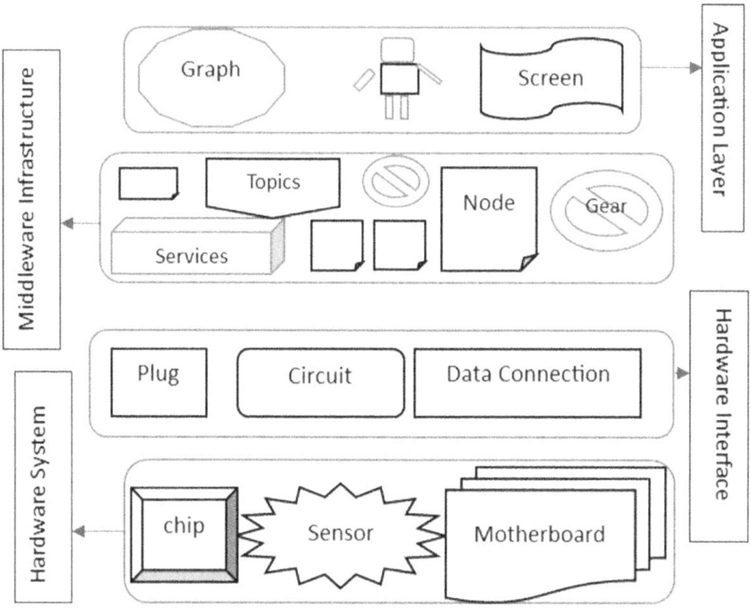

Figure 6.1 Robot Operating System (ROS) architecture.

Figure 6.1 illustrates a comprehensive layered structure of advanced technological systems focusing on robot or computation model design. The top of this architecture contains three components—"Application Layer", "Middleware Infrastructure", and "Kuboot Layer"—demonstrating how users access applications through middleware services. The Middleware Infrastructure executes data management and enables communication channel facilitation between the application-layer and hardware-layer operations. This level has elements Middleware "Subscribers" and Services to imply the possibility of the distributed systems implementation alongside service-oriented architecture.

The lowest tier, called hardware interface, works with hardware directly by using such components as Rod-ware Interactor and Hardware access. These permit simple hardware communications. The system that is located outside the main structure is the subsystem called Robot System, which unites with Consumer core assigning and Communication structure Interface to demonstrate how the system will support robotics and control systems and Internet of Things (IoT) applications.

The bottommost layer is the hardware interface layer, which is directly linked to physical hardware. It includes components like: Robot ware Interactor: specific module designated for managing distinct hardware interactions. (maybe "rod-ware" is a typo or abbreviation for something similar to "robot-ware" or a personal term?) Hardware Access is an item that facilitates low-level discussion of hardware resources.

Communication structure Interface: This could possibly relate to be an interface of combining or structuring components (is this "Combucture" supposed to be "Constructure" or "Component Structure"?), allowing us integrating robotics into other system components. Purpose and Design: The system appears to focus on modularity, and the system is scattered in a hierarchical and network-like organization. It also supports robotics, control mechanisms and IoT, making it a flexible modular framework capable of interfacing with a wide range of motor driven and hardware components, where hardware, software and networked devices runs in harmony. Figure 6.1 provides a visual representation of the organization and is a form of modular design, and it also shows how the elements of the system combine to form one cohesive structure. The system depicted in this figure is a visual representation of systems with hierarchical and interconnected relationships.

6.2.1 Integration of unmanned aircraft systems (UAS) with the ROS

The popularity of the ROS in the software architecture of UAS is explained by its modular structure, strong communication protocols, and a full set of functions [8]. The ROS allows developers of systems to build customizable UAS, which can be used between mapping and delivery missions and also conducts surveillance missions via its framework. UAS's control programs receive the following benefits from the ROS:

1. Code Modularity and Reusability: The ROS node model allows developers to divide large system capabilities of the UAS into small components. Autonomous nodes work concurrently in the flight control system and navigation processes as well as analysis of sensor data ROS reference packages enable developers to utilize the available common feature libraries that will enable them to abate their effort while lowering project schedules.
2. UAS Applications Are Validated by Simulation Tools such as Gazebo: Testers allow the authors to verify that the designs they created work by executing them in simulated three-dimensional environments that fully protect the system and reduce cost. The algorithms testing involves the developers to execute comprehensive simulated tests to test various components such as flight control pathways, environmental factors, and sensor outputs.
3. Full Documentation and Community Support: Community engagement offers much to the ROS platform since users create frequent software package releases and improve features that can be used in its

software infrastructure. Extensive records of the ROS connect with tutorials and support forums enabling novice programmers to take a co-operative avenue of progressing to create the UAS project.

The UAS architecture gains many benefits by integrating with the ROS, including community-supported development, simulation functionality, and device adaptation capabilities [8]. Developers may create scalable and dependable UAS solutions that can be tailored to a range of use cases, such as mapping, delivery services, and surveillance, by utilizing the ROS.

6.2.2 Comparison with non-ROS alternatives

A comparison between ROS-enabled Unmanned Aerial Vehicle (UAV) systems and their counterparts, like bespoke flight controllers (DJI SDK, PX4 without the ROS, or other proprietary systems), will assist the reader in grasping the value propositions of each approach quite clearly. ROS-based UAVs are great for research and prototyping due to their modularity, flexibility, and overwhelming community support.

Additionally, the open architecture of the ROS allows for easy implementation of AI, sensor fusion, and swarm intelligence, which requires more advanced robotics systems.

On the other hand, the ROS imposes certain limitations such as requiring excessive computational power and having latency problems in real-time execution scenarios. These limitations make it less desirable within the realms of mission-critical or high-speed applications. The flight controllers like PX4 without the ROS or DJI SDK have optimized implementations of systems offering low-latency processing and high reliability. These systems cater to commercial UAV operations where responsiveness and robustness are key factors. For instance, DJI SDK comes with high-level APIs that make development easier, unlike using barebones drone control, but do not permit deeper customization. PX4 is open source and more user-friendly, but, like any other embedded system, it is tailored toward embedded flight control and thus is the superior choice for real-time applications.

Figure 6.2 shows the UAV development path that describes multiple UAV approaches and applications from the unmanned aerial vehicle field. ROS-based UAVs start the development path because they offer open-source software and modular hardware that support both AI development and sensor combination.

The extensive user base of these UAVs enables high customization and provides multiple setup and maintenance challenges. The other options provide real-time processing capabilities through low-latency performance, which also guarantees industry-level stability. The systems deploy proprietary security but are default on ROS-based system flexibility. UAVs will dominate future commercial industry operations through delivery services combined with surveillance functions along with mapping operations and

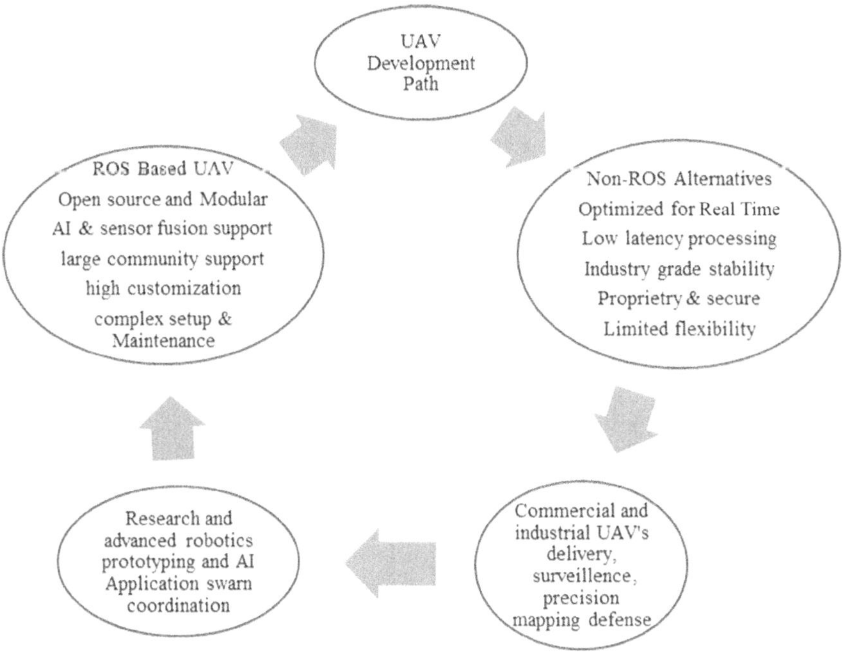

Figure 6.2 UAV development path.

defense monitoring activities because stability and efficiency are the main priorities.

The field of research, along with advanced robotics, depends largely on UAVs to develop AI-driven systems, coordinate swarming behaviors, and construct robotic prototypes. UAVs have progressed through multiple development approaches which actively optimize between system adaptability and functionality and purpose-specific needs.

6.3 COMMON UAS APPLICATIONS

Several different industries now benefit from advanced solutions developed through the evolution of unmanned aerial systems. The enhancements made possible through drone technology enable better search and rescue operations and technologized improvements for agricultural production. Existing UAS applications will benefit from technological developments as these capabilities create numerous innovative expansion areas. Unmanned aerial systems (UAS) show versatility because they serve several different sectors at once. The ROS identifies complex software frameworks that bring notable speed improvements to operations [9].

Some of the most popular uses for UAS are listed in the following.

6.3.1 Autonomous navigation and path planning

ROS-equipped UAS can use algorithms to navigate across difficult terrains on their own: real-time obstacle detection and protection functions are achieved through LiDAR in combination with camera sensors. The systematic method to discover efficient ways that decrease travel durations while conserving energy exists under path optimization. Vehicle reroutes automatically when a mission needs or environmental conditions change. This is known as automatic airspace system which is called dynamic rerouting. These applications are widely applied in environmental monitoring in the program of agricultural surveys and delivery services.

6.3.2 Sensor integration and data processing

To ensure acquisition of essential information, UAS are fitted with numerous sensors as follows [10]: cameras—to carry out visual inspections, videography, and aerial photography; LiDAR—3D mapping and high-density mapping; and multispectral and thermal sensors for precision farming and infrastructure measurements. The ROS users also receive convenient sensor integration opportunities along with automatic tools that can handle the results of the data collected. With this technology in place, it is possible to have real-time analytics that are able to help organizations make timely decisions.

ROS-based UAVs allow the incorporation of several sensors that generate complex tasks in the field of tactical control, which will make the drones respond to the various variations in the environmental conditions. When LiDAR is applied to a narrow environment, it shall also boost the degree of situational awareness and reliability of UAV when this information is combined with GPS positioning information and the same capability of camera perception of the distributed sensors. In order to develop a UAV operation based in the ROS, the autonomous AI elements must become the core of the computer. Deep learning methods of drones recognize targets almost instantaneously, making automatic target detection ability the reason for enhancing drones' performance in such unknown environments.

The UAV systems utilize learned optimal flight courses taken from the environmental feedback data to continue to enhance their flight operations using reinforcement learning algorithms. By using AI and applying it to the ROS, aerial mapping surveillance and search operation operations are better able to have autonomous decision-making systems than they would be without. Discussing about sensor fusion and AI-driven autonomy would be very useful to make this chapter more informative because it shows how UAV technology advances along its path.

6.3.3 Multiple unmanned aircraft systems

Operating in concert needs swarm coordination methods [11]. The ROS makes this possible by a UAV system that achieves efficient communication via publish–subscribe messaging and allocates distribution duties through distributed algorithms and safety controls to protect drone flight groups from aerial collisions. Swarm coordination technologies enable their deployment across different stages of mapping activities together with search and rescue activities and disaster emergency work. The integration between UAS adaptability and ROS functionality allows these systems to handle autonomous navigation and sensor-based data collection tasks as well as swarm coordination duties. The UAS technology revamps industrial procedures through performance boosts and offers options for discovery through higher efficiency levels.

6.4 CASE STUDIES AND APPLICATIONS

Modern technology advancements have become readily observable through the ongoing industrial implementation across all its functions. Practical problems studied through real-world applications demonstrate innovation's efficiency as illustrated in Figure 6.3. Each technological improvement adds more AI applications to UAS operations, leading to improved industrial output combined with long-term sustainability features that support higher living standards across society. The use of the ROS technology on digital platforms allows UAS applications to achieve important breakthroughs in several industries [12]. Scientific research documents substantial real-world deployments of ROS-based systems using case studies to demonstrate system adaptability.

Distributed Control Architectures: Time-critical systems in distributed UAS control operate because ROS 2 establishes methods for the management of multiple unmanned aerial vehicles [13]. In distributed control architectures, each UAS operates autonomously as a node while using the Data Distribution Service (DDS) protocol to maintain communication with central systems and peer systems. Multiple drones work together by sharing data during real-time search and rescue missions to achieve both maximum coverage and operational cooperation. The protection of UAV operations from cyberattacks relies on an essential secure communication which ROS-based frameworks address through algorithms for encryption and key exchange systems that protect node-to-node message transmission. Secure communication frameworks become crucial for both military missions and critical infrastructure observation operations.

The search and rescue operations attain the greatest coverage with better cooperation due to the application of multiple drones that communicate with each other in real time. Secure UAV Communication Frameworks: ROS-based frameworks achieve framework security through the combination of encryption algorithms and secure key exchange systems to ensure the safety of UAV communications between nodes during UAV operations that must be secured due to the presence of cyberattacks which are dangerous to UAV operations, particularly during military activities and cases of critical infrastructure monitoring. These frameworks have authentication protocols that provide safety safeguards to the accessibility of UAS networks [14].

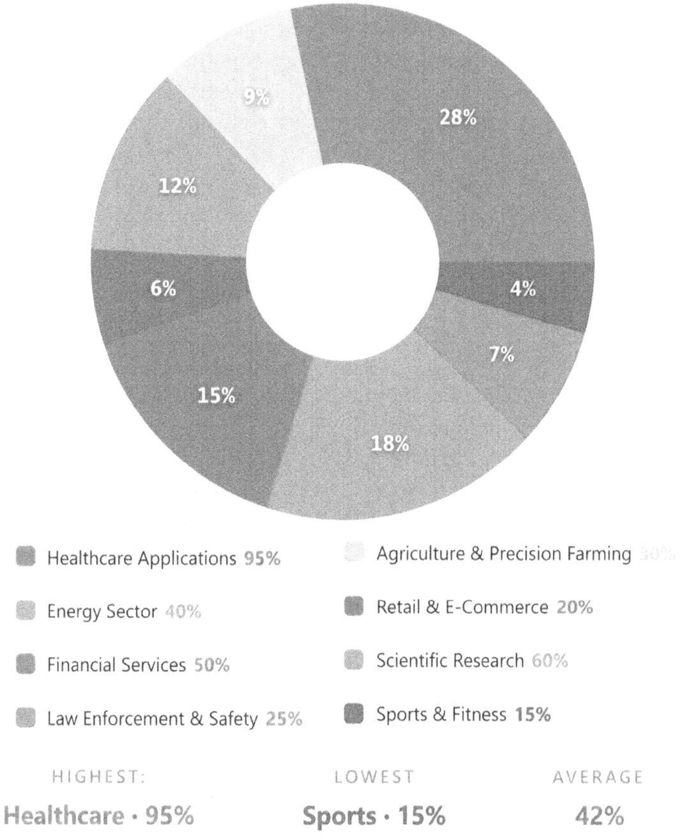

Figure 6.3 Applications of modern technologies.

Swarm Drone Flight Control: The research indicates that the ROS provides the option to make flight controllers to develop drone swarm operations based on nature-inspired swarm robotics concepts. Swarm robot systems deploy a number of drone systems that take advantage of natural swarm behavior dynamics in their cooperating activity. Allowing the control of drone units to be more possible due to the ROS framework features of drone movement coordination and service-based drone communication systems that contribute to the creation of an effective swarm flight controller to support the drone light performances and precision farming campaigns.

This analysis in Table 6.1 depicts multiple AI and technological approaches used across several sectors while emphasizing their data precision rates and operational performance. Healthcare: Healthcare researchers at a U.S. hospital obtained 95% accuracy with their AI diagnostic system

Table 6.1 Comparative study of case studies on the accuracy of different applications

Fields	*Case study*	*Application*	*Accuracy*
Healthcare Applications	AI-Assisted Diagnostics: An American hospital in the United States started using AI to identify early signs of cancer. Medical imaging data analysis using this system achieved 95% accuracy, which led to a 50% shorter diagnosis duration and predictive analytics for patient care.	• Remote monitoring through IoT devices. • Robotic surgeries for precision operations.	The system achieves 95% accuracy in analyzing imaging data, while human interaction might help refine the ultimate diagnosis.
Agriculture and Precision Farming	Drone-Assisted Crop Monitoring: Drones having multispectral sensors along with other aerial reconnaissance equipment monitored crop growth for a Brazilian farming group. Using this data allowed for improved optimization of irrigation and pesticide regulations reaching a 30% yield boost.	• Soil analysis for better planting strategies. • Livestock monitoring with radio frequency identification (RFID) tags. Farm operators control IoT-connected automated irrigation through smartphones and devices.	Moderate to high accuracy (dependent on drone sensors and data processing algorithms).
Energy Sector	Wind Turbine Inspections with Drones: A European wind energy company deployed drones that autonomously examined turbines while flagging wear and cracks without involving human staff. Drones helped the company decrease maintenance expenses by 40% while cutting down operational downtime by 40% too.	• Predictive maintenance in oil and gas pipelines. Drones use thermal imaging to inspect solar panel performance. • Smart grids for efficient energy distribution.	Structural inspection accuracy during defect detection is limited by the quality outcomes of drone image acquisition.
Transportation and Logistics	Autonomous Delivery Drones: An Asian logistics enterprise deployed autonomous drones to service remote package destinations during a pilot project. This initiative shortened deliveries by 70% along with a 20% reduction of expenses.	• The company practices fleet management utilizing GPS and IoT technologies. • Predictive analytics for vehicle maintenance. The system tracks orders in real time, which results in higher levels of customer satisfaction.	The system delivers over 90% navigational accuracy while demanding improvement in dynamic obstacle detection methods.

(Continued)

Table 6.1 (Continued) Comparative study of case studies on the accuracy of different applications

Fields	*Case study*	*Application*	*Accuracy*
Urban Planning and Development	Smart City Initiatives in Singapore: Through IoT- and AI-based smart city technologies, Singapore constructed a framework that delivers real-time traffic observation while decreasing energy usage and managing public services effectively. The system decreased traffic congestion levels by 30% while raising building energy efficiency by 25%.	• Real-time air quality monitoring. • Smart waste management systems. • Autonomous public transport systems.	Together with excellent traffic management and energy efficiency, the system faces constraints from real-time system performance.
Education and E-Learning	Virtual Reality (VR) in Medical Training: At a U.K.-based medical institute, students learned complex surgical procedures within a virtual reality simulation system, which improved their skill acquisition by 40%. Users achieved a 40% enhancement in their ability to learn from this training system.	• AI-based personalized learning platforms. • Virtual classrooms for remote education. Virtual educational models that use gaming methods help learners stay involved in their coursework.	Numerous teachers rely on simulations, but the precision of their work matters less than how much students participate and how useful the content is.
Environmental Conservation	Drone-Assisted Wildlife Monitoring: Conservationists across African territories deployed drones for both population measurements of elephants and migration surveillance and poaching prevention activities. The initiative cut poaching events by 60% during its implementation.	Seed-planting drones play a part in reforestation programs. The analysis of glacier melting uses satellite images as a monitoring tool. AI systems perform the analysis of environmental impact reports.	Wildlife monitoring using drones succeeds with accurate results above 80% but encounters limitations with terrain conditions and post-flight processing requirements.
Retail and E-Commerce	AI-Powered Recommendation Systems: Through artificial intelligence, an e-commerce giant studied consumer behavior to generate individual product suggestions, which raised their total sales by 20 percent.	• Augmented reality (AR) for virtual try-ons. • Chatbots for 24/7 customer service. • Automated inventory management systems.	Performing recommendations shows unmatched accuracy rates exceeding 90% by analyzing customer behaviors.

(Continued)

Table 6.1 (Continued) Comparative study of case studies on the accuracy of different applications

Fields	*Case study*	*Application*	*Accuracy*
Military and Defense	UAS for Surveillance: Defense agency implemented UAS platforms for border surveillance that effectively detected threats while shortening their response interval.	• AI for cyber threat analysis. • Autonomous vehicles for reconnaissance missions. • Predictive maintenance for military equipment.	Very high accuracy for threat detection (95%+); critical in time-sensitive missions.
Manufacturing and Industry 4.0	Predictive Maintenance in Factories: A German manufacturing plant deployed IoT sensors that monitored equipment inside its production facility. The system identified potential equipment issues in advance of actual failures to reach a 30% reduction in downtime, which resulted in higher productivity levels.	• Robotics for assembly line automation. • Digital twins for process optimization. Real-time quality evaluation depends on automation through artificial intelligence systems.	The system delivers fault predictions at high levels of accuracy (90%+), and its real-time tracking capabilities lead to better manufacturing efficiency.
Entertainment and Media	Streaming Platforms and AI Content Recommendations: AI analytics performed by a prominent streaming service generated personalized recommendations for content and achieved a 35% growth in user engagement.	• Virtual concerts and events in the metaverse. Generic AR attractions can be found within theme park premises. • Automated content creation tools.	The accuracy metrics exceed 90% in user preference prediction and the engagement statistics demonstrate algorithm effectiveness.
Financia Services	Fraud Detection with Machine Learning: Real-time machine learning algorithms deployed by a global financial institution detected fraudulent transactions, which resulted in a 50% reduction of financial losses.	• AI-driven credit scoring models. • Blockchain for secure transactions. • Robo-advisers for investment management.	Extremely high accuracy for fraud detection (over 95%) with continuous learning models.

(*Continued*)

Table 6.1 (Continued) Comparative study of case studies on the accuracy of different applications

Fields	*Case study*	*Application*	*Accuracy*
Scientific Research	AI in Drug Discovery: Through AI analysis, the pharmaceutical company reduced drug candidate identification time by 60% through its massive dataset evaluation.	• Climate modeling with supercomputers. The utilization of automated robots explores space. • Gene editing tools like Clustered Regularly Interspaced Short Palindromic Repeats (CRISPR) for advanced studies.	High accuracy (90%+) in large-scale data analysis, accelerating research outcomes.
Law Enforcement and Public Safety	Predictive Policing with Big Data: Through big data analysis, a city department of police successfully predicted crime areas and optimized resource placement which reduced criminal activity by 25%.	• Drones for crowd monitoring during large events. • AI-powered facial recognition systems. • Real-time emergency response coordination.	Moderate to high accuracy in crime prediction (80%-90%), with concerns over data bias.
Sports and Fitness	Wearable Technology in Athlete Performance: An athletic team employed wearables to track players' performance levels and control injury risks. The participation of wearable technology reduced the prevention of injuries by 15% while simultaneously enhancing team fitness.	• Wearable technology strongly assists with AI-driven virtual coaching platforms. • Real-time analytics for strategy planning. • AR for fan engagement during live events.	Wearable technology delivers performance forecasts and monitoring analytics at high accuracy levels exceeding 90% in real-time operation.

allowing earlier cancer detection while cutting evaluation duration to half. Predictive analytics linked with IoT-based remote monitoring enables physicians to perform robotic surgeries which improve patient care outcomes. Agriculture: Daily drone crop inspections throughout Brazil's farmland helped farmers raise their production levels by 30% by making accurate irrigation and pesticide applications.

The combination of soil pattern analysis with intense livestock observation raises operational effectiveness. Energy: The implementation of drone-based inspections for European wind turbines both lowered maintenance expenses and shortened turbine downtime by 40%. Between them, predictive maintenance and thermal imaging produce enhanced energy sector performance.

Transportation: Delivery drones operating autonomously throughout Asia cut both total delivery duration and expenses by 70% and 20%, respectively, through the implementation of GPS systems and IoT solutions for fleet optimization. Urban Planning: Smart city applications in Singapore achieved a 30% boost in traffic movement and a 25% improvement in energy performance through AI and IoT.

Other Sectors: AI systems demonstrate their growing importance by demonstrating effectiveness with 95%+ accuracy in finance fraud detection along with 90%+ accuracy in manufacturing predictive maintenance and 80%–90% accuracy in machine learning applications for crime prevention.

AI-based recommendation systems operating in e-commerce storefronts and entertainment services demonstrated more than 90% precision, thus boosting consumer interaction and purchasing numbers. AI, along with IoT and automation, introduces improvements to efficiency, reduces costs, and generates better decisions in various fields, but institutions are still challenged with inaccurate data, biased algorithms, and capacity limitations of systems.

6.5 ROS TOOLS AND RESOURCES IN THE DEVELOPMENT OF UAS

Multiple resources and tools offer an opportunity to create UAS [15] based on the ROS. The testing, visualization, and integration techniques turn out to be more efficient in UAS applications due to the toolkit provided by the ROS platform.

6.5.1 Simulation tools of UAS

The 3D simulation program Gazebo allows application developers to run their programs in real-life situations. It has, among the features, ability to test various UAS systems and virtual sensor reproduction and physical simulation algorithms. Gazebo interface to the ROS enables developers to test their workflows and algorithms prior to their experimental deployment on physical hardware. RViz: This visualization tool enables the developers to display the robot designs and sensor readings in addition to showing the status of the system. Flight paths are also represented, sensor outputs are displayed, and the level of subsystem conditions is represented as a vital tool of UAS development RViz.

6.5.2 Flight controllers

PX4: It is an open-source flight control software acting as an autopilot environment in which the ROS incorporates without any problems [16]. PX4 is characterized by the strong flight control system that entails package operations in addition to the stability and automated waypoint flights. The final stage of UAS system development and testing is the PX4 Facilitating PX4 that enables the integration of Gazebo-compatible simulation environment. The open-source autopilot ArduPilot is popular among industrial users to use ROS applications. The system possesses many functions that incorporate dynamic navigation into sensor fusion and has the ability to integrate several UAS platforms. The integration of the ROS into ArduPilot provides mission control with the ability to communicate with flight controller-ROS node in real time.

6.5.3 Development environments

Popular integrated development environments (IDEs) used as the development environment of the ROS include Visual Studio Code (with extensions to the ROS) and CLion [17]. The environments enhance the ROS development results with workspace administration systems and syntax, highlighting features and debugging abilities. Debugging Tools: There are two vital tools that allow monitoring of the state of the developer system: RQT (a framework of plugins) and ROS CLI (Command Line Interface) with the help of which developers check topics nodes and services. To solve complex UAS system problems, developers need these debugging tools. ROS-based UAS development can only proceed if developers use programming environments alongside flight controllers and simulation tools. Developers can generate trustworthy applications, scalable programs, and effective UAS systems using these tools for a variety of use scenarios.

6.6 CHALLENGES AND CONSIDERATIONS IN ROS-BASED UAS DEVELOPMENT

Using the ROS to develop a UAS drone offers benefits but brings technical difficulties that the team had not anticipated. According to our study, three main problems exist related to resource management within the computer system and security issues among the peripherals. The system must function better in all aspects of UAV performance for reliable unmanned aerial vehicle systems based on the ROS.

6.6.1 Lack of real-time processing capabilities

The ROS was designed for robotics research purposes, which implies that real-time processing limitations do not need to exist. The UAS controls its flight path and mission tasks by processing data in real time. As ROS 2 progressed, each update brought important middleware, but researchers have found missing methods for deterministic performance.

6.6.2 Communication and data protection security

Robots depend on basic security elements to send updated data with telemetry and sensor feedback through wireless communications. ROS ME (a custom framework or extension) is susceptible to data protection hazards when potential attackers manipulate unauthorized clients to gather information or disrupt network operations by jamming the network or masquerading as trusted hosts. ROS-driven UAV platforms must use robust security methods including decryption and access protection to decrease risks of being hacked.

6.6.3 Reviewing hardware combination problems together with data stream integration issues

Integrating the ROS with different flight controllers and components, such as LiDARs, GPS, cameras, and IMUs (inertial measurement units), in a UAV system proves difficult since they lack native support. Developers create personalized driver technology and interface layers to fit sensors together when regular solutions do not exist. Integration of multiple sensors into fusion systems creates special challenges when gathering data from all sensors at once.

6.6.4 Issues due to heavy resource consumption and minimal system power

Smaller UAVs designed for taking pictures and rescue tasks have limited onboard processing abilities. Outdoor robot software that uses three separate ROS modules needs too many processing resources for basic platforms to handle. In future research, more solutions need to be developed like choosing more basic ROS versions or optimizing procedures to simplify processing tasks on edge systems.

6.6.5 System defect and reliability problems

Protective and rescuing UAVs are supposed to be automated and, therefore, require a high-level defect tolerance technology to avert disasters.

ROS publish–subscribe communication messages between nodes may be long-running, and so may load slowly and fail without any indication. Engineers have to find alternative methods of data flowing across the system and establish protection systems to offset disasters. They should also establish procedures to monitor issues.

6.6.6 Compliance and regulatory issues

Applications of UAVs should be guided by the Federal Aviation Administration (FAA) and the European Union Aviation Safety Agency (EASA) standards and requirements, which mandate appropriate airspace regulations and safety controls to be applied in practices of radio communication. The ROS-based UAV systems require remote identification capabilities such as safe landing and spatial boundary control to meet the safety standards and government approvals.

6.6.7 Complexity of software and learning curve

The ROS provides good power and support, but it is not friendly to learning for some new developers of UAVs. This gap should be filled by experienced people by debugging distributed systems and managing software updates. These difficulties can also be averted through detailed documentation, standardized initiatives, and mechanized examination systems.

6.6.8 UAS devices

In field operations, UAS devices must ensure system survival and maintain stable flight control within a limited time frame. A key concern in this context is meeting strict timing requirements in ROS nodes that handle sensor data processing and flight control commands. In ROS 2, the DDS protocol offers a better real-time behavior as compared to ROS 1, which is weak in real-time applications.

6.6.9 System real-time management in different operating conditions

This involves much simulation, testing, and validation. The security concerns are another factor to consider in UAS operations, as they are key in supporting crucial applications.

6.6.10 Other difficulties

Network Protection: Data can be misused or breached as they go through nodes, which must be avoided. This can be prevented by other security measures like TLS (transport layer security) and encryption. The authentication and authorization comes as a result of allowing nodes of the authorized system and users to gain access to connections. Preventive steps against a potential harmful cyberattack are provided through security audits and frequent software updates.

6.6.11 Sensor and actuator diversities

Sensor and actuator diversity means that the system uses different kinds of sensors and actuators rather than just one standard type. Along with the onboard computer platform, this setup has not yet been fully tested or proven to work with UAS hardware through ROS.

Because of this, more work is needed to make sure the system can properly communicate with real hardware. In particular, the ROS drivers for these sensors and actuators need to be checked and validated, and it should also be confirmed that these drivers can actually be implemented for the specific hardware being used.

6.6.12 Resource restrictions

Since most of the UAS platforms use a limited processing power, ROS node processing requirements and memory utilization must be optimized in specific cases.

6.6.13 Special interfaces and middleware

These establish communication between proprietary UAS modules and non-ROS system platforms. Although the basis of the ROS is still held, the development of UAS solutions needs significantly more consideration of time constraints, security requirements, and hardware considerations, which will make solutions developable, reliable, and efficient. Therefore, constructors must make good use of the ROS during UAS activities by foreseeing solutions to such complications.

6.7 USING THE ROS WITH GAZEBO FOR UAV AND ROBOT SIMULATION

Combination of the ROS and Gazebo (an open-source 3D robotics simulator) is capable of building effective UAV and robot simulations. The joint use of the ROS and Gazebo gives users an advanced simulation platform to test robots and UAVs. Gazebo gives accurate 3D physics, while the ROS links different parts of the system via topic services and actions. The platform delivers accurate physical simulation, sensor replications, onboard intelligence features, and common computer hardware risk reduction.

To begin, you should install the ROS and Gazebo through the following commands:

```
sudo apt-get update
sudo apt-get install ros-noetic-desktop-full
sudo apt-get install ros-noetic-gazebo-ros
ros-noetic-gazebo-plugins
```

While the program is running, you can start Gazebo by issuing the following command:

```
roslaunch gazebo_ros empty_world.launch
```

Now, we will create a basic robot through URDF (unified robot description format). The my_robot.urdf file represents a basic robot with a plain rectangular body through the following content:

```
<?xml version="1.0"?>
<robot name="simple_robot">
  <link name="base_link">
    <visual>
      <geometry>
        <box size="0.5 0.5 0.2"/>
      </geometry>
      <material name="blue">
        <color rgba="0 0 1 1"/>
      </material>
    </visual>
  </link>
</robot>
```

To place the robot in Gazebo, run the following command:

```
rosrun gazebo_ros spawn_model -file my_robot.urdf -urdf
-model my_robot -z 1
```

When the robot exists inside Gazebo, we can utilize ROS topics for control. The following Python script sends velocity operations to the /cmd_vel topic to drive the robot at 0.5 m/s straight ahead:

```
#!/usr/bin/env python3
import rospy
from geometry_msgs.msg import Twist
def move_robot():
    rospy.init_node('robot_mover', anonymous=True)
    pub = rospy.Publisher('/cmd_vel', Twist, queue_size=10)
    rate = rospy.Rate(10) # 10 Hz
```

```
    move_cmd = Twist()
    move_cmd.linear.x = 0.5 # Move forward at 0.5 m/s

    while not rospy.is_shutdown():
     pub.publish(move_cmd)
        rospy.loginfo("Moving thc robot forward...")
        rate.sleep()

if __name__ == '__main__':
    move_robot()
```

Running this script with python:

```
rosrun my_robot_package move_robot.py
```

After running this command, the Gazebo simulation robot will consistently advance forward. So, it is required to add a LiDAR sensor into our simulation setup now.

Adding a Hokuyo LiDAR sensor to my_robot.urdf can be done by the addition of the following code:

```
<gazebo>
  <sensor type="ray" name="laser">
    <update_rate>10</update_rate>
    <ray>
      <scan>
        <horizontal>
          <samples>100</samples>
          <resolution>1</resolution>
          <min_angle>-1.57</min_angle>
          <max_angle>1.57</max_angle>
        </horizontal>
      </scan>
      <range>
        <min>0.1</min>
        <max>10.0</max>
      </range>
    </ray>
  </sensor>
</gazebo>
```

We can retrieve LiDAR data through the /scan topic. The following Python program uses a LiDAR sensor to log detected minimum distances:

```
#!/usr/bin/env python3
import rospy
from sensor_msgs.msg import LaserScan
```

```
def lidar_callback(msg):
    rospy.loginfo(f"Min distance: {min(msg.ranges)} meters")

def read_lidar():
    rospy.init_node('lidar_reader', anonymous=True)
    rospy.Subscriber('/scan', LaserScan, lidar_callback)
    rospy.spin()
if __name__ == "__main__":
    read_lidar()
```

Run the script using:

```
rosrun my_robot_package read_lidar.py
```

The program displays the exact readings that LiDAR detects from nearby targets. Now, analyze how Micro Aerial Vehicle ROS (MAVROS) helps ROS users control PX4-based drones through simulation.

First, install the required packages:

```
sudo apt-get install ros-noetic-mavros
ros-noetic-mavros-extras
```

To start the drone simulation, enter the following command:

```
roslaunch mavros px4.launch fcu_url:="
udp://:14540@127.0.0.1:14557"
```

Our simulation is ready when we use ROS services to operate the drone.

The following Python script uses the /mavros/cmd/takeoff service to set the drone's altitude at 10 meters:

```
#!/usr/bin/env python3
import rospy
from mavros_msgs.srv import CommandTOL

def takeoff():
    rospy.init_node('drone_takeoff', anonymous=True)
    rospy.wait_for_service('/mavros/cmd/takeoff')
    takeoff_service = rospy.ServiceProxy('/mavros/cmd/
takeoff', CommandTOL)

    altitude = 10.0 # Takeoff altitude
    response = takeoff_service(altitude=altitude)
    if response.success:
        rospy.loginfo("Takeoff successful!")
```

```
    else:
        rospy.logwarn("Takeoff failed!")
if __name__ == "__main__":
    takeoff()
```

Execute the takeoff script using:

```
rosrun my_drone_package drone_takeoff.py
```

The UAV moves upward to 10 m during the Gazebo simulation. This system enables developers to check their flight software without exposing real equipment to risks.

Users can use the ROS and Gazebo for safe and effective development tests on robots both on the ground and in the air. With AI-drivers and robot navigation systems and sensor input, applications of robotics are enhanced and can be utilized in the real environment. We will write autonomous flight controls in our future work to overcome the challenges and use AI learning techniques to improve AI robots.

6.8 FUTURE TRENDS IN THE ROS AND UAS

The determining element that drives the technologies in the development of robots and UAS is the ROS [18]. Current technological trends can also be viewed as future directions that are shaping ROS and UAS applications. UAS is more widely used in the fields of wireless communication, smart farming, and KubeEdge [19–23].

6.8.1 ROS 2 developments

ROS 2 is an incremental development over its predecessor since it is also capable of supporting the existing requirements:

a. Better Real-Time Performance: DDS middleware can ensure that user can have confidence in real-time operations in UAS determinism.
b. Architecture Support to Distributed Systems: Over-the-air updates (OTA) provide a support architecture that enables multiple UASs to coordinate their activities and thus provides distributed sensing and automation of swarm activities.
c. Enhanced Protection Capabilities: The encryption and authentication capabilities of the platform provides communication channels that ensure that operators of drones are free from concerns about the escalating cybersecurity threats of UAS operations.

6.8.2 AI implementation

The implementation of the AI technology will revolutionize the application of UAS because it will enable it to perform high-capacity autonomous operations:

a. ROS 2 Autonomous Decision-Making: The features of ROS mission optimization, obstacle avoidance, and path planning are made possible by the introduction of machine learning algorithms.
b. Computer Vision: UAS platforms are able to monitor their targets and scan objects with the help of AI-assisted vision systems, which makes the system able to completely examine its surrounding environment.
c. Predictive Maintenance: With AI assistance and sensor data, one can predict and fix hardware problems to increase reliability and minimize downtime.

6.8.3 Standardization efforts

Interoperability standards has become necessary as the fields such as defense and business are increasingly adopting the unmanned aerial systems technology, and hence, there is a possibility of more drones creeping into the scene:

a. Military Drone Standards: Special focus is given to the military operations in the context of streamlining the software development based on the interoperability of the system.
b. Cross-Platform Compatibility: Standardized APIs work with protocols to provide bridges that enable ROS-based systems to co-exist as one system with non-ROS platform.
c. Regulatory Compliance: Standard processes and global regulatory compliance help sensitive applications receive adoption and trust from customers. The future of the Real-time Operations and UAS will be defined by the further development of the real-time technology, artificial intelligence features, and standardization of framework models development. Such future trends will enable the development and interoperability of highly safe advanced systems, which leads to new military and commercial opportunities.

6.9 CONCLUSION

The UAS industry has been transformed by an open-source software known as the ROS, which allows the development of robots to be customizable, scalable, and modular. This chapter has touched on the fundamental functionality and architecture of the ROS such as the nodes, topics, services, and the parameter server, which all make computing and communication easy in UAS applications. ROS integration in UAS has resulted in innovations in autonomous navigation, sensor fusion, swarm coordination, and real-time data processing, and it has made a large-scale breakthrough in productivity in disciplines such as defense, logistics, and aerial surveys. Negative aspects of the ROS include the following: it is not effective in real-time processing, and it lacks security and hardware compatibility. Nevertheless, these limitations has been fixed in ROS 2, which is able to give a better platform to operate in real time, is more secure, and has AI-driven robotics. Through the integration of AI and machine learning, intelligent flight planning, autonomous decision-making, and predictive maintenance will be enhanced, and UAS capabilities that use the ROS will have a positive effect. The continued development, standardization, and compliance with regulations of the ROS will increase the industrial application of UAS in future. The ROS will continue to be a vital tool in developing intelligent, dependable, and efficient robotics in the coming days as the need to possess autonomous aerial system increases.

REFERENCES

1. P. Canelas, M. Tavares, R. Cordeiro, A. Fonseca and C. S. Timperley (2022). An Experience Report on Challenges in Learning the Robot Operating System, *IEEE/ACM 4th International Workshop on Robotics Software Engineering (RoSE)*, Pittsburgh, PA, pp. 33–38, doi: 10.1145/3526071.3527521.
2. E. A. Nurrohmah, I. K. Wibowo, M. M. Bachtiar and M. M. Lathief (2021). Improvement of the Processing Speed of The Robot's Vision System Using Robot Operating System. *International Electronics Symposium (IES)*, Surabaya, Indonesia, pp. 482–487, doi: 10.1109/IES53407.2021.9593989.
3. D. Yukhimets, M. Belousov and A. Gubankov (2024). Development of ROS-based Information-Control System for Hybrid Unmanned Underwater Vehicle. *International Conference on Ocean Studies (ICOS)*, Vladivostok, Russian Federation, pp. 152–156, doi: 10.1109/ICOS63634.2024.10775401.
4. A. Amandine, E. Antoine, J. G. Ramírez-Torres and G. Jean-Pierre (2024). Mobile Manipulation Based on ROS1/ROS2 Middleware: Software Architecture and Evaluation of ROS2 Planning Algorithms. *21st International Conference on Electrical Engineering, Computing Science and Automatic Control (CCE)*, Mexico City, Mexico, pp. 1–6, doi: 10.1109/CCE62852.2024.10771072.
5. A. M. Nataraja, R. Fernández-Pascual and A. Ros (2024). Enhanced System-Level Coherence for Heterogeneous Unified Memory Architectures. *IEEE International Symposium on Workload Characterization (IISWC)*, Vancouver, BC, pp. 273–283, doi: 10.1109/IISWC63097.2024.00032.
6. I. Davila, A. Valenzuela, R. Ortiz, J. Choi and F. Yumbla (2024). Software Architecture and Simulation Interface for Autonomous Underwater Vehicles. *IEEE Eighth Ecuador Technical Chapters Meeting (ETCM)*, Cuenca, Ecuador, pp. 1–6, doi: 10.1109/ETCM63562.2024.10746001.

7. A. Al Arafat, K. Wilson, K. Yang and Z. Guo (2024). Dynamic Priority Scheduling of Multithreaded ROS 2 Executor with Shared Resources. *IEEE Transactions on Computer-Aided Design of Integrated Circuits and Systems*, vol. 43, no. 11, pp. 3732–3743, doi: 10.1109/TCAD.2024.3445259.
8. R. Iwakami, B. Peng, H. Hanyu, T. Ishigooka and T. Azumi (2024). AUTOSAR AP and ROS 2 Collaboration Framework. *27th Euromicro Conference on Digital System Design (DSD)*, Paris, France, pp. 319–326, doi: 10.1109/DSD64264.2024.00050.
9. M. Bueno, I. Bernardino, S. Asif and P. Webb (2024). Unlocking the potential of Robot Manipulators: Seamless Integration Framework. *IEEE 20th International Conference on Automation Science and Engineering (CASE)*, Bari, Italy, pp. 2610–2617, doi: 10.1109/CASE59546.2024.10711476.
10. N. Shaukat et al. (2024). Trustworthy ROS Software Architecture for Autonomous Drones Missions: From Robo Chart Modelling to ROS Implementation. *20th IEEE/ASME International Conference on Mechatronic and Embedded Systems and Applications (MESA)*, Genova, Italy, pp. 1–7, doi: 10.1109/MESA61532.2024.10704818.
11. C.-H. Cheng, K.-T. Huang, Y.-F. Huang, Y.-X. Xu and K.-S. Liao (2024). Deep Learning Object Detection for Vehicle-to-Everything Systems with ROS 2.0 Distributed Communication Architecture. *International Conference on Consumer Electronics - Taiwan (ICCE-Taiwan)*, Taichung, Taiwan, pp. 527–528, doi: 10.1109/ICCE-Taiwan62264.2024.10674437.
12. D. Larkin, M. Novitzky, J. Kim and C. M. Korpela (2021). ATAK Integration through ROS for Autonomous Air-ground Team. *International Conference on Unmanned Aircraft Systems (ICUAS)*, Athens, Greece, pp. 1116–1122, doi: 10.1109/ICUAS51884.2021.9476676.
13. J. Jackson, D. Koch, T. Henrichsen and T. McLain (2020). ROSflight: A Lean Open-Source Research Autopilot. *IEEE/RSJ International Conference on Intelligent Robots and Systems (IROS)*, Las Vegas, NV, pp. 1173–1179, doi: 10.1109/IROS45743.2020.9341653.
14. M. Ivanytskyi, Y. Znakovska and Y. Averyanova (2023). Meteorological Information Access and Decision-Making for UAS Flight Planning. *17th International Conference on the Experience of Designing and Application of CAD Systems (CADSM)*, Jaroslaw, Poland, pp. 28–32, doi:10.1109/CADSM58174.2023.10076518.
15. Y. Xie, K. Wu, T. Zhao and X. Gao (2020). Research and Design of Dynamic Strategy Distributed Control Architecture in Power Internet of Things. *International Conference on Artificial Intelligence and Computer Engineering (ICAICE)*, Beijing, China, pp. 403–407, doi: 10.1109/ICAICE51518.2020.00085.
16. M. Albero, F. Blanes, G. Benet, J. E. Simo and J. Coronel (2006). Advanced Distributed Architecture for a Small Biped Robot Control. *6th IEEE-RAS International Conference on Humanoid Robots*, Genova, Italy, pp. 358–363, doi: 10.1109/ICHR.2006.321297.
17. N. Souli et al. (2024). Performance Evaluation of a Prototype UAV-Based Secure Communication System Employing ROS and Chaotic Communications. *International Conference on Unmanned Aircraft Systems (ICUAS)*, Chania-Crete, Greece, pp. 1034–1041, doi: 10.1109/ICUAS60882.2024.10556964.

18. U. Jahn et al. (2020). A Recommendation for a Systems Engineering Process and System Architecture for UAS. *IEEE 3rd International Conference and Workshop in Óbuda on Electrical and Power Engineering (CANDO-EPE)*, Budapest, Hungary, pp. 91–96, doi: 10.1109/CANDO-EPE51100.2020.9337752.
19. V. Aggarwal, et al. (2024). CleanComm: A Hybrid ACO-WOA-MOA Model for Sustainable Communication. *IEEE 4th International Conference on ICT in Business Industry & Government (ICTBIG)*, Indore, Madhya Pradesh, India.
20. P. Ajmani, N. Singh, and P. Verma (2022). Internet of Vehicles Taxonomy and Evaluation: Architectures, Protocols, and Issues. *5th IEEE International Conference on Contemporary Computing and Informatics (IC3I)*, Uttar Pradesh, India.
21. L. M. Saini, et al. (2024). KubeEdge for Scalable IoT Applications. *7th IEEE International Conference on Contemporary Computing and Informatics (IC3I)*, Vol. 7.
22. V. Sharma, B. Balusamy, G. Ferrari, and P. Ajmani (Eds.). (2024). *Wireless Communication Technologies: Roles, Responsibilities, and Impact of IoT, 6G, and Blockchain Practices* (1st ed.). CRC Press. doi: 10.1201/9781003389231.
23. P. Ajmani and P. Saigal (2023). *5G and IoT for Smart Farming. Applying Drone Technologies and Robotics for Agricultural Sustainability*. IGI Global, pp. 124–139.

Chapter 7

Image-processing applications in modern UAV systems

Pranav Paranjpe and Rajesh Bodade

7.1 INTRODUCTION

The advances in the unmanned aerial vehicle (UAV) technology have led to many industrial changes and upgraded the traditional aerial observations with better information-gathering skills. What had begun as simple remote-controlled flying devices have now become sophisticated aircraft with computing power and sensing technologies integrated into them.

The manipulation and analysis of visual data are image-processing approaches that allow for the extraction of new information, improvement of image quality, and transformation of images into analytical resources. UAV applications require image processing in order to generate relevant information from the gathered visual data, which enables them to perform object detection and tracking and produce detailed maps of terrain and structures.

The three main factors that make image processing important to UAV systems are as follows:

1. Enhanced Data Interpretation: The processing of visual image data provides useful information for applications such as crop health assessment, infrastructure inspection, and environmental monitoring.
2. Real-Time Decision Support: The system helps take real-time decisions through on-board processing in critical time-sensitive situations such as search and rescue operations and traffic control.
3. Automation Capabilities: Image processing improves automation capability of UAV systems so that they can perform automated obstacle detection and thus can execute missions independent of human operators.

UAV has experienced tremendous growth in its application range during the recent period, serving in fields such as precision agriculture, urban planning, disaster management, and environmental monitoring. The advancement of UAV technology has emerged from improved hardware features and

DOI: 10.1201/9781003598350-9

software processing methods, especially in image processing and computer vision.

Different image-processing methods improve capabilities of the UAV technology in diverse application domains as this chapter details. This chapter evaluates modern technological developments and performance measurements, and forecasts future developments in this field which is progressing aggressively.

7.2 UAV CLASSIFICATION AND CHARACTERISTICS

UAVs encompass a range of designations and technical properties that reflect the rapid developments of emerging technologies in the field. Modern UAV systems have unique operation characteristics that influence their functional capacity and range of applications. The core UAV classification model draws on multiple operational attributes to systematically define the functional capabilities and operational characteristics of each UAV category.

7.2.1 Size-based classification system

Size-based classification is one of the primary organizational frameworks that are applied in the classification of UAV vehicles. Large-scale operations involving large UAV platforms with wingspans of more than 10 meters are carried out in specialized surveillance purposes and large-range missions. Due to their outstanding endurance features, such platforms can surpass 24-hour persistent flight time. The large size of these platforms (5–10 m) leads to compromises in operational efficiency and deployment flexibility, which limits the use of these platforms in geological surveys, maritime protection missions, and others.

Small UAVs, ranging in size from 50 cm to 2 m, are the most versatile products used in the industry to carry out duties across civilian environments and commercial business sectors. These systems operate in the altitude of 100–500 m and have a flight endurance of 2–8 hours. Small-sized UAV platforms that range between 1 and 1.4 m in size have gained a popular use in urban applications and accurate agricultural purposes, [1] assert that mini-UAVs provide optimum performance in a limited environment and have sufficient capacity to carry heavier loads such as high-level sensors (Figure 7.1).

7.2.2 Operation-based classification system

In the operation-based system, the main elements that are considered are endurance, range, and mission capabilities. Operation based classification system operate with minimal cost and in short ranges that extend to 5 km

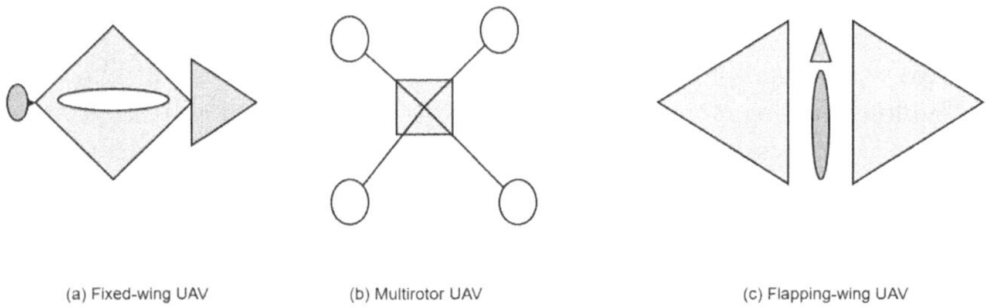

Figure 7.1 Examples of UAVs: (a) mini UAV (b) long-endurance UAV, and (c) flapping-wing UAV).

and achieve an operational time of 1–2 hours. Most urban surveillance operations and rapid response applications make use of these systems. Systems classified under the mid-range category operate between 10 and 30 km distance and sustain a flight time of 5–10 hours, which enables them carry out environmental observations and infrastructure maintenance.

Long-endurance category possesses the highest level of UAV operational capabilities such as a flight time of longer than 24 hours and an operational range of over 150 km. These systems utilize a high-power management technology and aerodynamics to extend their operational flight time without affecting their performance in missions (Figure 7.2).

7.3 IMAGE-PROCESSING METHODOLOGY FRAMEWORK

Image-processing algorithms are utilized in UAVs to perform identification activities where untamed visual images are transformed into processable data. This system includes several steps including preliminary data collection, sophisticated analysis, and interpretation. There is a comprehensive procedure that is used for processing UAV images and analyzing its structural components to derive necessary results.

7.3.1 Data acquisition and preprocessing

Visual data are acquired in the first stage through multiple sensor instruments before the analysis of its details. Key components are as follows:

1. Sensor Integration: Application requirements determine the selection and integration process for appropriate imaging sensors, which include RGB (Red Green Blue), multispectral, thermal, and hyperspectral sensors.
2. Image Enhancement: This process improves images by using applications of various techniques that reduce noise and increase the virtue of contrast and color correction.

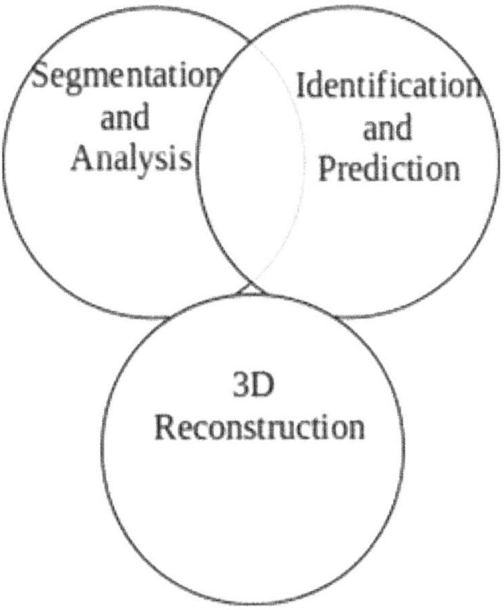

Figure 7.2 Relationship between UAV categories by techniques and applications (self-created).

3. Geometric Correction: The geometric correction method eliminates photograph imperfections brought about by camera lenses, UAV flight movements, and other atmospheric factors.
4. Image Registration: This involves matching different images or image sequences toward similar analysis and 3D reconstruction.

The study demonstrates that appropriate preprocessing mechanisms can help increase the quality of downstream analysis since their noise-cutting mechanisms can raise the segmentation accuracy by 15%–20% even in worse environmental circumstances.

7.3.2 Segmentation and analysis framework

Image segmentation procedures have adaptive algorithms that vary according to environmental conditions. According to [2], the development of thresholds in conjunction with machine learning practices brings the best results from the segmentation tasks of multi-level images. Technicians can employ morphological inspection operations to detect specific structural abnormalities and structural defects in infrastructures (Figures 7.3–7.5).

7.3.3 Advanced classification systems

The existing classification methods apply ensemble learning approaches to attain reliable performances across diverse operational environments. According to [3] if multiple classification methods are combined, an accuracy level of 96 percent can be achieved in detection applications of objects. The security applications can carry out a complex tracking and monitoring exercise by the addition of the temporal analysis features to the system.

The choice between different classification methods depends entirely on two factors: application demands and system processing power restrictions. Table 7.2 demonstrates popular machine learning methods used in UAV image processing and their performance, resource usage, and computational speed.

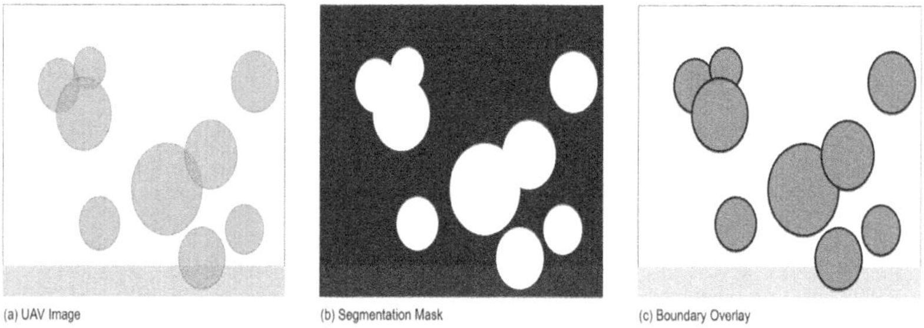

Figure 7.3 Segmentation result of trees in UAV images.

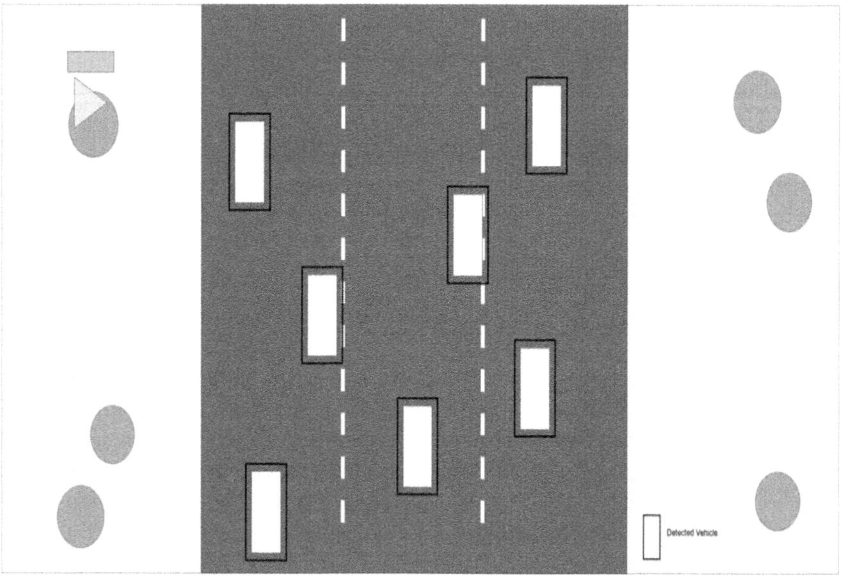

Figure 7.4 Car detection in drone images by cascade classifier.

Figure 7.5 Tobacco plant segmentation and recognition results.

7.3.4 Three-dimensional reconstruction methodologies

Advanced three-dimensional reconstruction methods led to the development of accurate mapping techniques and modeling procedures. Companies that use Structure from Motion (SfM) algorithms, as explained by [3], can document the structure of buildings with millimeter-level accuracy. Deep learning and multi-view stereo techniques work together to make real-time three-dimensional reconstruction processes work in continuously changing environments (Figure 7.6).

Table 7.7 provides a full breakdown of various 3D reconstruction methods based on their accuracy, processing time, resource use, and cost. The information given is still useful as a reference tool to help you choose a method for three-dimensional reconstruction based on your needs.

Figure 7.6 Example of 3D-reconstructed images from 2D drone images.

7.4 ADVANCED IMAGE-PROCESSING APPLICATIONS

Advanced image-processing systems have made a big difference in how UAVs are used in different operational areas. Modern systems have unmatched abilities to process and analyze data in real time, which helps in complicated decision-making across different operational settings.

7.4.1 Agricultural and environmental monitoring

Advanced image-processing technology proves highly efficient in the evaluation of crop health and yield forecasting within agricultural contexts. [4] proved that the utilization of deep learning algorithms and multispectral imaging could achieve the detection of crop diseases with an accuracy higher than 95% in agricultural fields. The combination of NDVI analysis and machine learning algorithms allows farmers to execute an accurate inspection of crop well-being by stage monitoring.

With the addition of hyperspectral analysis, thermal imaging has brought major developments in environmental monitoring systems. According to [3], an accuracy of 92% can be achieved in population monitoring through UAV thermal imaging with automated object detection processing. In many modern mapping applications to monitor forest canopies and geological features, highly accurate three-dimensional imaging results can be achieved by employing LiDAR systems along with traditional imaging technology.

Complexity of implementation, detection performance, and false positive tendency of agricultural monitoring applications are given in Table 7.4. Table 7.6 presents the evaluation metrics for environmental monitoring applications, highlighting precision levels, operational coverage area, and weather sensitivity as the primary assessment criteria.

7.4.2 Urban infrastructure and security applications

Modern urban monitoring systems have the capability to perform instantaneous object detection and tracking operations. According to [4], the deployment of convolutional neural networks detects objects at accuracy rates above 98% in traffic monitoring applications. Deep learning algorithms running in infrastructure inspection systems achieve detection at an accuracy rate of 94%.

Table 7.5 presents a comprehensive comparison between the applications of urban monitoring and infrastructure inspection, including typical performance, real-time capability, and reliability assessment. The data provided serve as a useful tool in selecting suitable methods for the purpose of meeting particular qualification needs and boundary limitations.

7.5 PERFORMANCE COMPARISON AND ANALYSIS (TABLES 7.1–7.8)

These tables represent different applications and methodologies for processing UAV images in detail. The search results show how the accuracy and processing speed relate to resource use when choosing algorithms for applications.

Table 7.1 Indicative Endurance classification for UAV systems [1,2]

Category	*Flight endurance*	*Operational range*	*Primary applications*
Very low cost	Very low	Very short	Recreational use, training, basic photography
Close range	Low to medium	Short	Local inspection, mapping, field scouting
Mid-range	Medium	Medium	Corridor mapping, regional surveillance, research
Long endurance	High	Long	Wide-area surveillance, environmental monitoring, defense/research applications

Table 7.2 Qualitative comparisons of machine learning algorithms for UAV image processing [3–7]

Algorithm	*Detection accuracy*	*Processing speed*	*Memory/compute usage*
YOLOv4	High	High	Medium
Faster R-CNN	High to very high	Low to medium	High
SSD	Medium to high	High	Low

Table 7.3 Qualitative comparison of image-processing techniques in UAV applications [6,8,14]

Technique	*Application domain*	*Typical performance*	*Processing complexity*	*Resource usage*
Global thresholding	Agriculture / simple vegetation separation	Medium	Low	Low
Adaptive thresholding	Forestry / non-uniform illumination scenes	Medium to high	Low to medium	Medium
Deep learning (CNN)	Urban monitoring / complex visual scenes	High to very high	Medium to high	High
SVM classification	Wildlife detection / crop-state classification	Medium to high	Medium	Medium
Morphological operations	Infrastructure inspection / mask refinement	Medium	Low	Low
Random forest	Environmental monitoring / land-use classification	Medium to high	Medium	Medium to high

Table 7.4 Qualitative comparison of agricultural monitoring applications [8–10]

Application type	*Processing method*	*Typical detection performance*	*False-positive tendency*	*Implementation complexity*
Crop disease detection	Multispectral analysis	High	Low	High
Plant count	CNN + watershed	High	Low	Medium
Growth stage analysis	SVM + color features	Medium to high	Low to medium	Medium
Weed detection	Random forest	Medium to high	Medium	Low to medium
Yield prediction	Deep learning	Medium to high	Medium	High

Table 7.5 Qualitative comparison of urban monitoring and infrastructure inspection results [6, 7,11–13]

Monitoring task	*Method used*	*Typical performance*	*Real-time capability*	*Operational reliability*
Traffic flow analysis	CNN	High	High	High
Building inspection	Edge detection + ML	Medium to high	Low to medium	High
Crowd monitoring	YOLO v4	High	High	High
Infrastructure defects	ResNet50	Medium to high	Low to medium	Medium to high
Construction progress	3D reconstruction	Medium to high	Low	Medium to high

Table 7.6 Qualitative environmental monitoring performance metrics [14–18]

Application	*Processing technique*	*Accuracy (%)*	*Coverage area (ha/hr)*	*Weather dependency*
Forest cover analysis	Segmentation + CNN	High	Medium to high	Medium
Wildlife tracking	Object detection	Medium to high	Medium	High
Water body mapping	Spectral or multi-band analysis	High	Medium to High	Medium
Land use classification	Random forest / CNN	Medium to high	High	Medium
Erosion monitoring	3D modeling / change detection	Medium to high	Medium	High

Table 7.7 Qualitative comparison of 3D reconstruction methods [19,20]

Method	*Typical geometric accuracy*	*Processing time*	*Resource requirements*	*Cost efficiency*
Photogrammetry	High	Medium to high	High	Medium to high
LiDAR integration	Very high	Medium	Very high	Low to medium
Structure from motion	High	Medium	Medium	High
Multi-view stereo	Medium to high	Medium to high	Medium to high	Medium

Table 7.8 Qualitative system requirements for different processing methods [6,7,11,12,19–21]

Processing method	*CPU requirements*	*GPU requirements*	*Memory requirement*	*Storage requirement*
Basic image processing	Medium	Low	Low to medium	Low
Deep learning	High	High	Medium to high	Medium to high
Real-time analysis	High to very high	High to very high	Medium to high	Medium to high
3D reconstruction	High	Medium to high	Medium to high	High
Multi-sensor fusion	Medium–high	Medium to High	Medium to high	Medium

7.6 CONCLUSION

The UAV technology with modern image-processing capabilities is bringing great new functionalities to different applications. This chapter dealt with the basic structures, methods, performance measurements, and the current status of image-processing applications using the UAV technology.

The Chapter produces several important specific discoveries:

1. Methodology Evolution: The methodology of research applied to UAV applications has evolved from simple techniques like thresholding and filtering to complex methods such as deep learning methods, which have improved the accuracy of performance and the stability of operations in difficult environmental conditions.
2. Application Diversity: The multiple domains that UAV-based image processing serves through specialized processing techniques include precision agriculture, environmental monitoring, inspection of urban infrastructure, and security applications.
3. Performance Trade-offs: Several factors help determine the performance results when it comes to choosing imaging techniques since there is no single method that offers maximum results in every operational condition.
4. Integration Challenges: Integration of advanced image-processing methods into UAV platforms becomes challenging due to power requirements, processing speed, and efficiency.

Specific research efforts for the future development of UAV-based image processing should concentrate on the following aspects:

1. Edge Computing Integration: Edge computing technologies must be integrated with UAV systems to provide on-board processing functions. This approach helps operations run smoothly without ground stations, thus enabling more self-piloted flights.
2. Multi-sensor Fusion: A better understanding of the environment can be achieved by combining information from different types of sensors, such as RGB, thermal, and LiDAR, than using a single sensor.
3. Temporal Analysis: Development of new algorithms for analyzing image sequences through time patterns can be useful for applications like change detection and predicting trends.
4. Automated Decision Support: Systems can take active decisions by integrating with automated decision support systems. These systems turn imagery output into useful operational recommendations.

Moving forward, advancements in hardware technology and improvements in algorithms will expand the capabilities of UAV-based image-processing applications. This will create new opportunities in science, business, and public service.

BIBLIOGRAPHY

[1] Ahmed, F., Mohanta, J.C., Keshari, A., & Yadav, P.S. (2022). Recent Advances in Unmanned Aerial Vehicles: A Review. *Arabian Journal for Science and Engineering*, 47, 7963–7984. https://doi.org/10.1007/s13369-022-06738-0

[2] Mohsan, S.A.H., Khan, M.A., Noor, F., Ullah, I., & Alsharif, M.H. (2022). Towards the Unmanned Aerial Vehicles (UAVs): A Comprehensive Review. *Drones*, 6(6), 147. https://doi.org/10.3390/drones6060147

[3] Bochkovskiy, A., Wang, C.-Y., & Liao, H.-Y.M. (2020). YOLOv4: Optimal Speed and Accuracy of Object Detection. arXiv:2004.10934. https://doi.org/10.48550/arXiv.2004.10934

[4] Ren, S., He, K., Girshick, R., & Sun, J. (2015/2016). Faster R-CNN: Towards Real-Time Object Detection with Region Proposal Networks. arXiv:1506.01497. https://doi.org/10.48550/arXiv.1506.01497

[5] Liu, W., Anguelov, D., Erhan, D., Szegedy, C., Reed, S., Fu, C.-Y., & Berg, A.C. (2016). SSD: Single Shot MultiBox Detector. In European Conference on Computer Vision. https://doi.org/10.1007/978-3-319-46448-0_2

[6] Tang, G., Ni, J., Zhao, Y., Gu, Y., & Cao, W. (2024). A Survey of Object Detection for UAVs Based on Deep Learning. *Remote Sensing*, 16(1), 149. https://doi.org/10.3390/rs16010149

[7] Chen, C., et al. (2023). YOLO-Based UAV Technology: A Review of the Research and Its Applications. Drones, 7(3), 190. https://doi.org/10.3390/drones7030190

[8] Zualkernan, I., Abuhani, D.A., Hussain, M.H., Khan, J., & ElMohandes, M. (2023). Machine Learning for Precision Agriculture Using Imagery from Unmanned Aerial Vehicles (UAVs): A *Survey. Drones*, 7(6), 382. https://doi.org/10.3390/drones7060382

[9] Guebsi, R., et al. (2024). Drones in Precision Agriculture: A Comprehensive Review of Applications, Technologies, and Challenges. *Drones*, 8(11), 686. https://doi.org/10.3390/drones8110686
[10] Tsouros, D.C., Bibi, S., & Sarigiannidis, P.G. (2019). A Review on UAV-Based Applications for Precision Agriculture. *Information*, 10(11), 349. https://doi.org/10.3390/info10110349
[11] Liang, H., Lee, S.-C., Bae, W., Kim, J., & Seo, S. (2023). Towards UAVs in Construction: Advancements, Challenges, and Future Directions for Monitoring and Inspection. *Drones*, 7(3), 202. https://doi.org/10.3390/drones7030202
[12] Choi, H.W., et al. (2023). An Overview of Drone Applications in the Construction Industry. *Drones*, 7(8), 515. https://doi.org/10.3390/drones7080515
[13] Spencer, B.F. Jr., Hoskere, V., & Narazaki, Y. (2019). Advances in Computer Vision-Based Civil Infrastructure Condition Assessment. *Engineering*, 5(2), 199–222. https://doi.org/10.1016/j.eng.2018.11.030
[14] Manfreda, S., McCabe, M.F., Miller, P.E., Lucas, R., et al. (2018). On the Use of Unmanned Aerial Systems for Environmental Monitoring. *Remote Sensing*, 10(4), 641. https://doi.org/10.3390/rs10040641
[15] Dainelli, R., Toscano, P., Di Gennaro, S.F., & Matese, A. (2021). Recent Advances in Unmanned Aerial Vehicle Forest Remote Sensing - A Systematic Review. Part I: A General Framework. *Forests*, 12(3), 327. https://doi.org/10.3390/f12030327
[16] Dainelli, R., Toscano, P., Di Gennaro, S.F., & Matese, A. (2021). Recent Advances in Unmanned Aerial Vehicle *Forest* Remote Sensing - A Systematic Review. Part II: Research Applications. Forests, 12(4), 397. https://doi.org/10.3390/f12040397
[17] Velez-Nicolas, M., Garcia-Lopez, S., Barbero, L., & Ruiz-Ortiz, V. (2021). Applications of Unmanned Aerial Systems in Surface Water and Groundwater Research: A Review. *Remote Sensing*, 13(7), 1359. https://doi.org/10.3390/rs13071359
[18] Sibanda, M., Mutanga, O., Chimonyo, V.G.P., Clulow, A.D., Shoko, C., Mazvimavi, D., Dube, T., & Mabhaudhi, T. (2021). Application of Drone Technologies in Surface Water Resources Monitoring and Assessment: A Systematic Review. *Drones*, 5(3), 84. https://doi.org/10.3390/drones5030084
[19] Deliry, S.I., & Avdan, U. (2021). Accuracy of Unmanned Aerial Systems Photogrammetry and Structure from Motion in Surveying and Mapping: A Review. *Journal of the Indian Society of Remote Sensing*, 49, 1997–2017. https://doi.org/10.1007/s12524-021-01366-x
[20] Berra, E.F., & Peppa, M.V. (2020). Advances and Challenges of UAV SFM MVS Photogrammetry and Remote Sensing: Short Review. *The International Archives of the Photogrammetry, Remote Sensing and Spatial Information Sciences*, XLII-3/W12-2020, 267-272. https://doi.org/10.5194/isprs-archives-XLII-3-W12-2020-267-2020
[21] Elamin, A., et al. (2022). UAV-Based Multi-Sensor Data Fusion for Urban Land-Use Classification Based on a Deep Convolutional Neural Network. *Remote Sensing*, 14(17), 4298. https://doi.org/10.3390/rs14174298

Part 3

Communication and security

Chapter 8

Mist computing in distributed UAV swarm control

Rolando Herrero and Martin Cadirola

8.1 INTRODUCTION

Internet of Unmanned Aerial Vehicle (I-UAV) networks involve swarms of UAVs that are used to offer low-latency and low-loss communication to support the transfer of sensor readouts to applications. Applications rely on *Machine Learning (ML)* techniques to extract knowledge [1]. This approach is consistent with applications of *modern Cyber-Physical Systems* (CPSs), where to extend battery life on UAVs, low-end processors are utilized to sense the environment and provide primary connectivity [2].

Extending battery life requires low power consumption, and low power consumption leads to low *Signal-to-Noise Ratios* (SNRs). Low SNR limits computational complexity and reduces transmission rates. Because of this, there is another level of computational complexity that is placed on dedicated resources in cloud or edge infrastructures. IoT and standard Internet protocols are utilized to transfer readouts from UAVs to this additional layer of dedicated computational resources. This situation is presented in Figure 8.1. It can be observed that readouts were transformed into knowledge and then provided to consumers as end users. The current scenario for the application being considered describes a case where each UAV sent readouts to the gateway, where multiple readouts were then sent to the application.

A low-latency alternative is mist computing, whereby readout processing is performed by UAVs in rotation. This topology of mist computing is shown in Figure 8.2. Accordingly, UAVs form a cluster through which they transmit the readouts to one of the UAVs, which plays the role of *Cluster Head* (CH) [3]. This UAV processes the readouts and extracts the knowledge to be forwarded to the end consumer. The CH is selected so that it maximizes the overall lifetime of the network. Note that this topology does not require any dedicated application to extract the knowledge. Even though latency is lower, a drawback of this approach is the increased energy consumption of the UAVs.

While I-UAV scenarios rely on IoT networking protocols designed for device-to-gateway connectivity, due to signal coverage, protocol stacks fall

DOI: 10.1201/9781003598350-11

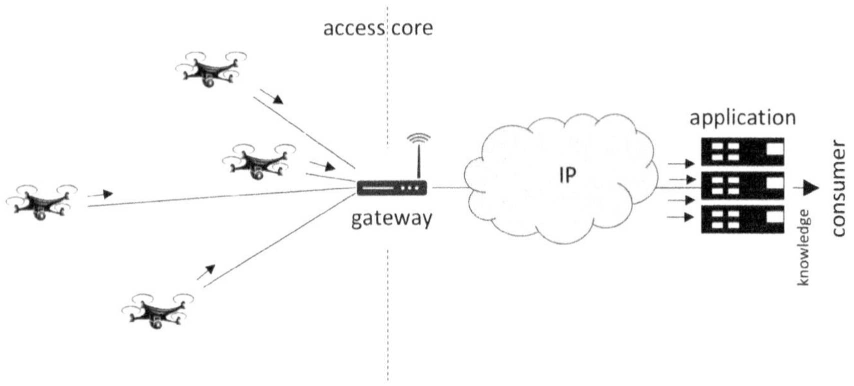

Figure 8.1 Cloud-based intelligence.

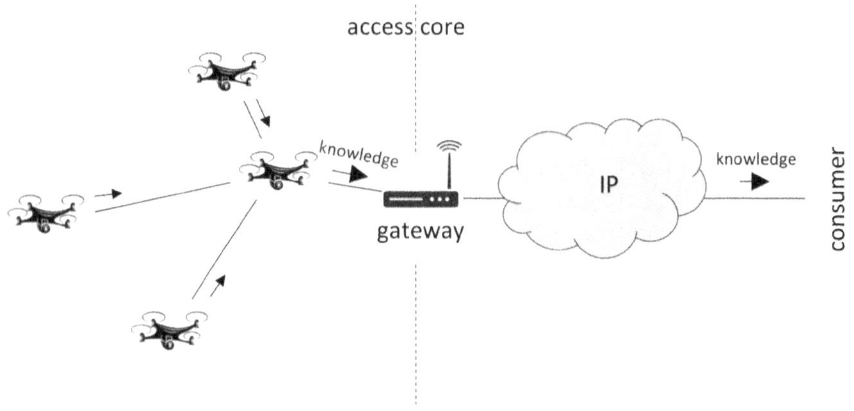

Figure 8.2 Mist-based intelligence.

into two main categories: *Wireless Personal Area Networks* and *Low-Power Wide-Area Networks.*

In this context, IoT networking protocols usually follow a layered scheme, which generally consists of five layers with each layer playing a specific role:

- The first layer, Physical Layer, interfaces the digital and analog domains so that access to the communications channel can be achieved.
- The second layer, Link Layer, provides inter-device connectivity between neighbors.
- The third layer, Network Layer, provides end-to-end connectivity between devices.
- The fourth layer, Transport Layer, manages the trade-offs between latency and loss to support diverse application requirements.

- The fifth layer, Application Layer, provides support for the actual application messages.

This five-layer architecture is commonly referred to as the Internet Protocol Suite, developed by the *Internet Engineering Task Force* (IETF).

In this chapter, the focus is on the LoRa-based protocol stack shown in Figure 8.3. LoRa is a physical- and link-layer technology that introduces an Low-Power Wide-Area Network (LPWAN) topology, supporting long-range connectivity between devices over distances of several kilometers. To support end-to-end IPv6 connectivity, LoRa layers are combined with an *IPv6 over Low-Power Wireless Personal Area Networks* (6LoWPAN) adaptation layer, providing wider *Internet Protocol version 6* (IPv6) connectivity to support IoT applications [4]. Regardless of the architecture (cloud or mist computing-based), the actual sensor readouts are encapsulated over the *Constrained Application Protocol* (CoAP) to provide a path from devices to knowledge extraction applications [5]. In this context, this chapter discusses the use of CoAP over LoRa, where the CH relies on *Representational State Transfer* (REST) transactions to retrieve readouts from the other UAVs in the cluster. Note that CoAP sessions are encapsulated over a combination of User Datagram Protocol (UDP) and Datagram Transport-Layer Security (DTLS), respectively, providing low latency and security.

The structure of this chapter is as follows: Section 8.2 presents a literature review and articulates the motivations for this work. Section 8.3 includes descriptions of the physical layer. Section 8.4 includes descriptions of the link layer. Section 8.5 describes the IPv6 and the IPv6 adaptation layer. Section 8.6 describes the transport and application layers. Section 8.7 describes the mist computing scheme. Section 8.8 describes an experimental framework used to validate the stack. Section 8.9 concludes this part of the report.

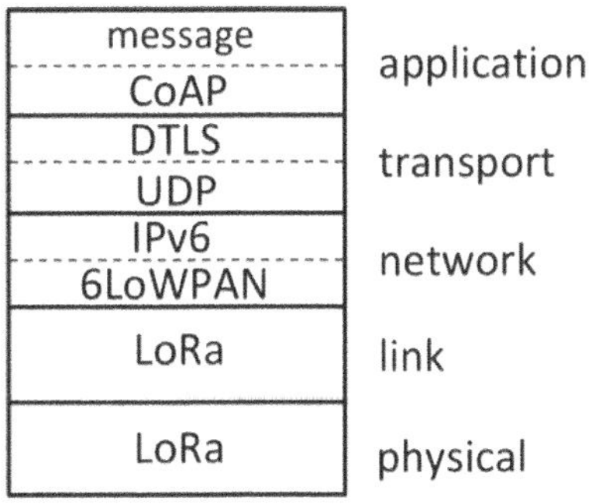

Figure 8.3 Protocol stack.

8.2 LITERATURE REVIEW AND MOTIVATION

Swarms of UAVs within the context of I-UAV to support wireless communications that enable the transmission of sensor readouts is a research area driven by significant attention within the last few years. In particular, a study [6] introduced a novel wireless communication system for UAVs by integrating LoRa for long-range, low-power wireless communication with the *Robot Operating System* (ROS) for robust coordination. The integrated approach is particularly designed for UAV-to-X applications, such as those in remote areas or disaster zones, where traditional communication networks may be not available or unreliable. By using the LoRa protocol, one can achieve long-range communication reliably. Besides, ROS provides a basic framework to manage complex behaviors of UAVs and efficient multiple-UAV coordination. By proving its viability and efficiency of the proposed system for various UAV-to-X applications, this work presents a successful implementation and testing of the proposed system on a real-world system. The main issue addressed in a recent work [7] is the challenge of ensuring reliable communication between a swarm of UAVs operating within the 5G ecosystem and guided by the 3GPP standards. This chapter therefore discusses the need for robustness of communications for swarms of UAVs and presents the recommendations regarding the design of ultra-reliable communication systems both from a perspective of software protocol stacks and Radio Frequency (RF) hardware. The authors discuss how the ideas could be demonstrated with the Easy-Swarm open-source platform utilizing LoRa radio and low-latency channel access protocol. Results on a real-world test with 10 UAVs and 20 robotic cars show that LoRa outperforms Wi-Fi in terms of reliability allowing for higher swarm density and longer coverage. Furthermore, LTE demonstrated the highest reliability and the lowest latency when network connectivity was strong. In a study [8], the authors present a collaborative framework that exploits cross-technology communications to achieve covertness constraints. The aim is to minimize the age of covert information (AoCI) and energy consumption by jointly optimizing data scheduling, power allocation, and offloading decisions by collaborating with UAVs. A multi-agent actor-critic algorithm that incorporates federated learning and attention mechanisms (cluster-MAAC-attention) is presented to solve the previous problem. Our simulation results show that our algorithm reduces the worst AoCI by 4 times and reduces the penalty by 20 times compared to existing schemes. A UAV telemetry system, designed as an independent Wireless Systems control module for installation in any UAV payload without subsystem interference, is used in a recent work [9] as the experimental object for the sounding rocket, where its components within the flight attitude measurement system of the sounding rocket guidance control system are explained, and actual launch data is analyzed and discussed. In a work [10], the authors present the development of a low-cost, long-range data logger payload for UAVs. UAVs are increasingly utilized for various applications, including observation and surveillance, both recreationally and commercially. However, due to their autonomous nature, operators lack real-time situational awareness of the drone's status during flight. To meet this significant need, this proposed system is designed to actively retrieve important mission data, including data points like altitude and atmospheric pressure, in real-time. This system, which includes an Arduino Uno, Arduino Nano, a 10-DOF IMU, a micro-SD card converter, a LoRa SX1278 module, and a 7.4V lithium-ion battery, utilizes the long-range functionality of the LoRa module in conjunction with the Arduino IDE and Node-RED IoT platform to send real-time data to the operator's base station. The proposed system is incorporated in an MJX Bug 16 Pro UAV for real-world experiments to collect data and evaluate performance. Similarly, a paper [11] presented the design and development of a LoRa-based repeater that can work with UAVs in order to provide a long-range communications capability in difficult spaces, like trees or in parts of the world where ground-based repeaters are limited by short range, high-cost, and potential negative impact to the environment (flora and fauna).

This innovative approach offers a more efficient and environmentally friendly solution. Furthermore, the system can be enhanced by integrating with other wireless technologies like Wi-Fi and Bluetooth, significantly increasing its versatility and applicability in various scenarios. In a research [12], the authors investigate the feasibility of using LoRa technology for *Search and Rescue* (SaR) missions in mountainous terrain. Their focus lies on characterizing the radio signal strength between a victim wearing a LoRa-equipped safety helmet and an UAV tasked with locating them. Recognizing the lack of existing data on such links in real-world mountain environments, the study presents measurements of excessive aerial *Path Loss* (PL) observed during UAV flights over a mountainous canyon. These measurements were then used to derive an equivalent model, crucial for implementing signal-strength-based localization techniques. The findings reveal that an angle-dependent fitting effectively describes the excessive aerial PL, and interestingly, no substantial seasonal variations in PL were observed despite the presence of icy canyon walls. Similarly, in a study [13], the authors present Transponder, an open-source, lightweight, and low-cost solution, for improving Search and Rescue (SaR) operations. Transponder is a drone-mounted system designed to analyze Wi-Fi beacons or probe requests in areas lacking network infrastructure. It utilizes LoRa communications and incorporates a Complex Event Processor to enhance data analysis, enrich information, and provide crucial first-aid details. The paper also includes a demonstration of Transponder's effectiveness in a realistic scenario. Finally, in another study [14], the authors introduce a novel approach for the Internet of Things (IoT) that leverages the integration of IoT with

UAVs. This innovative method establishes a "flying gateway" by deploying a UAV equipped with a LoRa-based gateway system. This system, comprising hardware components like Arduino, Raspberry Pi boards, and RAK 2445 boards for LoRa connectivity, enables the UAV to collect data from IoT nodes across extended areas. By flying over these nodes, the UAV acts as a mobile gateway, gathering and transmitting data to a LoRa server. This groundbreaking approach effectively extends the coverage of terrestrial LoRa gateways, making it possible to reach remote and rural areas previously inaccessible to traditional IoT infrastructure.

The chapter's work advances the current state-of-the-art specifically via the use of LoRa long-range readout delivery and the easy-to-use novel redundancy generation algorithm. The redundancy algorithm improves the robustness of the communication system against application-layer loss that contributes to predictability. Because the redundancy algorithm is generic, it can be adapted as an enhancement of any UAV swarm communication protocol.

8.3 PHYSICAL LAYER

LoRa is a powerful protocol stack that is optimized for *LPWAN* systems and capable of operation of over 10 years on a single battery charge [15]. The physical, link, and application layers architect LoRa as a lean stack that permits efficient device-to-device communication within the LoRa system. The network and transport layers are unnecessary as LoRa does not have native support for IPv6.

The LoRa stack is divided into different layers that are further broken down into sublayers covering different functionalities. Security is a fundamental consideration at both the link and application layers. The link layer secures communication by means of device authentication and AES-128 encryption. AES-128 relies on a unique 64-bit EUI-64 network key. The application layer also relies on AES-128 encryption for data confidentiality while in transmission. (AES refers to the *Advanced Encryption Standard*, and EUI refers to *Extended Unique Identifier.*)

The physical layer of LoRa uses *Industrial, Scientific, and Medical* (ISM) bands, i.e., 433 MHz, 868 MHz, and 915 MHz, respectively, and it uses *Chirp Spread Spectrum* (CSS) modulation. CSS allows LoRa to be power efficient and enables LoRa devices to communicate over long distances, generally more than 10 kilometers at lower data rates with the environmental conditions taken into account. The actual coverage will also depend on other environmental factors such as terrain and constructions.

A single gateway can cover very large areas, often hundreds of square kilometers [1]. The performance of the Gateway will change depending on how many particular devices are connected. For example, at a distance of 8 km and a rate of 250 *bits per second* (bps), it could support around 50 devices. At a distance of 2.5 km and a rate of 5 *kilobits per second* (kbps), it could support about 700 devices.

Data transmission rates differ by region and frequency band. In Europe, the 868 MHz band typically supports rates between 250 bps and 50 kbps, while the 915 MHz band in the United States accommodates higher rates from 980 bps up to 21.9 kbps. This difference arises mainly due to the number of subchannels (10 in the 868 MHz band versus over 64 in the 915 MHz band). Subchannel bandwidth varies with the spectral band, with uplink and downlink channels generally ranging from 125 to 500 KHz in both cases.

8.4 LINK LAYER

Long-Range Wide-Area Network (LoRaWAN), the networking mechanism for LoRa technology, defines both the media access algorithm and the network architecture. Unlike many WPAN technologies that rely on mesh networking for long-range communication, LoRaWAN employs a simpler approach.

While mesh networks thus allow for long-range communication, their intermediate nodes have to be coordinated for data aggregation, affecting their battery lifetime and transmission capacity. LoRaWAN targets low-power devices and thus optimizes for coverage over transmission rate. Direct communication between one device and multiple gateways increases redundancy and simplifies network management. It also eliminates complex handover mechanisms for mobile environments.

LoRa gateways act as a relay that receives data transmitted by LoRa devices through their LoRa interface and sends it to the server using their IP interface. This provides redundancy for the delivery of data, while the server performs the appropriate security verifications and discards the duplicate copies. In this way, LoRaWAN shifts most of the complexity to the server, making the gateways relatively lightweight embedded devices.

The spread spectrum approach allows LoRaWAN to achieve long-range communication with much lower transmission rates compared to technologies like IEEE 802.15.4, while channel interference can be mitigated by using multiple gateways.

LoRaWAN's *Media Access Control* (MAC) is based on a simplified version of ALOHA, a predecessor of *Carrier-Sense Multiple Access with Collision Avoidance* (CSMA/CA). It allows devices to transmit without listening to the channel beforehand, hence minimizing network synchronization and reducing power consumption. Considering that LoRa modulation creates very few collisions, ALOHA is also appropriate for LoRaWAN's star topology, where many devices send data to the same gateway [16].

Most importantly, gateways need to have sufficient capacity to support the throughput from many devices. This is realized through several strategies: multi-channel modulation supporting simultaneous transmissions, variable payload lengths, and adaptive data rates. Devices with better links—of a higher SNR—can send faster, which reduces channel access times and increases the overall network capacity. Another factor that affects transmission rate is the spreading factor of the CSS modulation. In general, devices closer to the gateways send faster, as the quality of the link is improved. Reducing the transmission rate significantly prolongs the battery life of devices.

LoRaWAN is highly scalable: networks can be deployed with a single gateway and easily expanded by adding more gateways as the capacity demand grows. LoRaWAN devices have three classes of operation (A, B, and C), each representing a different trade-off between battery life, latency, and throughput, depending on power consumption and interaction with the gateways.

Class A, which is supported by all devices, allows gateways to send downlink packets only just after receiving traffic from the device; thus, it provides the longest battery life. Class B supports scheduled downlink transmissions from gateways; thus, it represents a moderate power consumption and battery life. Class C, which allows continuous downlink communication from gateways except in periods of device transmission, thus offers the shortest battery life but the lowest latency.

8.4.1 Network layer

This section focuses on IPv6 and its limitations, followed by presenting 6LoWPAN as a solution for those limitations.

8.4.1.1 IPv6

Initial standardization of this protocol was through *Internet Protocol version* 4, or IPv4. However, IPv4 has one serious problem: its 32-bit addressing allows only approximately 4.3 billion possible unique addresses. This small pool has long since been depleted, and the workaround of implementing *Network Address Translation* (NAT) everywhere has been widely deployed [17].

NAT works by mapping several private IP addresses within a LAN to a single public IP address visible on the WAN. This is achieved by combining private IP addresses with unique transport-layer ports, such as UDP ports, which enable multiple devices to share a single public IP address. While NAT does expand the number of devices that can connect to the Internet, this technology introduces significant disadvantages.

NAT seems to violate the basic principle of layered architecture, whereby routers examining transport-layer information, in this case, port numbers, make routing decisions; this should not be the case since routing decisions must strictly rely on network-layer information, which is the IP address. In other words, leveraging transport-layer information to make decisions on routes heavily increases the processing overheads and complexity for the routers, which seriously impacts the efficiency of a network.

To solve the limitations of IPv4, the more sustainable solution was to simply increase the size of the IP addresses. With 128-bit IPv6 addresses, there is an overwhelmingly larger address space, there is no longer a need to use NAT, and there will be sufficient addresses for the foreseeable future.

Here is a huge improvement with IPv6, as a 128-bit address space greatly increases the number of addressable devices when compared to a 32-bit IPv4 address. This increases the number of addresses available for the increasing number of devices that will be connected to the Internet.

IPv6 addresses are interestingly written in a different way. The addresses contain eight 16-bit hexadecimal values, which are separated by colons. Each value may drop leading zeros for readability. Moreover, any number of 16-bit values of consecutive zeros may be represented with a double colon (: :). An example of an IPv6 address is 2001:0:0: 0:0:10: 21:10, which can be re-written as 2001::10:21:10.

IPv6 presents a new header format, which is shown in Figure 8.4. The header has some important fields:

- Version (4 bits): It is always set to 6, unequivocally identifying it as an IPv6 packet.
- Traffic Class (8 bits): It carries Quality of Service (QoS) information, allowing for differentiated treatment of various types of traffic based on their priority or sensitivity.
- Flow Label (20 bits): It is used to identify datagrams that are part of a common session flow in order to improve processing and optimization.
- Payload Length (16 bits): It specifies the number of bytes in the following data payload.
- Next Header (8 bits): This specifies the protocol that was used to encode the data in the payload, which could be something like UDP, or potentially some higher level protocol.
- Hop Limit (8 bits): This works as a counter, which is decremented as it traverses the network. This helps to avoid infinite loops so that eventually they are dropped if they cannot reach their destination in a reasonable number of hops.
- Source and Destination (128 bits): it is a unique 128-bit address of the device initiating the communication and the device that is the destination for the packet.

If the Hop Limit reaches zero, the packet will be dropped from the network. After the IPv6 header, the payload containing data is handled by the destination device.

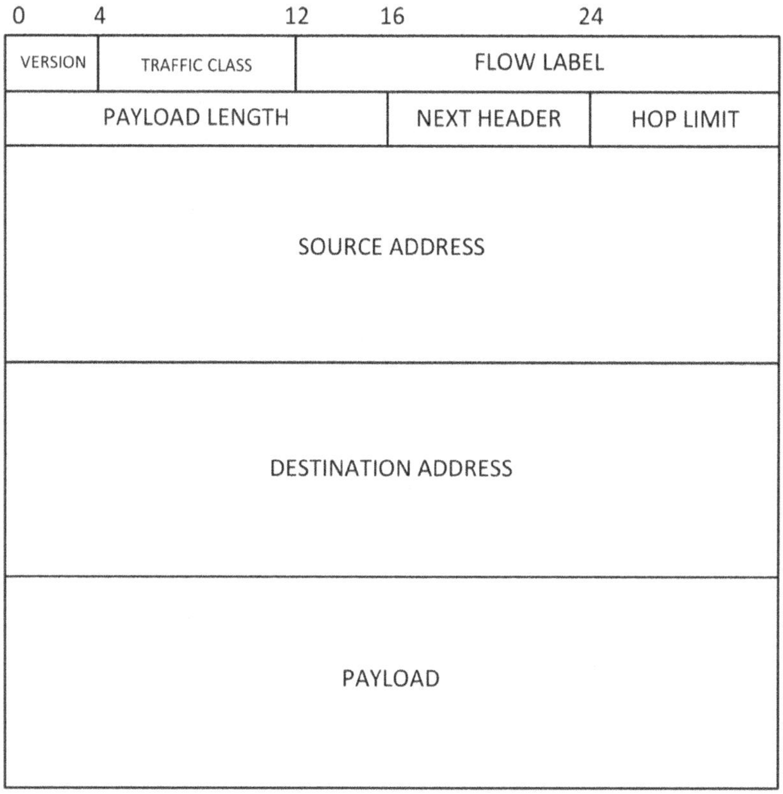

Figure 8.4 IPv6 structure.

8.4.1.2 6LoWPAN

The IPv6 is not directly compatible with most of the physical and link-layer technologies used in IoT. To bridge this gap, 6LoWPAN and similar mechanisms operate between the network and link layers. They adapt and translate IPv6 datagrams into a format suitable for transmission over *Low-Power Lossy Networks* (LLNs) [18].

In essence, 6LoWPAN natively provides IPv6 to constrained and low-power devices while efficiently optimizing overhead. This IPv6 connectivity also offers connectivity to any other host and router on the Internet. Even though it was initially developed for IEEE 802.15.4, most of the features in 6LoWPAN have been generalized for other physical- and link-layer technologies such as LoRa. In practice, the term "6LoWPAN network" commonly refers to an LLN that utilizes 6LoWPAN principles.

The 6LoWPAN architecture, depicted in Figure 8.5, consists of multiple WPANs. These WPANs serve as access stub networks, wherein the network's traffic usually stays and is not typically routed into another network.

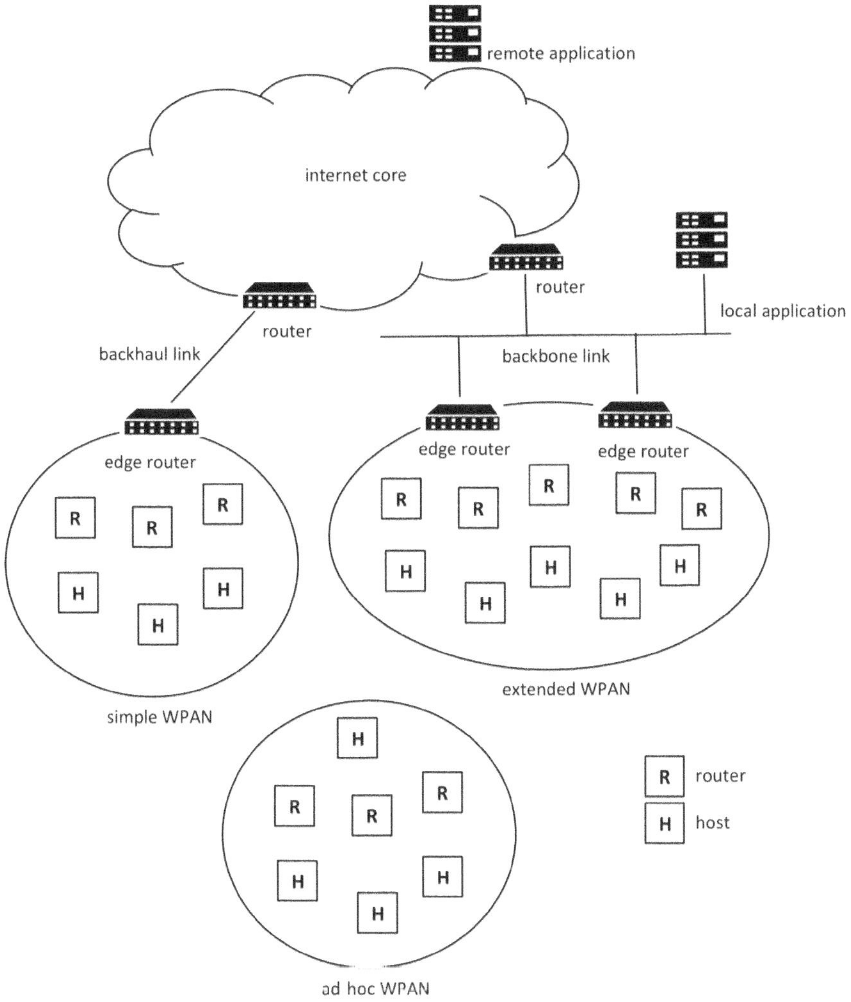

Figure 8.5 6LoWPAN topologies.

All nodes in a WPAN have IPv6 addresses with the same prefix, which is generally acquired through RA messages. There are three categories of WPANs:

- Simple WPANs: Connect to the IP core through edge routers.
- Ad hoc WPANs: Locally connect devices without access to the IP core.
- Extended WPANs: Connect to the IP core via multiple-edge routers along a backbone link.

An edge router, positioned between the access and core networks, serves as a traditional IoT gateway. It manages traffic entering and exiting the

WPAN by performing 6LoWPAN adaptation, *Neighbor Discovery* (ND) for intra-link communication, and other functions such as IPv4-to-IPv6 translation for communication with entities in the IP core.

Devices within a WPAN can operate as either hosts or routers, depending on the source and destination addresses of the datagrams. A sensor may act as a host when generating readouts and as a router when forwarding traffic from other devices in a mesh network. Most devices possess a single interface (wireless) for both transmitting and receiving datagrams, even when acting as routers. In this case, routers forward datagrams on the same interface they receive them, extending network coverage. This approach differs significantly from traditional Internet routers, which typically forward traffic received on one interface to another.

Within WPANs, core communication adheres to standard IPv6 rules. Each device is uniquely identified by its IPv6 address and can send and receive datagrams. However, constrained devices have limited processing capabilities, restricting the types of transport and application protocols that are suitable. While UDP is generally appropriate in this environment, *Transmission Control Protocol* (TCP) and *Hypertext Transfer Protocol* (HTTP) are often unsuitable due to their computational complexity and excessive resource consumption.

As indicated before, while primarily designed for IEEE 802.15.4, 6LoWPAN can be adapted to function over other link-layer technologies with minor modifications. Essential requirements for these link-layer protocols include (1) framing capabilities, (2) unicast transmission, and (3) unique addresses that enable the derivation of unique IPv6 addresses through *Stateless Address Autoconfiguration* (SAA).

Given that IPv6 fragments must be at least 1,280 bytes, 6LoWPAN incorporates its own fragmentation mechanism to adapt datagram transmission to link layers with small *Maximum Transmission Units* (MTUs). To minimize the number of fragments and enhance efficiency, link-layer frames should ideally have larger payloads, with a minimum of 60 bytes. This is crucial because, for a given packet loss rate, a higher number of fragments increases the likelihood of datagram loss within the network.

6LoWPAN compression can significantly improve efficiency by compressing IPv6 and UDP headers, maximizing link-layer payload size and minimizing the number of fragments. While not strictly necessary, incorporating error detection and correction mechanisms within the adaptation layer for other link layers can enhance reliability. Moreover, the adaptation layer must support security features such as encryption and authentication.

8.4.2 Transport and application layers

Interaction with IoT devices is done by means of REST message exchanges. Specifically, REST requests and responses support connectivity between clients and servers. There are many session-layer protocols that support REST Application Program Interfaces (APIs) including the well-known

HTTP and the aforementioned CoAP. From all of them, it is this latter protocol that best works with 6LoWPAN. CoAP is an extra-lightweight session-layer protocol that was designed for the transmission of sensor and actuation traffic over LLNs. CoAP was standardized by the IETF as RFC 7252 in 2014. CoAP is expected to become in the coming years the default mechanism for the transmission of device data in the context of access IoT network. CoAP applications range from smart grid and building automation to asset tracking and Industry 4.0. CoAP also includes built-in mechanisms for resource discovery and management.

Figure 8.6 shows the CoAP's architecture that comprises two key sublayers. The lower message sublayer interfaces with the UDP transport layer, a choice necessitated by the need to support constrained devices that may not be capable of handling the overhead of TCP.

In addition, to mitigate some of UDP's limitations, the message sublayer offers two operational modes:

- Confirmable: Requests and responses require acknowledgments from the receiving end to ensure successful delivery, introducing a degree of reliability akin to TCP.
- Non-Confirmable: This "fire-and-forget" mode transmits messages without expecting acknowledgments, suitable for situations where reliability is not critical.

The upper request/response sublayer handles the construction and interpretation of application-level messages, enabling seamless communication between CoAP clients and servers.

Figure 8.7 shows how an application interacts with a sensor. The application sends a CoAP GET message to the sensor that requests temperature

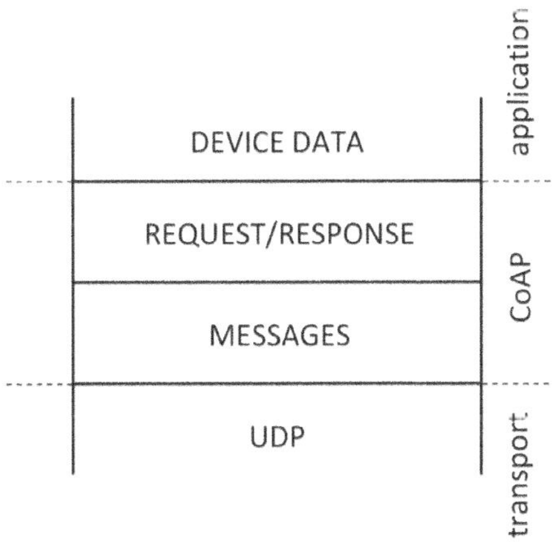

Figure 8.6 CoAP layered structure.

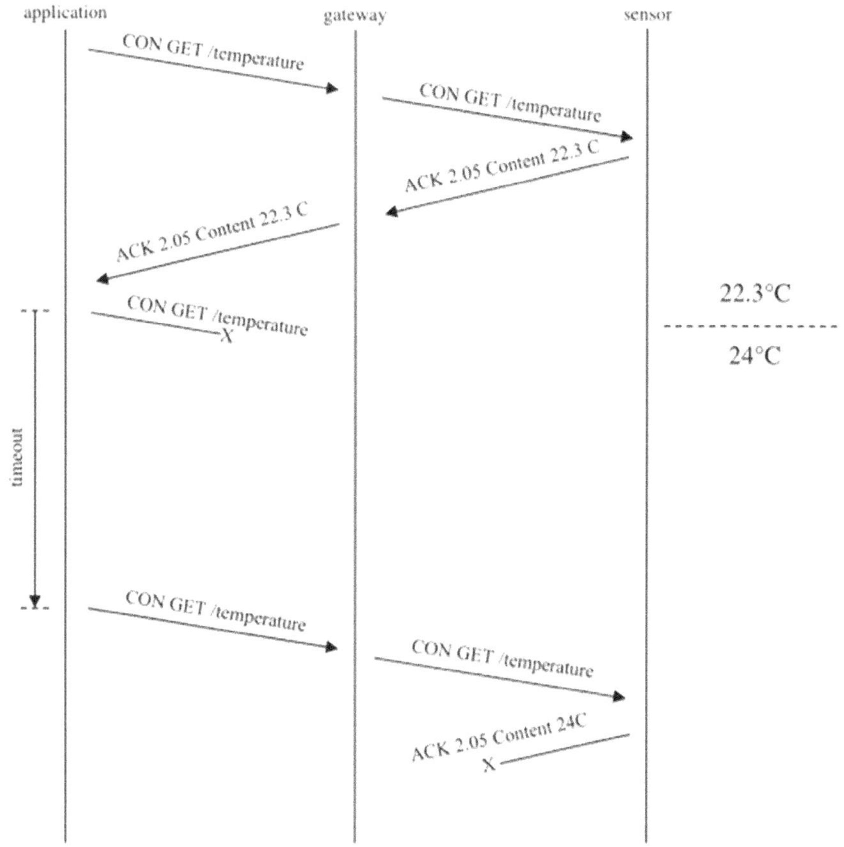

Figure 8.7 REST message flow.

measurements with a *Uniform Resource Locator* (URL) describing what is needed, which in this example is "/temperature" [19].

These messages usually go through a gateway device acting as a pass-through for the sensor to connect to the application. The gateway passes roughly the same CoAP message to the application and regenerates the necessary information for the lower-level 6LoWPAN and LoRa layers. Moreover, in complex configurations, two sessions might exist: the application could send a message to the gateway and receive it from the sensor via the gateway.

The sensor, upon receiving a CoAP GET message, will send one message with the temperature reading in a CoAP 2.05 Content response. This response is also an acknowledgment of the request.

Confirmable CoAP will include its own retransmit functionality (e.g., if a message was lost in transit, the sender will automatically retransmit it at exponentially longer intervals) based on RFC 7252.

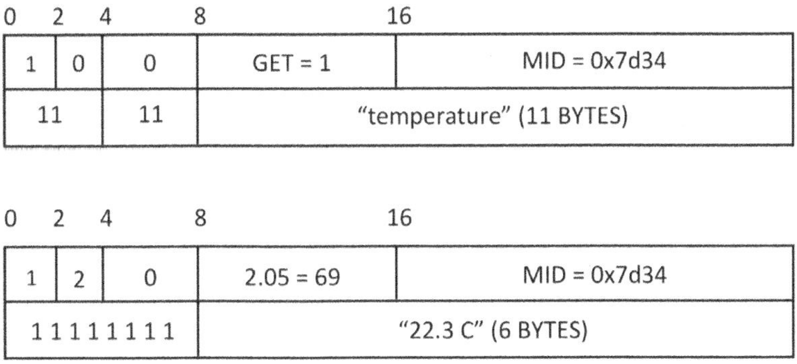

Figure 8.8 CoAP packing.

Figure 8.8 depicts the structure of CoAP requests and responses. Each message has 2-byte *Message Identifier* (MID). In a request, the URL is included as representing the resource being accessed, and in response to that request, the readouts are included, represented in text.

CoAP requests and responses nominally occupy 17 and 12 bytes, respectively. When the protocol stack is included, requests are 31 bytes and responses will occupy 26 bytes.

In the best-case transmission time, it would take about 44.28 milliseconds to transmit a request and 37.14 milliseconds to transmit a response.

8.4.3 Mist computing infrastructure

This part of the study considers the details of quantifying the effect of both non-redundant and redundant mist computing compared to cloud computing for UAV communications. We propose a mathematical model to account for redundancy, representing application-layer network packet loss as a function of physical-layer packet loss. We will use the estimate of application-layer packet loss to suggest redundancy levels.

Figure 8.9 shows the channel model used in this section to model the bursty packet loss that characterizes wireless communication. In this Gilbert–Elliott channel model, two states are considered: a good state with a low packet loss probability (e_B) and a bad state with a high packet loss probability (e_B) [20]. The two parameters p and α in this model control the transitions between the good and bad states. In essence, these two parameters determine the packet loss behavior of the network and the extent of its burstiness. For simplicity, it can be assumed that $e_G = 1$ and $e_B = 0$, which implies no packet loss in the good state and full packet loss in the bad state.

The CH UAV transmits multiple copies of a GET request, which are responded to with multiple copies of readouts from another UAV within the cluster as shown in Figure 8.10. This UAV dynamically determines the

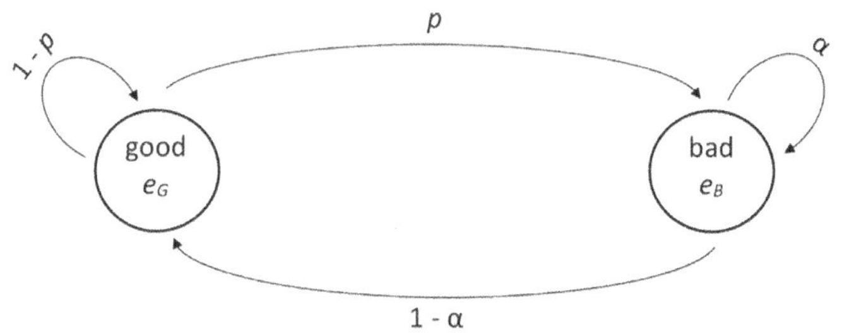

Figure 8.9 Wireless channel model.

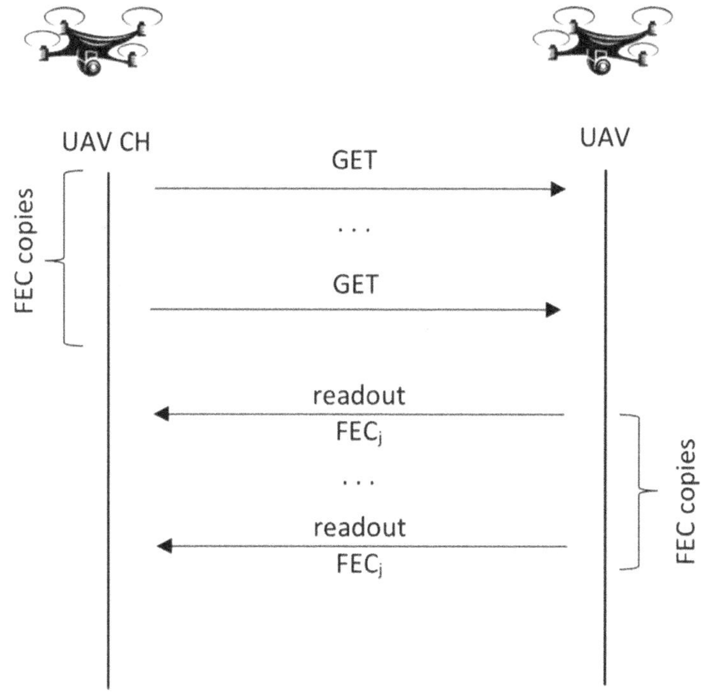

Figure 8.10 UAV CH to UAV message flow.

number of responses to send based on network-layer loss and burstiness. This *Forward Error Correction* (FEC) level is included in the response sent back to the CH UAV, enabling the latter to transmit the appropriate number of request copies [21].

The probability of success is given by $P_{success} = P_{response|request} \times P_{request} = P_{response} \times P_{request} = P_{msg}^2$, where P_{msg} is the probability of successful transmission given by $P_{msg} = P_{msg|good} \times P_{good} + P_{msg|bad} \times P_{bad}$, but because $P_{msg|bad} = 1 - e_B = 0$

and $P_{msg|bad} = 1 - e_G = 1$, then $P_{msg} = P_{good}$. Therefore, the probability of success results $P_{success} = P_{good}^2$.

P_{good} is the probability that the channel is good. This probability results as $P_{good} = \frac{1-\propto}{1-\propto+p}$. Therefore, for a successful transaction, the corresponding probability is given by

$$P_{success} = \left[\frac{1-\propto}{1-\propto+p}\right]^2 \tag{8.1}$$

where p and $\propto$ are network packet loss and network loss burstiness probabilities, respectively. When multiple n copies are transmitted, the success probability follows a Binomial distribution

$$P_{success}^n = \sum_{i=1}^{n}\binom{n}{i}\left[\frac{1-\propto}{1-\propto+p}\right]^{2i}\left[\frac{p}{1-\propto+p}\right]^{2(n-i)} \tag{8.2}$$

obtained from Equation (8.1).

The algorithm based on Equation (8.2) to estimate the FEC level is shown below:

1. Let $\hat{n} = 1$.
2. Compute the estimators using *Maximum Likelihood Estimation* (MLE) as $\hat{\propto} = \frac{J_{H,H}}{J_{H,H}+J_{H,L}}$ and $\hat{\propto} = \frac{J_{L,H}}{J_{L,H}+J_{L,L}}$ where $J_{H,H}$, $J_{L,H}$, $J_{H,L}$ and $J_{L,L}$, respectively.
3. Compute $P_{success}^{\hat{n}}$ as follows:

$$P_{success}^{\hat{n}} - \sum_{i=1}^{n}\binom{\hat{n}}{i}\left[\frac{1-\hat{\propto}}{1-\hat{\propto}+\hat{p}}\right]^{2i}\left[\frac{\hat{p}}{1-\hat{\propto}+\hat{p}}\right]^{2(\hat{n}-i)}$$

   where $\hat{p}$ and $\hat{\propto}$ result from Step 1.
4. If $P_{success}^{\hat{n}} < \epsilon$, then let $\hat{n} = \hat{n} + 1$, if $\hat{n} \leq Q$, and go back to Step 3.
5. $\hat{n}$ specifies the FEC level.

Note that ϵ represents a probabilistic threshold of intended successful transmission rate. Similarly, Q represents the maximum possible FEC level. The computational complexity of this algorithm is given by the worst-case scenario given by the iteration of the $P_{success}^{\hat{n}}$ computation. This later complexity is given by $O(\hat{n})$, which combined with the iterations leads to an overall complexity of $O(Q^2)$.

8.5 EXPERIMENTAL FRAMEWORK

This section introduces an experimental framework to test the effects of cloud computing, mist computing, and mist computing with controlled redundancy, as indicated by the algorithm in Section 8.7. This scenario is implemented using the well-known Kapanga Firebird framework [22]. Kapanga Firebird is a protocol virtualization toolset that enables the emulation of protocol stacks within layered architectures. Figure 8.11 illustrates the user interface of Kapanga Firebird, demonstrating the configuration of four UAVs.

Moreover, Kapanga Firebird supports the emulation of Unmanned Aircraft Systems by allowing the deployment of UAVs through the specification of flightpaths, power constrains, computational capabilities, and engine characteristics. In this context, Kapanga Firebird is instrumental to the evaluation of the performance of IoT networks in UAV-based topologies. The configuration and scripting used for this emulation can later be deployed on UAVs carrying a Kapanga Firebird agent for use in real-life scenarios. In this context of this research work, this is done on Ecotronics Ventures' *Autonomous Vehicle for Environmental Research* (AVE-R) platform.

AVE-R is a robust unmanned aerial system developed and tested in Maryland, featuring a substantial 10 kg payload capacity. Initially

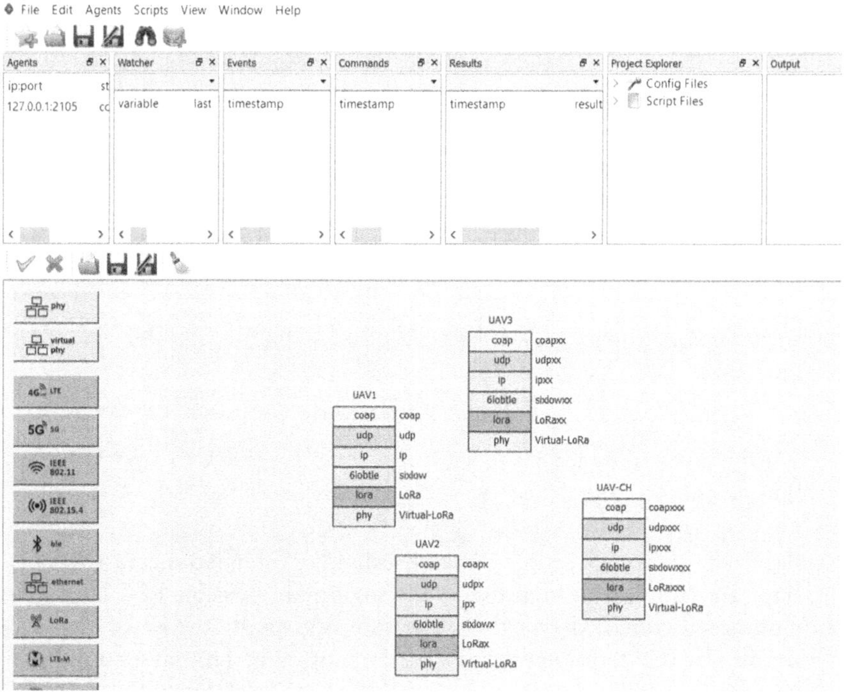

Figure 8.11 Kapanga firebird user interface.

demonstrated during NASA's SCOAPE-II campaign in June 2024, AVE-R played a crucial role in measuring air quality pollutants near oil platforms in the Gulf of Mexico. The findings from this deployment have been published in the American Geophysical Union and American Meteorological Society journals. Beyond its environmental sensing capabilities, AVE-R serves as a versatile research platform for unmanned communications including swarm deployments. Figure 8.12 shows a UAV from the AVE-R platform.

Figure 8.13 shows topology under consideration for the emulation of the proposed IoT communication technologies. Specifically, it consists of four UAVs with a circular flightpath of radius 200 meters with centers deployed in a grid pattern as shown in the figure. All four UAVs are located and separated by 90° flying with an angular velocity of $\Omega = {}^{36^\circ}\!/_{s}$. UAVs receive CoAP requests to transmit environmental temperature readouts every 20 ms. In the cloud computing scenario, these requests come from an application in the cloud. For the mist computing scenario, the requests come from the CH UAV shown in the figure. Several 300-second test cases are considered for different levels of network packet loss ($p = \{0, 0.01, 0.02, 0.03, 0.05, 0.08, 0.1, 0.12, 0.15\}$) as well as low and high network packet loss burstiness ($\propto = \{0.3, 0.9\}$, respectively).

When comparing mist to cloud computing, predictions are correspondingly performed either at the UAV CH or at an application in the cloud. In all scenarios, we assume no network impairments between the gateway and the ultimate knowledge consumer. This assumption is justified because this connectivity generally relies on traditional wired Internet protocols. All impairments are instead applied to the constrained wireless IoT network that connects UAVs to the gateway.

Figure 8.12 AVE-R UAV.

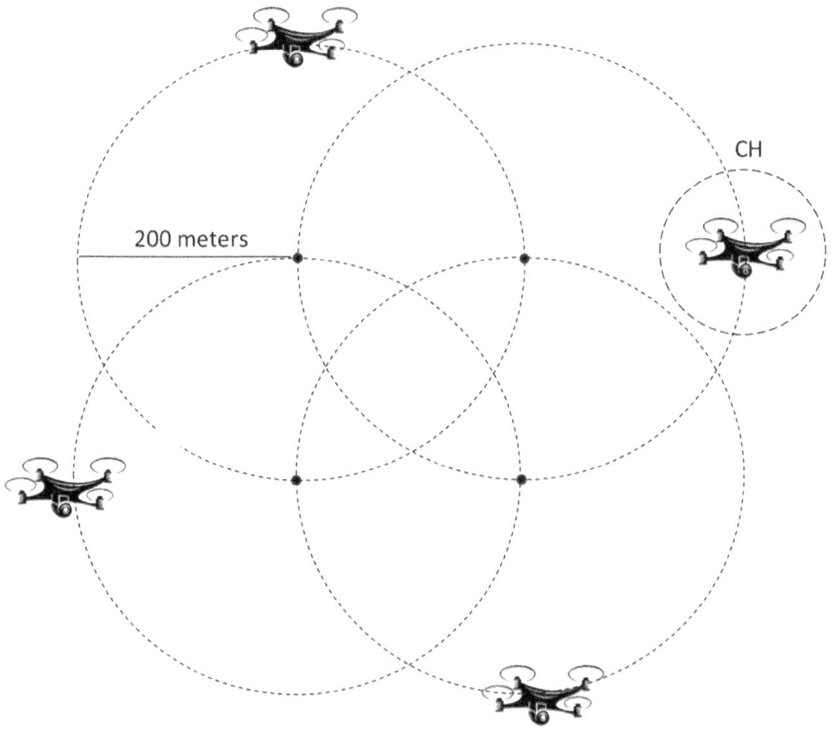

Figure 8.13 Kapanga firebird emulation topology.

In this context, prediction success is defined as the probability that all four readouts are received by the CH (for mist computing) or the application (for cloud computing). Given the assumption of no loss between the CH and the application, predictability can be assessed solely at the CH in the context of mist computing.

Figures 8.14 and 8.15 show the predictability as a function of the packet loss probability for both low ($\propto = 0.3$) and high ($\propto = 0.9$) network packet loss burstiness when the redundancy algorithm in Section 8.7 is used. Note that the predictability drastically improves when the algorithm is used for both low and high network packet loss burstiness. A comparative analysis of the two schemes reveals a substantial average performance improvement when the algorithm is applied. This improvement is more pronounced in scenarios with high network loss burstiness (241.83%) compared to low burstiness (135.77%). However, the algorithm's effect on the system's prediction accuracy is minimal. In terms of predictability, there is a larger decline without the algorithm in low-loss burstiness scenarios compared to with it: 71.72% versus 59.18%.

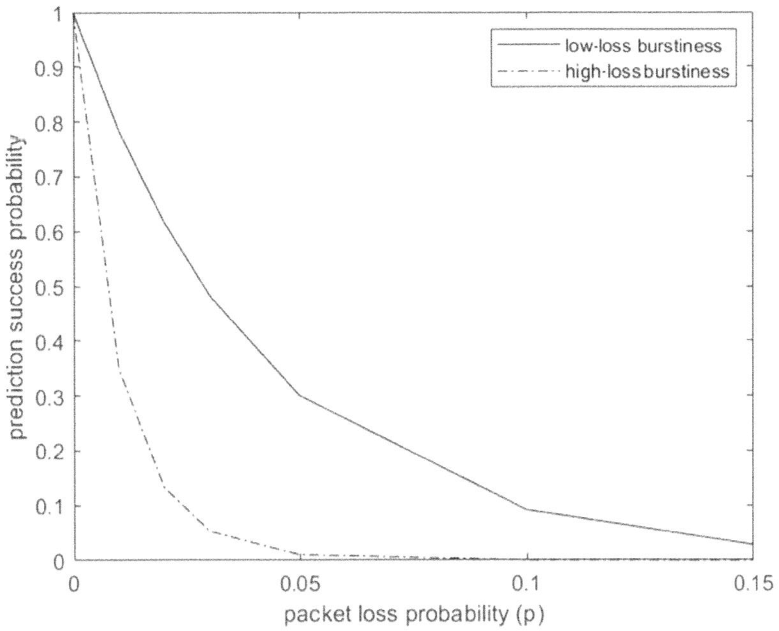

Figure 8.14 Predictability without redundancy.

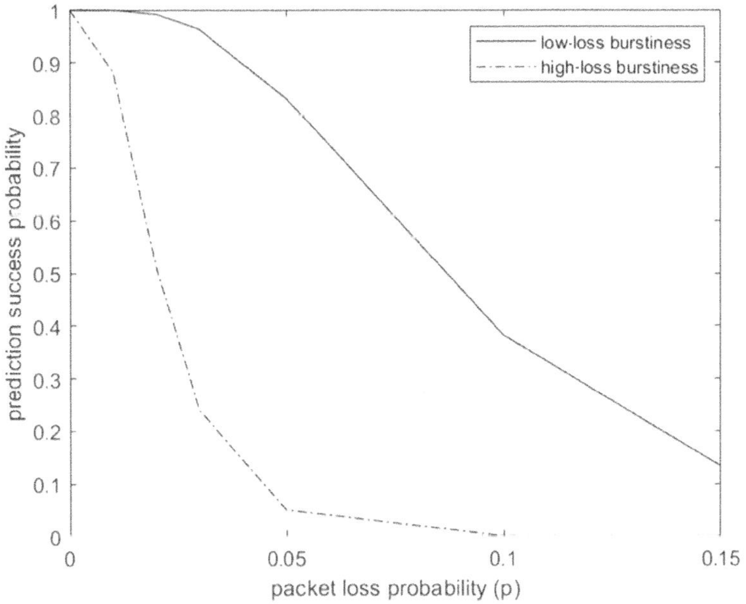

Figure 8.15 Predictability with redundancy (algorithm).

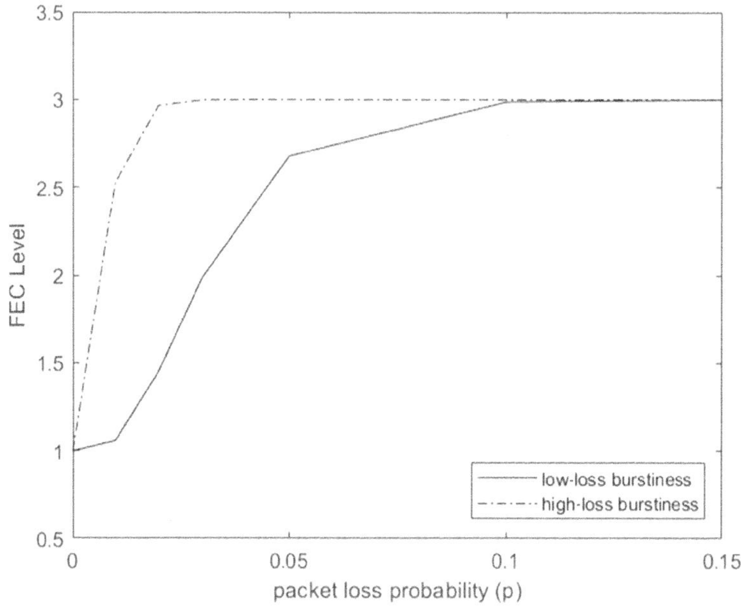

Figure 8.16 FEC level.

Figure 8.16 shows the average FEC level as a function of the packet loss probability for low and high network packet loss burstiness. Note that this average FEC level is obtained for a target application-layer loss of $\epsilon = 0.9$ and a maximum FEC level of $Q = 3$. As expected, as loss probability and loss burstiness increases, the average FEC level under consideration also increases.

Note that when comparing cloud computing to mist computing, the most crucial factor is the reduction in latency during the decision-making process. Specifically, when knowledge is extracted within a CH (compared to a cloud application), any subsequent actuation benefits from lower latency. In this context, the *Round-Trip Time* (RTT) for this Kapanga Firebird configuration is approximately 38 milliseconds.

This single-cluster scenario can be extended to support multiple clusters. In this context, the CHs associated with each cluster can communicate directly with the application to perform predictions. The number of clusters is constrained by overall throughput and latency requirements.

8.6 CONCLUSIONS AND FUTURE WORK

This chapter introduced an I-UAV networking topology supporting mist computing. By employing a novel FEC algorithm, this topology enhances the system's predictability.

To this end, we considered three scenarios of increasing levels of predictability:

- Cloud Computing: Traditional cloud-based processing.
- Mist Computing: Processing performed closer to the data source, offering reduced latency.
- Mist Computing with FEC: This builds on the previous section on mist computing with the proposed FEC to improve upon it and mist computing.

In ideal conditions, where computational complexity and energy consumption are not restricting factors, mist computing is naturally superior to cloud computing due to the minimized latency.

In addition, when considering mist computing, the proposed FEC has a meaningful benefit to the application-layer *Packet Delivery Ratios* (PDRs), therefore improving the overall predictability of the system.

For the future work, there are some aspects that warrant further investigation as follows:

- Integration of WSN Routing Mechanisms: Exploring the applicability of routing protocols developed for *Wireless Sensor Networks* (WSNs) to efficiently propagate data readouts from UAVs to applications.
- Diverse Topologies Support: Exploring the performance and adaptability of this proposed system in more complex and realistic scenarios involving dense swarms of UAVs with diverse communication patterns and topologies.
- Inter-Cluster Communication: Although inter-cluster communication is not normally necessary for this project, as the predictions will occur either at the CH depending on readouts that have been obtained from UAVs within the cluster or at a cloud application that collects aggregated readouts from all CHs, I will explore effective methods for inter-cluster communication.

Ultimately, this chapter provided an introduction to research respecting a meaningful contribution in the area of I-UAVs, more specifically as it relates to the topologies of UAVs and IoT networking.

APPENDIX

Acronym	*Meaning*
6LoWPAN	IPv6 over Low-power Wireless Personal Area Networks
AES	Advanced Encryption Standard
API	Application Program Interface

(Continued)

Acronym	Meaning
bps	bits per second
CH	Cluster Head
CoAP	Constrained Application Protocol
CPS	Cyber-Physical System
CSMA/CA	Carrier-Sense Multiple Access with Collision Avoidance
CSS	Chirp Spread Spectrum
DTLS	Datagram Transport-Layer Security
EUI	Extended Unique Identifier
FEC	Forward Error Correction
HTTP	Hypertext Transfer Protocol
IETF	Internet Engineering Task Force
IoT	Internet of Things
IP	Internet Protocol
IPv4	Internet Protocol version 4
IPv6	Internet Protocol version 6
ISM	Industrial, Scientific, and Medical
I-UAV	Internet of Unmanned Aerial Vehicles
kbps	kilobits per second
LAN	Local Area Network
LLN	Low-Power Lossy Networks
LoRa	Long Range
LoRaWAN	Long-Range Wide-Area Network
LPWAN	Low-Power Wide-Area Network
MAC	Media Access Control
MID	Message Identifier
ML	Machine Learning
MLE	Maximum Likelihood Estimation
MTU	Maximum Transmission Units
NAT	Network Address Translation
ND	Neighbor Discovery
PDR	Packet Delivery Ratio
PL	Path Loss
QoS	Quality of Service
RA	Router Advertisement
REST	Representation State Transfer
ROS	Robot Operating System
RTT	Round-Trip Time

(Continued)

Acronym	*Meaning*
SAA	Stateless Address Autoconfiguration
SaR	Search and Rescue
SNR	Signal-to-Noise Ratio
TCP	Transmission Control Protocol
UDP	User Datagram Protocol
URL	Uniform Resource Locator
WAN	Wide-Area Network
WPAN	Wireless Personal Area Network
WSN	Wireless Sensor Network

BIBLIOGRAPHY

1. R. Herrero, *Fundamentals of IoT Communication Technologies*, Springer-Nature, 2021.
2. Y. Chen and T. Kunz, "Performance evaluation of IoT protocols under a constrained wireless access network," in *International Conference on Selected Topics in Mobile Wireless Networking*, Cairo, 2016.
3. A. Shinde and R. Bichkar, "Energy efficient cluster based secured data aggregation using genetic algorithm for WSN," in *3rd Asian Conference on Innovation in Technology (ASIANCON)*, Pune, 2023.
4. C. Bormann, Z. Shelby, S. Chakrabarti and E. Nordmark, "RFC 6775: Neighbor discovery optimization for IPv6 over low-power wireless personal area networks (6LoWPANs)," *IETF*, Wilmington, DE, 2012.
5. C. Bormann, Z. Shelby and K. Hartke, "RFC 7252: The constrained application protocol (CoAP)," *IETF*, Wilmington, DE, 2015.
6. M. Karatzia, P. Kolios and G. Ellinas, "Implementing mission-critical UAV swarm coordination through the integration of LoRa and ROS frameworks," in *International Conference on Information and Communication Technologies for Disaster Management (ICT-DM)*, Dubai, 2023.
7. Z. Yuan, J. Jin, L. Sun, K.-W. Chin and G.-M. Muntean, "Ultra-reliable IoT communications with UAVs: A swarm use case," *IEEE Communications Magazine*, vol. 56, no. 12, pp. 90–96, 2018.
8. X. Xia, H. Esmat, B. Lorenzo and D. Goeckel, "UAV-enabled covert cross-technology communication in heterogeneous IoT networks," in *IEEE 100th Vehicular Technology Conference (VTC2024-Fall)*, Washington, DC, 2024.
9. Y.-H. Ma, K.-H. Lee, T.-H. Shieh, C.-L. Yeh and H.-S. Lin, "Applying LoRA communication technology to uav telemetry system and payload design," in *IEEE International Workshop on Electromagnetics: Applications and Student Innovation Competition (iWEM)*, Taoyuan City, 2024.
10. M. Irfan bin Edi, N. Emileen Abd Rashid, N. N. Ismail and K. Cengiz, "Low-cost, long-range unmanned aerial vehicle (UAV) data logger using long range (LoRa) module," in *IEEE Symposium on Wireless Technology & Applications (ISWTA)*, Kuala Lumpur, 2022.

11. N. Khera, M. Chowdhary and A. Saxena, "Development of LoRa enabled UAVs for defence application," in *1st International Conference on Innovative Sustainable Technologies for Energy, Mechatronics*, Dehradun, 2024.
12. G. M. Bianco, A. Mejia-Aguilar and G. Marrocco, "Measurements and modeling of radiohelmet-UAV LoRa links in a mountain canyon," in *16th European Conference on Antennas and Propagation (EuCAP)*, Madrid, 2022.
13. A. Calabrò and E. Marchetti, "Improving SAR ops using Wi-Fi and LoRa on UAV," in *IEEE International Conference on Pervasive Computing and Communications Workshops and Other Affiliated Events (PerCom Workshops)*, Pisa, 2022.
14. A. Moheddine, F. Patrone and M. Marchese, "UAV and IoT integration: A flying gateway," in *26th IEEE International Conference on Electronics, Circuits and Systems (ICECS)*, Genova, 2019.
15. L. Alliance, "LoRaWAN 1.1 Specification," Fremont, CA: LoRa Alliance, 2017.
16. A. Lavric and V. Popa, "Internet of Things and LoRa low-power wide-area networks: A survey," in *International Symposium on Signals, Circuits and Systems (ISSCS)*, Iasi, 2017.
17. R. Graziani, *IPv6 Fundamentals: A Straightforward Approach to Understanding IPv6*, Pearson Education, 2012.
18. P. T. A. J. Hui, "Compression format for IPv6 datagrams over IEEE 802.15.4-based networks," *IETF*, Wilmington, DE, 2011.
19. J. Arkko, C. Jennings and Z. Shelby, "RFC 9039: Uniform resource names for device identifiers," *IETF*, Wilmington, DE, 2021.
20. R. Herrero, "Analysis of IoT mechanisms for media streaming," *Internet of Things*, vol. 9, Article 100177, 2020.
21. R. Herrero, "Modeling and comparative analysis of Forward Error Correction in the context of multipath redundancy," *Telecommunication Systems: Modelling, Analysis, Design and Management*, vol. 65, pp. 783–794, 2017.
22. *Kapanga Firebird*, Ecotronics Ventures LLC, https://www.ecotronics.com/, 2024.

Chapter 9

MAVLink communication framework

From basic operations to secure deployment

Iraq Ahmad Reshi, Sumaya Akhtar, and Hrisikesh Roy

9.1 INTRODUCTION TO MAVLINK PROTOCOL [1-2]

MAVLink (Micro Air Vehicle Link) is an open-source, lightweight communication protocol, which is commonly adopted by UAV systems to transfer telemetry, command, mission and status data from the drone to the ground station (Figure 9.1). It currently comes in two major versions, MAVLink 1, which is a compact packet-based communication protocol used for basic communication, and MAVLink 2, which adds extended support for messages, backward compatibility, extension fields and optional packet signing for greater security. MAVLink can be used with low bandwidth links like telemetry radio, wireless network, and UART serial communication .

The protocol lets messages go both ways, which makes it easy for the different parts of unmanned systems to talk to each other. A drone can send telemetry data, like GPS coordinates and battery status, to a ground control station (GCS). The GCS can then send commands to the drone, like "take off" and "land." MAVLink also lets vehicles talk to each other, which is very important for drone swarms, as well as for subsystems

MAVlink full security stack

Layer	Protection
Hardware	Secure Boot(HSM/TPM) Hardware Crypto Accelerations Tamper-Proof Key Storage
Transport Layer	Certificate-Pinning DTLS 1.3 /QUIC WireGuard vpn
Application Layer	MAVLink 2.0 Secure Sessions (AES-256-GCM +ECDSA/Falcon) Anti-Reply Counters
Key Management	Key Rotation Policies HSM-Based CA for PKI Post-Quantum Key Hybridization

Figure 9.1 Layered security stack for secure MAVLink deployment.

DOI: 10.1201/9781003598350-12

like flight controllers and companion computers to talk to each other. Every part of the network has a system ID that uniquely identifies the vehicle and a Component ID that identifies the subsystem. This system keeps communication organized when there are more than one vehicle [3].

MAVLink sends and receives information about UAVs using a standard format. MAVLink also has a number of pre-defined messages that allow users to quickly and easily communicate with UAVs. For example, GPS_RAW_INT sends real-time GPS coordinates back and forth during flight, ATTITUDE sends real-time position and altitude information, and BATTERY_STATUS sends battery information. MAVLink 2 supports creating new messages with a unique message ID to allow for a wide range of customization options, including sensor or camera control commands and custom payloads. Users can also use control messages like COMMAND_LONG and SET_POSITION_TARGET_LOCAL_NFS to issue commands to their UAV from a remote location.

Other messages, such as MISSION_ITEM and MISSION_REQUEST_LIST, also allow for waypoint navigation and mission planning.

Being able to communicate across many different types of communication channels is one of the major benefit of using MAVLink. The most popular type of MAVLink communication occurs using serial (UART) connections, which is regularly found in flight controllers and telemetry radios. For more complicated UAV setups, where Wi-Fi or Ethernet is needed, MAVLink messages is usually sent via User Datagram Protocol(UDP) or Transmission Control Protocol (TCP), particularly when running on Raspberry Pi or other companion computers. Short-range typically can be accomplished via Bluetooth, while long-range can be accomplished with RFD900x or SiK radios. This flexibility allows MAVLink to be effectively used across different robotic and UAV platforms [1,2].

MAVLink uses a heartbeat mechanism for sustaining robust communication. By default, the HEARTBEAT message is sent once every second. The vehicle type (quadcopter, rover, etc.), flight mode (GUIDED, AUTO, etc.), and system status (ACTIVE, STANDBY) are among the crucial system details it contains. The system deems the connection lost and initiates fail-safe protocols in autonomous systems if no heartbeat is detected. For safe UAV operations, this feature guarantees real-time link health monitoring

A parameter protocol in MAVLink allows for dynamic autopilot configuration changes. A ground station can obtain particular parameters (like Proportional–Integral–Derivative (PID) gains and flight modes) by using messages like PARAM_REQUEST_READ. PARAM_SET permits in-flight modifications in the interim. To validate the update, the autopilot responds with PARAM_VALUE. This system is frequently used in Mission Planner and QGroundControl to adjust vehicle performance without flashing the firmware again.

MAVLink also provides organized mission and geofence management for autonomous operations. MISSION_ITEM is used to define waypoints, and MISSION_REQUEST_LISTisusedtoobtainthevehicle'smissionplan.FENCE_POINT messages control geofencing, a safety feature that restricts flight areas.

These features play a critical role for swarm robotics, delivery UAVs, and automated survey drones, where precise navigation and operational limits are critical.

MAVLink is widely used in many industries, and therefore, it has proven to be a versatile protocol. Using this protocol, drone operators receive real-time GPS, altitude, and battery information via telemetry data produced by the drone. The received information can be used for many types of activities including agricultural surveying and search and rescue operations. To conduct autonomous missions, operators can also send waypoint information to drones using MAVLink protocols. The use of cameras, activating the cameras, and modifying gimbals while in flight are also uses of MAVLink. In addition to supporting companion computers for artificial intelligence and machine learning applications, MAVLink also enables communication among multiple drones in swarm robotics. As discussed earlier, MAVLink is highly useful in modern unmanned systems due to its lightweight, flexible, and interoperable communication design .

9.2 MAVLINK PROTOCOL BASICS [1-2]

The lightweight, message-based MAVLink protocol (Micro Air Vehicle Link) allows for communication between the UAV, the GCS, and the onboard system. There are three basic parts to each message: a payload (the actual data, e.g., GPS and sensor values), a header (the meta-data, e.g., system ID and message ID), and an integrity checksum which is the optional signing feature of MAVLink 2.0.

MAVLink is defined using dialects, which are XML files that define the set of messages available to MAVLink users. There are also proprietary XML files, such as ardupilotmega.xml and px4.xml, which define additional/custom messages that have been developed specifically for use with those particular platforms. Additionally, common.xml is a widely used dialect that provides support for general-purpose messages, enabling developers to develop messages beyond the defined dialects without breaking support for the standard messages.

The two major versions of the MAVLink protocol are: version 1, which contains standard messages and basic framing capabilities, and version 2, which adds support for security features, larger payloads, and the ability to define optional fields. However, version 2 still supports version 1 messages. All forms of telemetry data associated with the UAV (position and attitude), the management of parameters (PID values, thresholds), control of the UAV (take-off and landing), and the mission management process (upload and manage autonomous mission files) are supported by the MAVLink message set. Therefore, MAVLink provides an efficient, flexible, and extensible communication system for UAV and robotics systems. The basic message structure and the associated transmission, reception, acknowledgment, and error-handling states are illustrated in Figure 9.2.

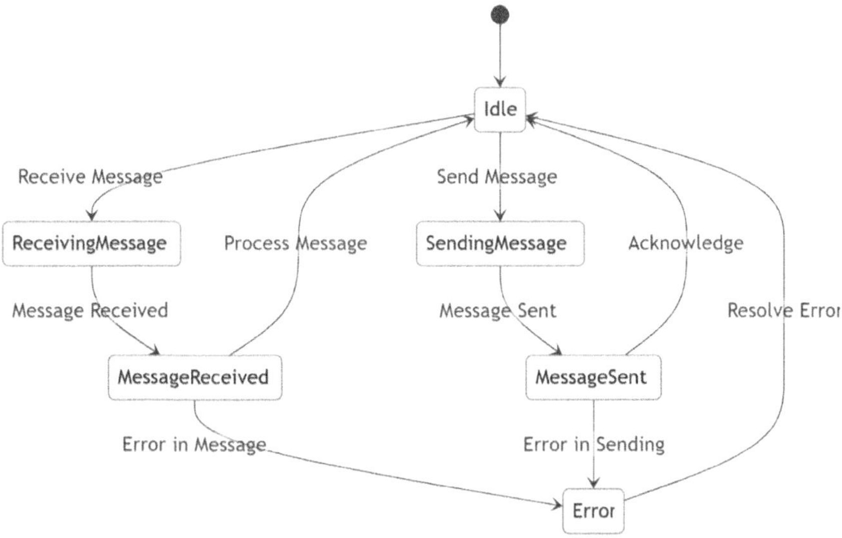

Figure 9.2 State diagram of MAVLink message transmission, reception, acknowledgement, and error handling.

9.3 POPULARITY OF MAVLINK IN DRONES AND ROBOTICS [1-2]

The ability to effectively communicate using the MAVLink protocol has allowed it to become the de facto standard for the drone industry. The MAVLink protocol is able to communicate over low-bandwidth/high-latency links commonly found in UAV's telemetry. The compact binary nature of the MAVLink protocol transmits information at faster speeds with less overhead than traditional text-based formats such as Json and Xml.Microcontrollers. The MAVLink protocol is also fully compatible with various robotics frameworks (such as the Robot Operating System (ROS)) and commonly used autopilot platforms (PX4 and ArduPilot), allowing it to be utilized in applications ranging from ground robotics to UAVs.

The MAVLink protocol provides a standardized format for all the key functions commonly associated with drones, such as GPS, Attitude, Battery Information, Mission Upload, Arm/Disarm, Take-off, and Calibration. As a result, developers no longer need to write their own code to implement the same functions. The MAVLink protocol is supported by a number of Ground Control Software applications, including QGroundControl and Mission Planner.

9.4 SETTING UP A MAVLINK COMMUNICATION SYSTEM

An important consideration when designing or deploying an unmanned aerial vehicle, unmanned ground vehicle, or any other unmanned system is establishing a reliable MAVLink communication framework.

MAVLink stands for "Micro Air Vehicle Link." It provides the means for communicating between drones and GCS. This section describes how to establish a system of MAVLink communications. The section highlights software tools used for establishing communication, the necessary hardware connections to support this communication, and a particular example of how to do this using a Pixhawk flight controller with Mission Planner or QGroundControl.

9.4.1 Hardware interfaces: how the devices connect?

MAVLink supports a variety of hardware connections. The most common type of serial connection between the Pixhawk and other devices (such as telemetry radios and Raspberry Pi) is via UART (Universal Asynchronous Receiver/Transmitter). The easiest way to connect to a laptop from a Pixhawk is through a USB port; however, there are also other ways to connect using networked protocols, such as UDP and TCP, particularly if using a companion computer with either ethernet or Wi-Fi. In summary the various connections are

- USB = Micro USB cable connection to a PC;
- UART/Serial = Telemetry modem and/or Raspberry Pi connection;
- UDP/TCP = Long-range or Wi-Fi connections over networks such as ethernet.

9.4.2 Software libraries: tools to work with MAVLink

To work with MAVLink messages, a suitable software library associated with the message protocol is used. MAVLink libraries are available for use include:

MAVLink C Library: This library is built into the firmware of many flight controllers (e.g., PX4 and ArduPilot).

Pymav link (Python): This library allows you to write some basic scripts for sending commands and reading data from a flight controller.

MAVProxy: This is a command-line tool that allows you to monitor MAVLink messages in real time while flying a drone.

Rust/JavaScript/Swift Bindings: On mobile devices and in a new platform, you can build custom apps that use MAVLink messages. If you are new to MAVLink, python and pymavlink, it will be helpful for you to learn and test MAVLink messages.

9.4.3 Example setup: Pixhawk and Mission Planner/QGroundControl

Connection setup of Pixhawk to a GCS using Mission Planner / QGroundControl.

Step 1: Hardware Connection

- Option 1 (USB): Connect the Pixhawk to your laptop with a USB cable. This powers it up and creates a virtual COM port.
- Option 2 (Telemetry): Use two 915 MHz (or 433 MHz) telemetry modules: One connects to the Pixhawk's TELEM1 port. The other connects to your laptop via USB.
- Option 3 (Wi-Fi/UDP): Use a Raspberry Pi connected to the Pixhawk to send MAVLink data to the GCS over Wi-Fi.

Step 2: Install the Software

- Download and install Mission Planner (for Windows) or QGroundControl (for all platforms).
- Optional: Install Python, and then install MAVLink tools with:

~~~
``bash
pip install pymavlink mavproxy
~~~

Step 3: Connect to the GCS

- Open Mission Planner or QGroundControl.
- If using USB, select the correct COM port and baud rate (usually 115200).
- If using telemetry, select the telemetry COM port with a baud rate of 57600.
- If using UDP/Wi-Fi, QGroundControl will automatically detect incoming MAVLink on port 14550.

Step 4: Confirm the MAVLink Connection

- Look for a successful HEARTBEAT message. This means your drone is connected.

9.4.4 Advanced: add a companion computer (optional)

For more advanced projects, you can connect a Raspberry Pi to the Pixhawk and run custom code. This setup allows you to:

- Process images
- Run AI algorithms
- Send custom MAVLink messages

Here's a sample MAVProxy command from the Pi to send MAVLink data to a GCS:

~~~
```bash
mavproxy.py --master=/dev/serial0--baudrate 57600--out=
udp:192.168.1.100:14550
~~~

This means:

- master is your connection to the Pixhawk.
- out is where you're sending data (GCS IP and port 14550).

9.5 MAVLINK MESSAGE HANDLING [1-2]

Working with the message-based MAVLink protocol necessitates effective message encoding, decoding, and handling. XML dialects like common.xml and ardupilotmega.xml are used to define each message. MAVLink libraries, like pymavlink or the MAVLink C library, encode structured values, like altitude or GPS coordinates, into small binary packets when a GCS or companion computer transmits data. Real-time communication is easy to implement since these packets are automatically decoded back into usable data objects on the receiving end. Telemetry parsing is among the most frequently performed tasks. Once a connection is made, the drone streams messages including GLOBAL_POSITION_INT (latitude, longitude, altitude), BATTERY_STATUS (voltage, current, percentage), and ATTITUDE (roll, pitch, yaw). With pymavlink, developers can simply filter and log particular message types, enabling real-time state monitoring and intelligent decision-making.

Arming, take-off, landing, and mission uploads are among the commands that MAVLink can transmit. Usually, commands are sent using COMMAND_LONG or MISSION_ITEM_INT, and the autopilot confirms success by sending COMMAND_ACK messages in response. Structured mission messages can be used to define waypoints and mission tasks in a similar manner. Overall, building autonomous or semi-autonomous drone control systems with little code is simple thanks to MAVLink's structured messaging, dependable encoding/decoding, and high-frequency communication. Figures 9.3 and 9.4 show the functional workflow and state transitions of MAVLink message handling.

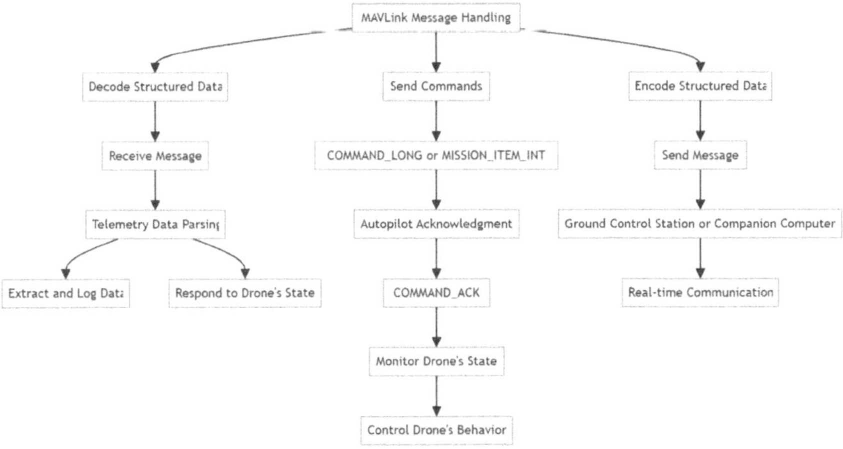

Figure 9.3 Functional workflow for MAVLink message handling.

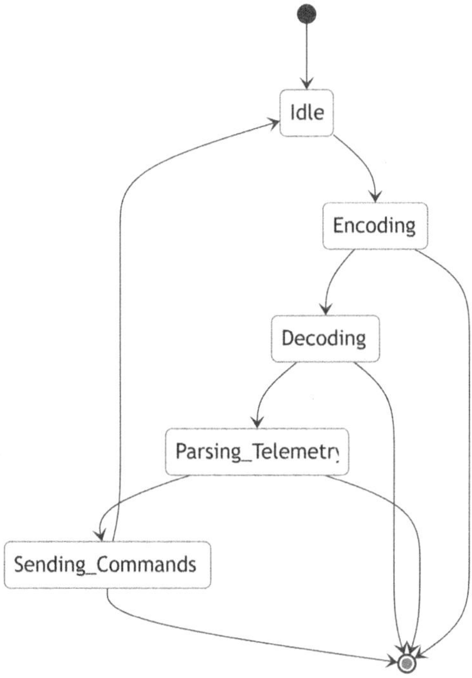

Figure 9.4 State-flow representation of MAVLink encoding, decoding, telemetry parsing, and command transmission.

9.6 MAVLINK AND GCS

Any Unmanned Aerial System (UAS) must have a GCS in order to monitor telemetry, plan missions, switch flight modes, send commands, and adjust system settings. All of this depends on MAVLink, which serves as the communication channel between the GCS and the flight controller (such as Pixhawk). MAVLink 1.0 and 2.0 are fully supported by contemporary platforms, which helps explain its success with Mission Planner, QGroundControl, and MAVProxy.

A feature-rich Windows GCS for ArduPilot is **Mission Planner**. It provides advanced parameter tuning, firmware flashing, log analysis, map-based mission planning, system health monitoring, and Software in the Loop (SITL) simulation. All of its functions are powered by MAVLink messages, which are connected via USB, telemetry radios, or network links.

QGroundControl is a user-friendly, cross-platform ground control station used with PX4 and ArduPilot-based systems. It supports both desktop and mobile devices and provides tools for mission planning, vehicle setup, telemetry monitoring, and real-time flight control. It supports camera control, sensor setup, log viewing, 3D mission planning, and MAVLink 2.0 functions like message signing. MAVLink connections over UDP are automatically detected by QGC, which simplifies and facilitates field-friendly wireless use. A comparative analysis of the GCS platforms is presented in Table 9.1.

Table 9.1 Comparison of Mission Planner, QGroundControl, and MAVProxy.

Feature/criteria	*Mission planner*	*QgroundControl*	*MAVProxy*
Primary platform	ArduPilot	PX4 (also support ArduPilot)	ArduPilot
User interfaces	Graphical	Graphical (GUI, Touch optimization)	Command-line
Operating system	Windows only	Windows, macOS, Linux	Windows, Linux, Raspberry Pi
Telemetry interfaces	USB, Serial, Telemetry, UDP	USB, Serial, UPD, TCP	USB, Serial, UDP
Mission planning	Yes (drag-drop with details option)	Yes (visual and intuitive)	Scripted mission upload supported
Parameter tuning	Full support with advance option	Full support with graphical slider	Supported via CLI
Firmware flashing	Yes (ArduPilot only)	Yes (PX4 and ArduPilot)	No
Scripting/ Automation)	Limited scripting support	Minimal scripting	Advance scripting in python
Extensibility/ Plugins	Community plugins available	Moderate (QML/ Qt-based plugins)	Highly modular
Companion computer integration	Limited	Good	Excellent
Video streaming	No native support	Yes (RTSP, MAVLink video)	Limited, requires an additional module
Offline mapping	Supported	Supported	Not supported natively
Use case suitability	Best for in-depth control/tuning	Best for mobile field use and px4 drone	Best for scripting, DevOps and swarms
Ideal users	Engineers, system tuners, researchers	Field pilots, px4 developers	Advanced users, researchers, developers
License	Open source (GPL)	Open source (Apache 2.0)	Open source (GPL)

A typical GCS operational workflow, from monitoring and mission planning to flight completion, is illustrated in Figure 9.5.

9.7 MAVLINK USE CASES [1-2]

The original purpose of MAVLink was to allow communication between drones and their GCSs via telemetry. While MAVLink was designed for

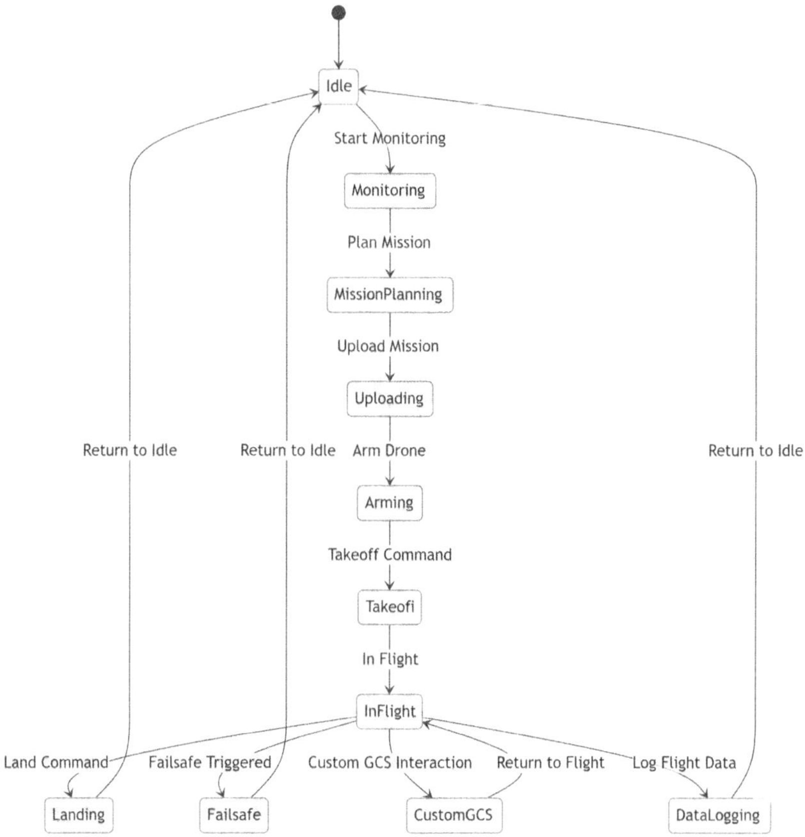

Figure 9.5 Typical GCS mission workflow.

single vehicle-to-GCS communications, the flexibility of the MAVLink message architecture makes it suitable for more advanced applications, including robotics, Internet of Things (IoT), and imaging/computer vision. This section will discuss these advanced applications of MAVLink beyond the traditional usage for UAV flight controls, with examples from the areas of swarm robotics, cloud-connected UAVs, and IoT/computer vision integrations.

9.7.1 Swarm robotics with MAVLink

The coordination of several drones or unmanned systems that collaborate either autonomously or semi-autonomously is the main focus of swarm robotics. For each vehicle in the swarm to communicate with the GCS, MAVLink is essential. Every drone is assigned a distinct system_id, which enables the GCS or a central controller (such as a Raspberry Pi or Jetson Xavier) to send broadcast or individual MAVLink commands.

In typical swarm setups:

- MAVLink routing is managed by MAVProxy or custom Python scripts that use pymavlink.
- Multiple drones can receive commands like ARM, TAKE-OFF, or MISSION_START simultaneously or sequentially from a single GCS.
- For inter-drone communication, like position syncing or formation commands, custom MAVLink messages/MAVLink dialects can be developed.

9.7.2 MAVLink in computer vision systems

MAVLink can transmit, receive, and act upon object detection data from internal or external processors in vision-based systems. For instance, a drone equipped with a camera and an edge AI module (such as the Jetson Nano or Raspberry Pi + Coral TPU) can instantly identify things like people, cars, or QR codes. The position, class, and confidence score of the object are sent to the flight controller in a custom MAVLink message by the system after it has been detected.

For instance:

- A VISION_POSITION_ESTIMATE can be used to send the drone's vision-based position and attitude estimate to the autopilot for external navigation or non-GPS positioning.
- A custom OBJECT_DETECTED message (via a MAVLink dialect) might include bounding box coordinates, timestamps, and labels.

This enables the autopilot to react in real time by changing its course, hovering for inspection, or initiating an event like sounding an alarm or dropping a payload. Companion computers that execute the computer vision pipeline and generate MAVLink messages in accordance with it facilitate integration. AI output and flight behavior are usually connected via MAVSDK and pymavlink, which makes MAVLink an effective intermediary between perception and action.

- Cloud dashboards, mobile apps, or automation systems can subscribe to this data for visualization, alerts, or control.

9.8 MAVLINK SECURITY

As the utilization of drones increases throughout sensitive sectors such as Agriculture, Delivery, Defense, and Surveillance, securing communications between the GCS and drones becomes critical. MAVLink 1.0's design was for speed and minimal bandwidth, and therefore, it is vulnerable to man-in-the-middle (MITM), spoofing, and eavesdropping attacks.

An attacker could intercept the telemetry, inject fake commands, or even take control of an unsecured drone.

To counteract these vulnerabilities, MAVLink 2.0 uses message signing to provide users with the capability to guarantee the authenticity and integrity of messages. A signature consisting of 32 bytes and created from the message's shared secret key, link ID, and timestamp is included with each message. Invalid or unsigned messages are rejected, thus preventing unauthorized command injection (however, this does not prevent the ability for an attacker to read or capture messages) without adding any significant processing overhead [3,4].

9.9 SECURITY LIMITATIONS OF MAVLINK

Performance and ease of use were the original goals of MAVLink's design, not security. The absence of default encryption is a major drawback. MAVLink messages are sent as plain text, which means that they are open to manipulation or eavesdropping since they do not contain encryption. Service interrupts due to an attacker who intercepts telemetry data from UAVs, sends malicious commands, or pretends to be a trusted source, compromise the safety and control of UAV operations.

The inadequate authentication techniques used in MAVLink's early iterations are another significant drawback. Older iterations of the MAVLink protocol went according to their designs with limited checksum techniques (CRC), which can easily be duplicated by individuals wishing to deceive others This means that a malicious user could impersonate a drone or a ground station. The inability to properly validate the identity of users, drones, or ground control stations increases the vulnerability of UAV systems to unauthorized access, hijacking, and malicious control attempts through illegitimate communication channels.

In addition, the existing state of MAVLink sequence numbers and timing information makes it possible for an individual to conduct replay attacks by sending former MAVLink messages again for the purpose of achieving results contrary to what was originally intended with the MAVLink messages sent. As a result of lacking user mutual authentication, it may be possible for a third party who intercepts a MAVLink message to modify the MAVLink message before it reaches its intended recipient, leading to a number of potential attacks against MAVLink communication systems. Flooding a MAVLink system with dummy messages may also create an opportunity for an individual to conduct denial-of-service attacks. These various vulnerabilities illustrate the importance of ensuring stronger cryptographic protections for all MAVLink protocol communications [5,6]. The exposure of plain text MAVLink traffic to a man-in-the-middle attack is illustrated in Figure 9.6.

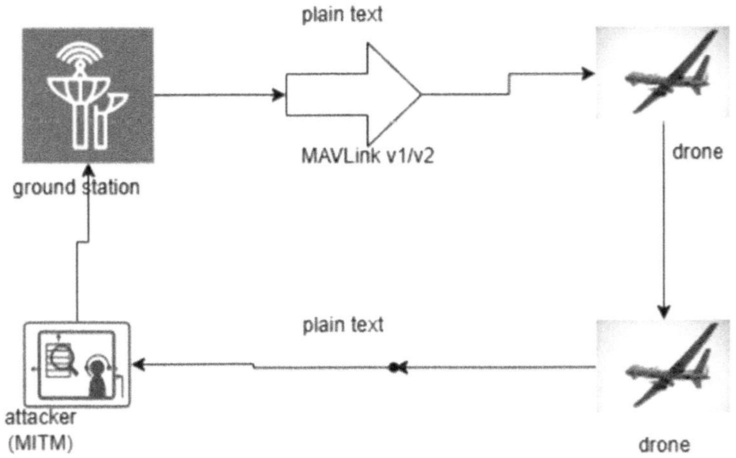

Figure 9.6 Exposure of unencrypted MAVLink communication to a man-in-the-middle attack.

9.9.1 Symmetric cryptography

Symmetric cryptography utilizes a single key for both the encryption and decryption of data. In order for both parties to successfully exchange information securely, they must each have the same secure key and keep it private. Symmetric cryptography is preferred because it is faster than asymmetric cryptography, as well as more effective when working with very large amounts of data [7,8]. The basic symmetric-key encryption and decryption process is shown in Figure 9.7.

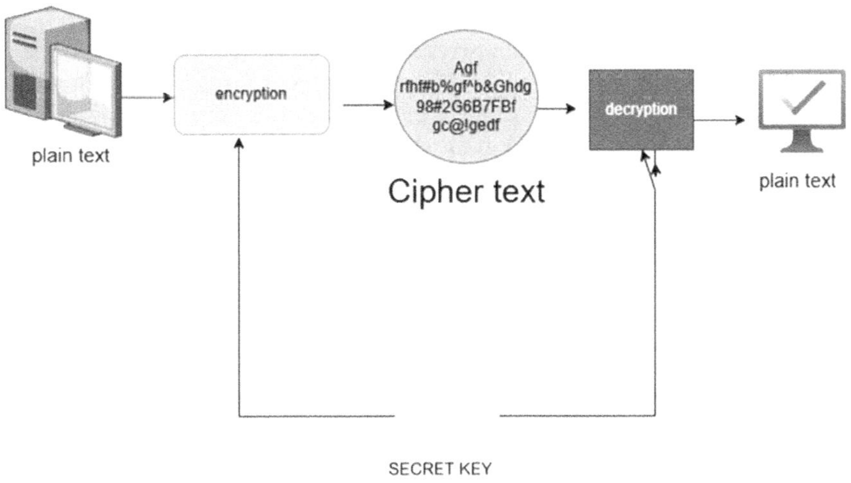

Figure 9.7 Symmetric-key encryption and decryption process.

9.9.2 Symmetric cryptographic solutions

Symmetric encryption is well suited for real-time UAV applications because it provides fast encryption and decryption with low computational overhead. MAVLink 2.0 supports built-in message signing for improved communication security. This mechanism uses a pre-shared symmetric secret key, link identifier, timestamp, and signature field to verify that messages originate from a trusted source and have not been modified or replayed. However, MAVLink 2.0 signing provides authenticity, integrity, and replay protection, but it does not provide data confidentiality. Therefore, encryption must be added separately if privacy of MAVLink payloads is required.AES-256 in Galois/Counter Mode (AES-256-GCM) is a suitable choice for securing MAVLink communication, as it provides both confidentiality and message authentication. In AES-GCM, an authentication tag is generated along with the ciphertext to verify message integrity and authenticity, helping protect UAV communication against tampering and unauthorized modification. HMAC (Hash-based Message Authentication Code) with SHA-256 is another authentication technique. HMAC uses a shared secret key and a hashing function to verify the message's source and integrity. HMAC is lightweight and efficient for verifying that messages have not been tampered with, even though it does not guarantee confidentiality.

However, symmetric approaches do not support forward secrecy and have problems with key distribution. A representative pre-shared-key MAVLink secure session is shown in Figure 9.8.

9.9.3 Asymmetric cryptography

Asymmetric cryptography, also known as public key cryptography, is a technique that encrypts and decrypts messages using a pair of related keys: one public key and one private key. By doing this, the message is shielded from unwanted access or use. A cryptographic key that can be used to encrypt a message is called a public key. It can only be decrypted with the private key of the intended recipient. Only the person who created the key has access to a private key, also referred to as a secret key.

To send an encrypted message, one must obtain the public key of the intended recipient from a public directory. They encrypt the message using that key before sending it. With private key, the recipient can then decrypt the message. If the sender signs a message using their private key, anyone with the sender's public key can verify the signature. This confirms the sender's identity and ensures that the message has not been modified [7,8]. The public/private-key encryption and decryption process is illustrated in Figure 9.9.

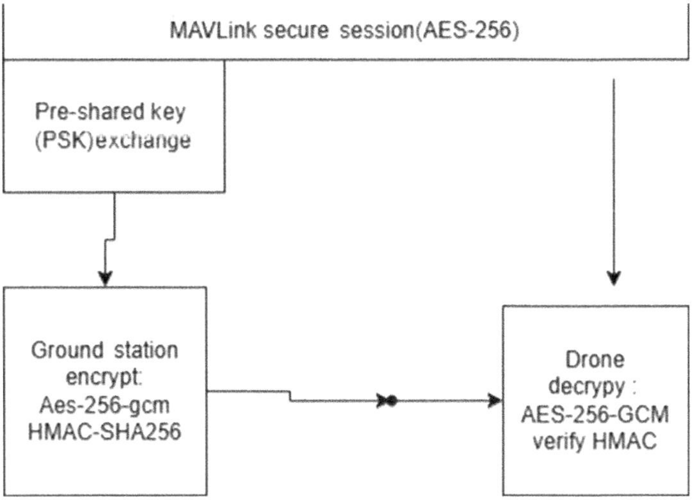

Figure 9.8 Illustrative pre-shared-key MAVLink secure session using AES-256-GCM with integrity verification.

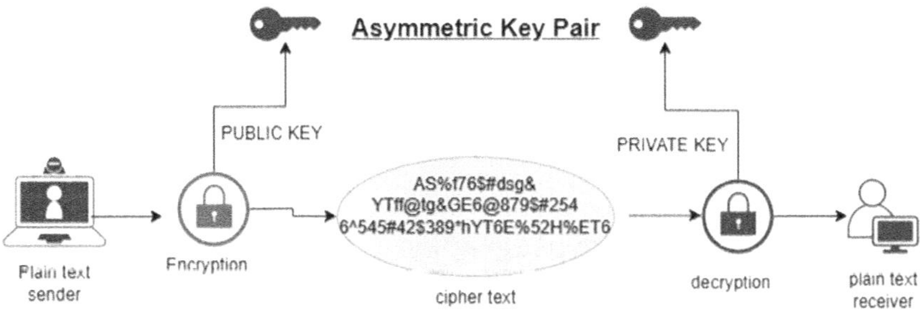

Figure 9.9 Public-key encryption and decryption using an asymmetric key pair.

9.9.4 Asymmetric cryptographic solutions

With public–private key pairs, asymmetric cryptography offers robust security. Digital signatures in UAV communications frequently use Ed25519 or ECDSA (Elliptic Curve Digital Signature Algorithm). The drone or GCS can sign messages with a private key. After that, the recipient can use the matching public key to confirm the signature. This procedure guards against spoofing and impersonation by ensuring message authenticity and preventing repudiation. However, unless optimized, asymmetric methods are frequently not the best for high-frequency telemetry and require more processing power [7,8].

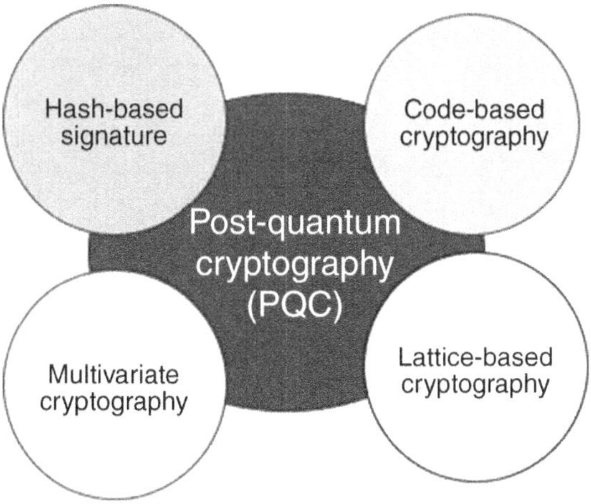

Figure 9.10 Major families of post-quantum cryptography.

By confirming identities via a reliable Certificate Authority (CA), certificate-based authentication—such as utilizing X.509 with TLS or DTLS—improves security. In order to create a secure, mutually authenticated channel, the drone and GCS exchange certificates during pairing. If a certificate is compromised, it can be revoked. However, this method is less appropriate for small, resource-constrained drones because the handshake process increases latency and demands more processing power.

9.9.5 Post-quantum cryptography

"Post-quantum cryptography" (PQC), or "quantum-resistant" or "quantum-safe" cryptography, includes cryptographic algorithms designed to withstand attacks from both classical and future quantum computers. These algorithms are crucial because many of the encryption techniques we currently use could be compromised by strong quantum computers. PQC includes creating new cryptographic algorithms based on mathematical puzzles that are believed to be challenging for even quantum computers to solve. These algorithms seek to protect private information from potential risks associated with quantum computing [9]. The major families of post-quantum cryptography are summarized in Figure 9.10.

9.10 COMPARATIVE ANALYSIS OF CRYPTOGRAPHIC SOLUTIONS

Symmetric encryption such as AES-256, combined with message authentication codes such as HMAC, provides an efficient security option for resource-constrained drones. However, its overall quantum resistance depends on how the encryption keys are generated and exchanged.

Asymmetric algorithms (ECDSA) require additional resources for processing power, but the level of authentication and non-repudiation provided by their use is quite strong though not quantum resistant. Future protection from quantum threats may be achieved through the use of post-quantum cryptographic (PQC) methodologies such as Kyber and Dilithium; however, there are limitations with respect to performance and bandwidth requirements [7–9]. Tables 9.2 and 9.3 summarize the comparative features and attack resistance, respectively.

Table 9.2 Comparative features of cryptographic solutions for MAVLink communication.

Feature	*Symmetric (Aes-256)*	*Asymmetric*	*Post-quantum*	*Hybrid (AES + ECDSA)*
Key type	Single shared type	Public/private key pair	Public/private key pair	Combined symmetric + asymmetric
Encryption speed	Fastest	Slow	Slowest	Balanced
Quantum resistance	Vulnerable to Grover algorithm	Vulnerable to Shor's algorithm	NIST-Standardized	Depends on components
Best for	Telemetry streams	Secure pairing/GCS auth	More secure connection	High security for UAV
Hardware support	Widespread acceleration	Limited MCU support	Experimental	AES+ECC common
Key distribution	Challenging (pre-shared)	PKI	PKI	Automatic + pre-shared
Use example	Real-time sensor data	Drone-GCS secure handshake	Military UAVs (long time)	Commercial drones

Table 9.3 Comparison of attack-resistance properties of cryptographic solutions.

Attack Type	AES-256	ECC/RSA	Kyber	Hybrid (ECC+Kyber)
Spoofing	No (without MAC)	Yes	Yes	Yes
Replay Attack	Needs nonce/ timestamp	Needs nonce/ timestamp	Needs nonce/ timestamp	Needs nonce/ timestamp
MITM	No	Yes (with signatures/ certificates)	Yes (with authentication)	Yes (with authentication)
Quantum Decryption	Partial (128-bit quantum security)	No	Yes	Yes

9.11 THE FUTURE OF MAVLINK IN ADVANCING DRONE COMMUNICATION

Originally developed as a simple telemetry protocol for micro air vehicles in a university research environment, MAVLink has evolved into an organized and efficient communication standard for transmitting and receiving information in the growing drone industry. Today, it supports a wide range of military, commercial, industrial, and research UAV applications. Like many communication protocols, MAVLink continues to evolve in response to emerging technologies and changing user requirements. This section examines how MAVLink is being adapted for the next decade of robotic and aerial innovation by addressing the growing need for autonomous systems, real-time data processing, secure communication, and network-integrated robotics.

9.11.1 Integration with emerging communication protocols and robotics frameworks

MAVLink's compatibility with other protocols like ROS 2 [10]: Robot Operating System DDS: Data Distribution Service, RTPS: Real-Time Publish-Subscribe is becoming more and more necessary as UAVs are integrated into larger robotic and networked systems. While frameworks like DDS and ROS 2 are designed for distributed, publish–subscribe systems typically used in fleets, SLAM-based navigation, or cloud robotics, MAVLink works well for direct command-and-control applications.

Projects MAVROS and MAVLink MicroRTPS serve as translators to close this gap, allowing the publication of MAVLink messages as ROS topics/DDS streams. This makes it possible for MAVLink drones to communicate with different robots, edge nodes, or even intelligent infrastructure. MAVLink drones could become a crucial component of autonomous systems if a future MAVLink-native ROS 2 plugin provides smooth communication in multi-agent, cross-platform robotic networks.

9.11.2 5G and edge AI: enabling real-time intelligence

5G [11] is one of the most important technologies for MAVLink's future. 5G enables UAVs to operate well beyond conventional radio ranges thanks to its significantly reduced latency (as low as 1 ms), increased bandwidth, and support for network slicing. Secure 5G networks can now transmit MAVLink messages, allowing:

- Real-time telemetry from drones hundreds of kilometers away.
- Low-latency video and control command streaming.

- Remote ground control operation through cloud-hosted centers.
- On-demand mission planning with over-the-air updates.

Drones can now process visual or environmental data onboard with companion computers like the Jetson Nano, Raspberry Pi with Coral TPU, or Qualcomm RB5 thanks to the development of edge AI. Machine learning models for anomaly detection, route planning, and object detection can be executed by these systems. Then, MAVLink acts as a language to convey AI decisions to the flight controller, such as "hover at target," "avoid zone," or "track object at X, Y, Z." This connection in autonomous UAV systems, makes it the decision-making channel.

Developers are also looking into custom MAVLink messages for AI feedback, including:

- OBJECT_TRACKING_STATUS
- TARGET_DETECTION_INFO
- ML_INFERENCE_RESULT

In order to guarantee compatibility across various vendors in vision-guided or AI-directed flights, these messages could be standardized in future languages.

9.11.3 Security and regulation: toward a trustworthy MAVLink ecosystem

When drones are used in sensitive domains such as defense, logistics, and urban monitoring, safety, security, and regulatory compliance must remain key priorities. Future MAVLink-supported systems are therefore expected to include stronger security and compliance features, such as transport-level encryption for MAVLink-over-TCP using TLS, blockchain-based identity and telemetry verification, and robust public key infrastructure for mission authorization. MAVLink-based systems may also evolve to support international standards and regulations related to airworthiness, UAS traffic management, and Remote ID. Additional improvements may include integrated digital signatures, encrypted communication, verified operator identity, and GPS timestamp-based location references for trusted and auditable drone operations [12].

REFERENCES

1. ArduPilot Development Team, "MAVLink Basics," ArduPilot Developer Documentation. Available: https://ardupilot.org/dev/docs/mavlink-basics.html.
2. MAVLink Development Team, "MAVLink Developer Guide." Available: https://mavlink.io/en/.

3. Allouch, A., O. Cheikhrouhou, A. Koubaa, M. Khalgui, and T. Abbes. 2019. "MAVSec: Securing the MAVLink Protocol for Ardupilot/PX4 Unmanned Aerial Systems." In *Proceedings of the 15th International Wireless Communications & Mobile Computing Conference* (IWCMC), Tangier, Morocco, pp. 621–628, doi: 10.1109/IWCMC.2019.8766667.
4. Khan, N. A., N. Z. Jhanjhi, S. N. Brohi, A. A. Almazroi, and A. Ali Almazroi. 2022. "A Secure Communication Protocol for Unmanned Aerial Vehicles." *Computers, Materials & Continua* 70 (1): 1–15. https://www.techscience.com/cmc/v70n1/44408/html.
5. Amorim, A., M. Taylor, T. Kann, W. L. Harrison, G. T. Leavens, and L. Joneckis. 2025. "Enforcing MAVLink Safety & Security Properties Via Refined Multiparty Session Types." In A. Dutle, L. Humphrey, and L. Titolo (eds.), *NASA Formal Methods (NFM 2025), Lecture Notes in Computer Science*, vol. 15682, Springer, Cham. doi: 10.1007/978-3-031-93706-4_1.
6. Hamza, M. A., M. Mohsin, M. Mohsin, and S. M. K. Abbas. 2024. "MAVLink Protocol: A Survey of Security Threats and Countermeasures." In *Proceedings of the 4th International Conference on Digital Futures and Transformative Technologies (ICoDT2)*, Islamabad, Pakistan, pp. 1–8, doi: 10.1109/ICoDT262145.2024.10740195.
7. Menezes, A. J., van Oorschot, P. C., and Vanstone, S. A., 1996. "*Handbook of Applied Cryptography*". CRC Press, Boca Raton, FL, USA.
8. Stallings, W. 2023. "*Cryptography and Network Security: Principles and Practice*". 8th ed. Pearson, Hoboken, NJ, USA.
9. Bernstein, D. J., Buchmann, J., and Dahmen, E. (eds.), 2009. "*Post-Quantum Cryptography*". Springer, Berlin, Germany.
10. Robot Operating System (ROS), "*ROS: Home.*" Available: https://www.ros.org.
11. Damigos, G., Lindgren, T., Sandberg, S. and Nikolakopoulos, G. 2023. "Performance of Sensor Data Process Offloading on 5g-Enabled UAVs." *Sensors* 23 (2): 864.
12. Shakhatreh, H., Sawalmeh, A.H., Al-Fuqaha, A., Dou, Z., Almaita, E., Khalil, I., Othman, N.S., Khreishah, A. and Guizani, M. 2019. "Unmanned Aerial Vehicles (UAVs): A Survey on Civil Applications and Key Research Challenges." *IEEE Access* 7, 48572–48634.

Chapter 10

Secure horizons

Future trends in drone communication

Sumaya Akhtar, Iraq Ahmad Reshi, Ummer Iqbal Khan, Prerna Ajmani, and J. Joshua Thomas

10.1 INTRODUCTION

The drone technology is evolving rapidly and affecting many industries like package delivery, agriculture monitoring, military surveillance, and emergency response work. As drone networks have become more advanced, the security of the underlying drone communication becomes significantly more important. The security of today's drone communication networks is primarily based on cryptographic protocols to establish the confidentiality, authenticity, and integrity of the data that are exchanged or transmitted from drone to drone, drone to ground control station, or drone to satellite links. Current drone security protocols continue to primarily rely on standard cryptography which includes symmetric cryptography for real-time data streams and asymmetric public-key systems for key exchange and authentication. The emergence of practical quantum computing capability poses a serious threat to the foundational assumptions of these existing cryptographic security implementations within drone networks [1,2].

Drones operate in different environments as compared to traditional networks. This is due to unique limitations including limited computational ability, limited bandwidth, intermittent connectivity, and requirement of real-time processing. Therefore, application of cryptographic protocols requires additional considerations. Drone networks demand cryptographic security solutions that operate at very low latency, adapt to highly dynamic and rapidly changing network conditions, and provide strong resilience while maintaining robust security [3].

Drone communication currently relies on classical public-key cryptographic mechanisms, such as Rivest-Shamir–Adleman (RSA), Diffie–Hellman, and Elliptic Curve Cryptography (ECC), for authentication, key exchange, and secure communication. The strength of these schemes is primarily based on computational difficulty of solving problems that include integer factorization and discrete logarithms. However, quantum algorithms can solve these problems efficiently as compared to classical methods. Consequently, the long-term security of existing drone communication

DOI: 10.1201/9781003598350-13

systems cannot be guaranteed by classical algorithms, thus creating a pressing need for post-quantum cryptography (PQC).

The importance of implementing PQC within drone networks lies mainly in the fact that security is of paramount importance in the operations of such networks. UAVs today are widely deployed in military operations, surveillance, rescue efforts, and even autonomous swarm missions; hence, there is a need for secure and efficient communication between them. Moreover, the fact that drones operate as a connected swarm makes it crucial that any form of security breach to one device would mean compromising the entire network. Additionally, many drones remain in operations for long periods, making them susceptible to future quantum attacks.

10.2 CURRENT CRYPTOGRAPHIC LANDSCAPE IN DRONE SECURITY

Various cryptographic challenges arise in drone communication, differentiating it from conventional networks. This is because it operates under conditions where computing capability is low, along with constrained bandwidth and time. Such factors result not only in performance issues but also in reduced security margin. Moreover, the characteristics of drone networks including mobility and ad hoc network topologies further complicate the security of communication. Therefore, it is important to reconsider and improve existing cryptography solutions or develop new, lightweight, and resilient security solutions that address the operational challenges of drone networks.

The authentication and key establishment in drone networks are primarily based on public-key cryptography. The classical public-key cryptographic mechanisms are primarily based on the computational hardness of mathematical problems such as integer factorization and discrete logarithm problem.

The public-key cryptographic systems based on RSA derive their security strength from the integer factorization problem. The RSA involves the computation of modulus N based on two large prime numbers p and q:

$$N = p \times q$$

The integer factorization problem is to determine p and q from the publicly known value modulus N. This is a computationally hard problem and forms the basis of RSA security [4].

The computational challenge is a significant issue when attempting to break RSA-based security. The brute-force method of factorizing the large modulus $N = p \times q$ is impractical as it requires an impractically large number of operations. General Number Field Sieve (GNFS) is an advanced factorization method that offers substantially improved efficiency, but it still requires considerable computational resources and time [4].

Discrete logarithm problem (DLP) is the problem of determining an unknown exponent (x) from the values (g) and $(g \wedge x)$ in a finite cyclic group. While computing $(g \wedge x)$ from a given (g) and (x) is computationally efficient, recovering (x) from (g) and $(g \wedge x)$ is believed to be computationally infeasible when the group is sufficiently large. This hardness forms the basis of cryptographic schemes like the Diffie–Hellman key exchange [5], DSA, and ECC. ECC is another widely used public-key cryptography in drone systems due to its efficiency and smaller key sizes. The computational hardness of ECC is based on the elliptical DLP. The Elliptic Curve Discrete Logarithm Problem (ECDLP) is defined as

$$Q = kP$$

where P and Q are points on an elliptic curve and k is an integer. The computational infeasibility is determining k from P and Q.

The above-stated mathematical problem is at the core of some of the most common cryptographic mechanisms implemented in drones' communication systems. Some examples are the Diffie–Hellman key exchange scheme that can be utilized by drones for establishing secure communications, DSA that enables authentication and verification of the commands issued to drones, and ECC-based mechanisms that ensure efficient cryptographic protection for drones.

Therefore, both RSA encryption and ECC-based public-key cryptography systems derive their security from the computational complexity of breaking them, considering the size of their security parameters. In the case of RSA, its security relies on the difficulty of integer factorization, while ECC-based cryptosystems are based on the complexity of DLP [6,7]. The more the bits used to represent the keys, the more the time an attacker requires to break those using classical means. The computation induced by larger key sizes and the associated power demands may conflict with the latency, processing, and battery-life requirements of autonomous flight operations and swarm missions. Although ECC offers comparable security with smaller key sizes than RSA, it is still based on mathematical assumptions that are vulnerable to future quantum attacks.

Because of the low battery capacity, computing capacity, and real-time demands of drones, it is impossible to increase the size of keys in classical encryption indefinitely. In classical systems, a larger size of keys means better protection. However, the advent of quantum computing creates a whole new approach in which such an idea is no longer enough. Quantum algorithms can crack the mathematical equations on which most classical cryptography schemes are based in polynomial time. It completely changes the perspective in regard to UAV security.

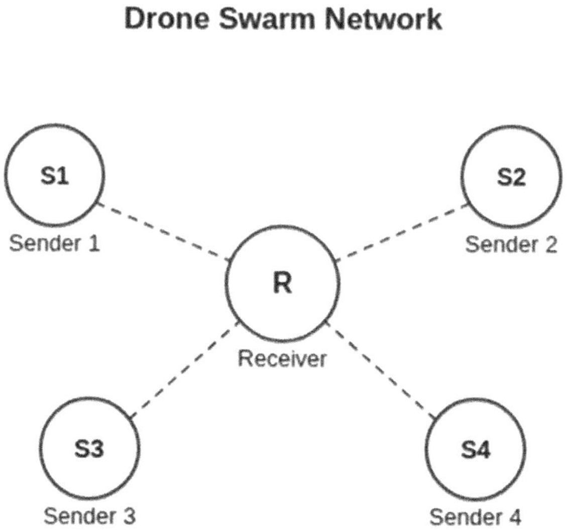

Figure 10.1 Drone Swarm Network Architecture

There is an urgent requirement for lightweight cryptographic schemes that are quantum resistant and at the same time can satisfy all the demands made upon UAV swarms.

Figure 10.1 presents the drone swarm network architecture considered in this study. The network consists of a central receiver node (R) and multiple sender drones (S1, S2, S3, and S4). Each sender drone establishes a secure communication link with the receiver to facilitate key exchange and encrypted data transmission within the swarm.

10.3 QUANTUM COMPUTING THREATS TO DRONE COMMUNICATION SECURITY

The emergence of quantum computing brings new threats to drone secure communications and dramatically change how UAV networks can be secured. Quantum computers use quantum bits, which can exist and function in superposition states, allowing for parallel computation across multiple solution paths simultaneously [8]. This ability translates to quantum systems providing an exponential computational advantage to solve certain mathematical problems on which current drone security systems rely upon.

Quantum computing poses a significant threat to traditional drone communication security because it exploits quantum mechanical phenomena, such as superposition and entanglement, to perform certain computations more efficiently than classical computers. Quantum algorithms have the capability to solve integer factorization and discrete logarithms in a much faster way than their classical counterparts.

Peter Shor's quantum algorithm poses the fastest, most severe, immediate threat to the security of classical cryptographic systems [1]. Shor's algorithm suggests that integer factorization–based RSA may be compromised by quantum computers in polynomial time. Similar concern also extends to the use of ECC-based implementations as the ECDLP can also be solved by efficiently using quantum algorithms.

In the discussed scenario, a quantum-capable adversary can comprise the network operations by injecting unauthorized commands, issue false instructions, intercept mission-critical information, or disrupt multi-drone coordination.

Grover's quantum search algorithm provides a quadratic speedup for brute-force attacks against symmetric cryptographic systems used in drone communications [2]. Although this speedup is less dramatic than that of Shor's algorithm, it still significantly reduces the security margins of symmetric encryption used for real-time drone data transmission.

Current drone communication systems often use the Advanced Encryption Standard (AES) with 128-bit keys to encrypt sensor data, video streams, and control commands. The Grover algorithm is known to lower the effective security strength of AES-128 to about 64 bits of classical security, which makes it less appropriate in protecting against quantum attacks in the future. In terms of providing long-term security to drones, where the mission would require extended observation or military activity, a quantum vulnerability like this calls for increasing the size of the symmetric key, such as switching from a 128-bit key to a 256-bit key to attain similar security levels.

10.4 HARVEST NOW, DECRYPT LATER ATTACKS ON DRONE COMMUNICATION

The quantum problem for drone communications isn't just a problem for the future. One of the most dangerous and well-developed attack techniques, namely, "Harvest Now, Decrypt Later" (HNDL) or "Store Now, Decrypt Later", is a threat to UAVs in today's operational reality. The approach involves an opponent capturing and storing encrypted communications sent between drones, although they are not decipherable without the current computing power. The opponent keeps these captured data with the idea that in the future, with powerful-enough quantum computers, they will be able to break the classical encryption and have access to the original plaintext. The HNDL threat is not merely theoretical. Public institutions such as intelligence agencies and cybersecurity firms have openly admitted that state-sponsored actors are already massively collecting encrypted information in preparation for its decryption. In 2022, the U.S. National

Security Agency (NSA) published a specific warning regarding the ongoing attempts by adversaries to intercept encrypted traffic prior to the development of quantum computing, and the need for national security systems to make a quantum transition to quantum-resistant algorithms as soon as possible [9]. Likewise, in 2023, the U.S. Cybersecurity and Infrastructure Security Agency (CISA), along with its partners from the United Kingdom, Australia, Canada, and New Zealand, issued an advisory confirming that the HNDL threat exists and the organizations that have systems with long data sensitivity lifespans should consider the shift to PQC as an operational priority and not a future goal. This is a readily applicable threat for drone networks that transmit classified Intelligence, Surveillance, and Reconnaissance (ISR) data, or sensitive military coordination or long-term infrastructure monitoring records. The documented evidence of adversarial interception of UAV communication links is indicative of the practical feasibility of HNDL attacks against aerial platforms. The obvious takeaway for HNDL is that if drone communications are being intercepted today, then they will be intercepted, and if they were recorded and stored, they will be decrypted retroactively when adversaries come to have access to quantum computing resources. When a Unmanned Aircraft System (UAS) is attacked using the HNDL model, a unique and significant imbalance is created in the security situation of drone communication. Typically, in classical cryptographic attacks, the adversary must break the encryption before the data being protected becomes useless. In HNDL, however, the adversary completely separates the interception and decryption operations. It is not necessary for an adversary to break the handshakes or session key exchanges of drones that are using RSA or ECC today. With low storage cost, the captured ciphertext and handshake records can be kept for an indefinitely long period of time: The cost of storing data has dropped significantly over the past few decades. Once a quantum computer becomes available with cryptographic capabilities, even years-old sessions can be opened [10]. This "retroactive exposure" differs from the threat model that classical cryptographic systems are designed against. When it comes to drone communication systems, the HNDL threat applies to a number of categories of data that are particularly sensitive. One of the first places where key exchange records can be found is at the risk of being compromised. For protocols that rely on the RSA or elliptic curve Diffie–Hellman, the process of negotiating keys generates public values which be captured by a passive attacker. If a future quantum computer running Shor's algorithm is able to recover the session key from these recorded public values, then the entire session can be recovered and the information decrypted. Second, the information about a drone's operations may be disclosed if its authentication credentials, certificates, or identity tokens are decrypted over time during the drone registration or mission authorization process or when associating with a ground station. Third, the ISR payload data such as high-resolution imagery, sensor readings, geolocation, and acoustic or signals intelligence gathered during longer missions can have intelligence value for years. Fourth, coordination messages and formation parameters from multiple UAVs could provide clues to operational tactics, routes, and mission goals. The HNDL threat covers the entire communication chain that includes key negotiations and mission data payloads.

The evolution of the quantum computing is a crucial factor to evaluate the HNDL risk for drone systems. The estimates, which are very variable, suggest that a quantum computer of size capable of breaking current 2048-bit RSA or 256-bit ECC will be possible in the next 10–15 years, and may be achieved as early as the 2030s [11]. Many of the military, border surveillance, and infrastructure monitoring UAV platforms will be in operation for over a decade, and some classified data collected by these platforms will be sensitive for decades, so the exposure period for HNDL is not trivial. Today's deployment of an RSA- or ECC-protected communication may still be in use when the first cryptographically relevant quantum computers go on-line, and data gathered from it today could be decrypted before the drone is retired.

In this regard, the difference between forward security and long-term data confidentiality is relevant. The HNDL model adds a new form of vulnerability even in drone communication systems that use ephemeral key exchange and provide forward security against classical attacks. If the ephemeral key exchange is based on elliptic curve Diffie–Hellman, the recorded ephemeral key exchange would lead to a quantum adversary to be able to recover the ephemeral session key and decrypt session contents. A perfect forward secrecy is not effective against a quantum adversary that can solve the ECDLP, since a compromise of the long-term private key would lead to compromise of past sessions, allowing classical attackers to conduct such sessions. It indicates that even the most secure contemporary-type drone communication systems are vulnerable to HNDLs attacks, when they use classical key exchange approaches.

Looking at the HNDL threat for communication with drones means facing a different scenario than just upgrading encryption after quantum computers. By definition, HNDL attacks pre-extract data before decryption can take place, and therefore any communication sent today using classical cryptography is susceptible to an attack by an adversary who is recording it. The best way to prevent this is to use quantum-resistant key exchange systems before they can be intercepted by drones, rather than after. This motivates to adopt post-quantum key-encapsulation methods like those based on the Learning with Errors problem to be addressed now, instead of a future upgrade. While symmetric encryption with a sufficiently large key size is believed to be secure against Grover's algorithm at reasonable security margins, what makes the symmetric data encryption layer vulnerable is mostly the classical asymmetric key exchange, asymmetric authentication mechanisms, etc., that it is used with, and not the symmetric encryption itself. Mitigating the asymmetric components in a UAV communication architecture with quantam-resistant elements is thus the most promising short-term solution to HNDL risk [9,12].

The HNDL threat thus serves as a mental connection between the quantum computing vulnerabilities and the PQC solutions. It shows that the need for moving to quantum-resistant communication mechanisms for drones is not a requirement of current-day quantum computers. The harvest stage of an HNDL attack has already commenced, as there are actors motivated and capable of collecting and storing thousands of encrypted drone communications. This fact alone means that the PQC is not just a promising research avenue, but it is also an essential security issue for drone networks that transport data of any real long-term sensitivity.

10.5 QUANTUM-RESISTANT APPROACHES FOR DRONE COMMUNICATIONS

Quantum-resistant cryptographic approaches are essential for drone communication networks as they are increasingly used in sensitive applications including surveillance, defense, disaster management, agriculture, logistics, and smart-city operation. Conventional public-key cryptography mechanisms like RSA and ECC are vulnerable to quantum attacks. Thus, PQC [13] is considered very important in drone communication security.

10.5.1 Lattice-based cryptography

Lattice-based cryptography is one of the most prospective post-quantum schemes for drone networks that relies on hard problems like Learning with Errors and Shortest Vector Problem. It can be used for encryption, digital signatures, and key exchange mechanisms. Lattice-based cryptography is implemented using mathematical operations, such as modular arithmetic, matrix, and polynomial calculation, thus making it efficient for embedded systems. It can be used for securing sensor data, command message, and also key establishment between drones and the ground control station [14–17].

10.5.2 Hash-based signatures

Hash-based signature is an PQC mechanism that can be used in authenticated communication. The security of hash-based signatures primarily relies on the strong properties of cryptographic hash functions such as collision resistance and pre-image resistance. Hash-based signatures are resistant to quantum attacks since they do not rely on number-theoretic assumptions. However, the key management of most hash-based systems is quite complex due to the use of one-time or multi-time signature generation. Stateless techniques like SPHINCS+ [18] lighten the burden of key management. However, its implementation has to be considered carefully with respect to drone networks.

10.5.3 Code-based cryptography

Code-based cryptography is another robust post-quantum candidate. These cryptosystems are built on the hardness of decoding certain error-correcting codes. McEliece [19] type of cryptosystems were one of the first and extensively studied techniques that offer reasonable security along with faster encryption/decryption compared to lattice-based and hashmac-based cryptography. However, their extremely large public keys can be a hurdle in UAV communication systems, where bandwidth is a serious constraint and also storage is limited.

10.5.4 Multivariate cryptography

Multivariate cryptography is primarily used for digital signatures and relies on the difficulty of solving a system of multivariate quadratic polynomial equations over finite fields. This cryptography could be quite ideal for drone authentication, command authentication, and swarm coordination, since it is capable of generating and verifying signatures quite quickly. An additional benefit of multivariate cryptography is the potential for producing relatively short signatures that are efficient in terms of computational requirements. However, most previously proposed multivariate signatures were cracked using cryptanalysis; hence, one must select cryptographically secure schemes that have undergone thorough analysis [20].

10.6 LATTICE-BASED CRYPTOGRAPHY FOR DRONE COMMUNICATIONS [14–17]

One of the most promising fields for post-quantum security in drone communication networks is lattice-based cryptography. In contrast, the security of lattice-based schemes is based on the hardness of problems formulated over high-dimensional lattices. These problems are currently thought to be intractable for classical computers and quantum computers alike, and thus could be appropriate for protecting unmanned aerial vehicle communication in the quantum era. Figure 10.2 illustrates the lattice-based key exchange protocol employed in a drone swarm environment. The protocol consists of three phases: key generation at the receiver drone, encryption by the sender drone using the receiver's public key, and decryption at the receiver to recover the shared secret. The scheme leverages lattice-based

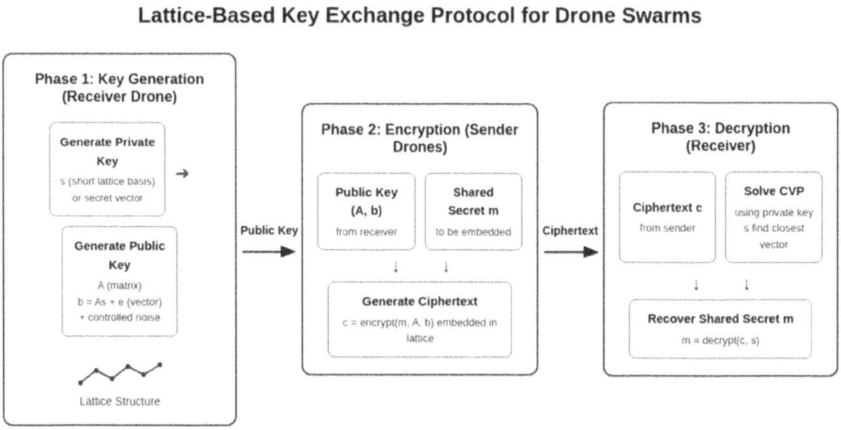

Figure 10.2 Lattice Based Key Exchange Protocol For Drone Swarms

cryptographic primitives to establish a secure shared key while providing resistance against quantum attacks.

A lattice may be regarded as a set of points in multi-dimensional space with a discrete structure. In cryptographic systems, these structures are used to build digital signatures, key-encapsulation schemes, and encryption schemes. Lattice-based cryptography can be used in drone networks for secure key exchange, authentication, data confidentiality, and the secure communication of mission-critical information.

10.6.1 Mathematical basis of lattices

A lattice is generated by taking all integer linear combinations of a set of linearly independent basis vectors. If the basis vectors are represented as

$$B = \{b_1, b_2, \ldots, b_n\}$$

then the lattice generated by the basis B can be written as

$$L(B) = \{\ z_1 b_1 + z_2 b_2 + \ldots + z_n b_n\ :\ z_i \in Z\ \}$$

where $z_1, z_2 \ldots z_n$ are integers and Z denotes the set of all integers.

The determinant of a lattice, also called as the volume of the fundamental parallelepiped, is written as $\det(L) = |\det(B)|$.

The determinant gives an idea of how densely lattice points are arranged in space. The choice of lattice dimension, basis, modulus, and noise distribution strongly influence both security and efficiency.

The choice of cryptographic algorithms for drone networks must confirm to the following major requirements:

1. The algorithm must be resilient against both classical and quantum attacks.
2. The algorithm must be suitable for embedded UAV hardware with limited processing power, memory, communication bandwidth, and battery capacity.

10.6.2 Learning with errors

The modern lattice-based cryptographic schemes are primarily based on the Learning with Errors (LWE) problem. LWE is based on the basic idea to hide a secret vector inside a noisy linear equation.

Let A be a public matrix, s be a secret vector, and e be a small error or noise vector. The public value can be represented as follows:

$$b = As + e \bmod q$$

where q is a modulus and e is sampled from a small noise distribution.

Without the error term, the equation becomes

$$b = As \bmod q$$

This equation may be solved using a linear algebra. However, when the error term e is added, recovering the secret vector s becomes computationally difficult.

The LWE assumption states that the following pair

$$(A, As + e)$$

is computationally difficult to distinguish from a random pair (A, u) where u is chosen randomly.

This hardness property is central to lattice-based encryption and key-encapsulation mechanisms. In drone communication, it allows a UAV and a ground control station to establish a shared secret even when an attacker can observe all public communication over the wireless channel.

10.6.3 Important lattice problems

The security of lattice-based cryptography mainly depends on the hardness of several well-known lattice problems.

10.6.3.1 Shortest vector problem

The shortest vector problem requires finding the shortest non-zero vector in a lattice. For a lattice L, the problem is to find a vector v in L, where v is not zero, such that

$$\|v\| = \lambda_1(L)$$

where $\lambda_1(L)$ represents the length of the shortest non-zero vector in the lattice:

$$\lambda_1(L) = \min \|v\|, \quad \text{for } v \in L \text{ and } v \neq 0$$

10.6.3.2 Closest vector problem

The closest vector problem involves finding the lattice point closest to a given target vector. Given a lattice L and a target vector y, the objective is to find v where

$$v = \arg\min \|y - x\|, \quad \text{for } x \in L$$

This problem is closely related to decryption in lattice-based cryptographic systems.

10.6.3.3 Shortest independent vectors problem

The shortest independent vectors problem requires finding several short and linearly independent vectors in a lattice. For a lattice L of dimension n, the goal is to find linearly independent vectors, $v_1, v_2, \ldots, v_n \in L$, such that the maximum length among them is minimized.

10.6.4 Lattice-based key encapsulation

Lattice-based key encapsulation is used for secure key establishment. A simplified lattice-based key-encapsulation mechanism can be represented using the following three algorithms:

$$(pk, sk) \leftarrow KeyGen()$$

$$(c, K) \leftarrow Encaps(pk)$$

$$K \leftarrow Decaps(sk, c)$$

Here, pk is the public key, sk is the private key, c is the ciphertext or encapsulated key material, and K is the shared secret.

The shared secret K is then passed through a key derivation function to produce a session key:

$$Ksession = KDF(K)$$

This session key can be used with authenticated encryption to protect real-time drone communication:

$$C = Enc_{Ksession}(M)$$

$$M = Dec_{Ksession}(C)$$

where M is the original message and C is the ciphertext.

For example, a drone and a ground control station may use a lattice-based key-encapsulation mechanism to establish *Ksession*. This key can then protect telemetry, mission commands, sensor readings, and navigation updates using efficient symmetric encryption.

10.7 NATIONAL INSTITUTE OF STANDARDS AND TECHNOLOGY (NIST) STANDARDIZATION AND THE FUTURE OF DRONE COMMUNICATION SECURITY

The NIST standardization [13] of PQC is major progression for the future of secure drone communication. The transition based on NIST-standardized PQC algorithms is an important step in preparing drone networks for the quantum era.

NIST has finalized a set of PQC standards: ML-KEM for key establishment, ML-DSA for digital signatures, and SLH-DSA as a hash-based digital signature alternative. Figure 10.3 depicts PQC algorithms considered for Securing Drone Communications.

A practical drone security architecture may use PQC in the following ways:

1. **ML-KEM** can be used for quantum-resistant session key establishment in drone networks.
2. **ML-DSA** is a digital signature mechanism and can be used for the authentication of drone identities, firmware updates, mission commands, and control messages.
3. **SLH-DSA** is a hash-based signature for high-assurance applications.

Figure 10.3 Post Quantum Cryptographic Algorithms

10.7.1 ML-KEM for drone key exchange

ML-KEM [21] is the primary NIST-standardized key-encapsulation mechanism for post-quantum secure key establishment. It enables two communicating parties to establish a shared secret key over an insecure channel. This shared key can then be used with symmetric encryption algorithms such as AES or ChaCha20-Poly1305 for protecting drone communication traffic. NIST specifies three ML-KEM parameter sets: **ML-KEM-512**, **ML-KEM-768**, and **ML-KEM-1024**, offering increasing security strength with corresponding performance trade-offs.

However, ML-KEM is not a direct replacement for all existing key exchange protocols without design changes. Traditional ECC-based schemes such as Elliptic Curve Diffie–Hellman (ECDH) are interactive key agreement mechanisms, whereas ML-KEM is a key-encapsulation mechanism. Therefore, protocol designers must carefully redesign the handshake flow, message formats, certificate handling, and session key derivation process when integrating ML-KEM into UAV networks.

In practical drone deployment, **ML-KEM-512** may be considered for lightweight or low-power drones where performance is critical, **ML-KEM-768** may be appropriate for general-purpose secure drone communication, and **ML-KEM-1024** may be used for high-security missions such as defense, border surveillance, and critical infrastructure monitoring.

10.7.2 ML-DSA for drone authentication

ML-DSA, derived from the CRYSTALS-Dilithium [22] submission, is the primary NIST-standardized post-quantum digital signature algorithm. It is designed for generating and verifying digital signatures that remain secure even against adversaries equipped with large-scale quantum computers.

In drone communication systems, ML-DSA can be used for authenticating drones, ground control stations, mission servers, firmware packages, software updates, and command messages. This is especially important because unauthorized commands, false navigation updates, and malicious firmware can lead to mission failure or physical compromise of UAVs.

ML-DSA is particularly suitable for drone authentication because it provides a practical balance between security and performance. Signature verification is an important operation in UAV networks, since drones may frequently need to verify commands, certificates, swarm membership messages, and route updates. Efficient verification allows drones to authenticate received messages without creating excessive processing delays.

10.7.3 SLH-DSA for conservative drone security

SLH-DSA, based on SPHINCS+, is a stateless hash-based digital signature algorithm standardized by NIST. Unlike ML-DSA and ML-KEM, which are based on lattice problems, SLH-DSA relies on hash function–based security assumptions. This makes it valuable as a conservative alternative, especially in applications where algorithmic diversity is important. NIST describes SLH-DSA as a backup digital signature method in case weaknesses are later found in ML-DSA.

10.8 FUTURE DIRECTION FOR QUANTUM-SAFE DRONE COMMUNICATION

The future of drone communication security will likely involve a gradual transition from classical public-key cryptography to post-quantum and hybrid cryptographic systems. In the short term, hybrid approaches that combine ECC with ML-KEM or combine Elliptic Curve Digital Signature Algorithm (ECDSA) with ML-DSA may be used to maintain compatibility while improving quantum resistance. In the long term, fully post-quantum drone communication frameworks are expected to become necessary for mission-critical UAV systems.

A practical post-quantum drone security architecture should include the following:

- **Quantum-resistant key exchange** using ML-KEM.
- **Quantum-resistant authentication** using ML-DSA or SLH-DSA.
- **Efficient symmetric encryption** using AES-256 or ChaCha20-Poly1305 after post-quantum key establishment.
- **Secure firmware and command signing** using post-quantum signatures.
- **Hybrid migration support** for compatibility with the existing drone infrastructure.
- **Hardware-aware optimization** for embedded processors, flight controllers, and low-power communication modules.

The adoption of NIST-standardized post-quantum algorithms will therefore play a central role in building secure, future-ready drone communication systems. For UAVs used in defense, disaster management, critical infrastructure inspection, smart cities, agriculture, logistics, and public safety, PQC will become an essential security layer for protecting commands, telemetry, payload data, and autonomous coordination against both present and future adversaries.

10.9 CONCLUSION

The emergence of quantum computing poses a significant challenge to the classical cryptographic mechanisms currently used in drone communication systems. As UAVs become increasingly important in civilian, commercial, and military applications, ensuring secure and reliable communications is essential. PQC offers a practical path toward protecting drone networks against future quantum-enabled threats. However, the successful adoption of PQC must account for the resource constraints, dynamic topologies, and real-time requirements of UAV environments. By integrating efficient quantum-resistant cryptographic solutions, drone networks can achieve long-term security, resilience, and operational reliability in the quantum era.

REFERENCES

1. P. W. Shor, "Polynomial-time algorithms for prime factorization and discrete logarithms on a quantum computer," *SIAM Journal on Computing*, vol. 26, no. 5, pp. 1484–1509, 1997. Available: https://doi.org/10.1137/S0097539795293172
2. L. K. Grover, "A fast quantum mechanical algorithm for database search," in *Proceedings of the 28th Annual ACM Symposium on Theory of Computing*, Philadelphia, PA, USA, 1996, pp. 212–219. Available: https://doi.org/10.1145/237814.237866

3. K.-Y. Tsao, T. Girdler, and V. G. Vassilakis, "A survey of cyber security threats and solutions for UAV communications and flying ad-hoc networks," *Ad Hoc Networks*, vol. 133, Aug. 2022, Art. no. 102894. Available: https://doi.org/10.1016/j.adhoc.2022.102894
4. A. K. Lenstra and H. W. Lenstra Jr., Eds., *The Development of the Number Field Sieve*, Lecture Notes in Mathematics, vol. 1554. Berlin, Germany: Springer, 1993. Available: https://doi.org/10.1007/BFb0091534
5. W. Diffie and M. E. Hellman, "New directions in cryptography," *IEEE Transactions on Information Theory*, vol. 22, no. 6, pp. 644–654, Nov. 1976. Available: https://doi.org/10.1109/TIT.1976.1055638
6. A. J. Menezes, P. C. van Oorschot, and S. A. Vanstone, *Handbook of Applied Cryptography*. Boca Raton, FL, USA: CRC Press, 1996. Available: https://cacr.uwaterloo.ca/hac/
7. W. Stallings, *Cryptography and Network Security: Principles and Practice*, 8th ed. Boston, MA, USA: Pearson, 2020.
8. P. Kaye, R. Laflamme, and M. Mosca, *An Introduction to Quantum Computing*. Oxford, UK: Oxford University Press, 2007. Available: https://global.oup.com/academic/product/an-introduction-to-quantum-computing-9780198570493
9. NSA, CISA, and NIST, "Quantum readiness: Migration to post-quantum cryptography," Aug. 2023. Available: https://www.nsa.gov/Press-Room/News-Highlights/Article/Article/3498776/nsa-cisa-and-nist-publish-quantum-readiness-guide/
10. P. Rosenzweig, "The alarming trend of cybersecurity breaches and failures in the U.S. government continues," Heritage Foundation Issue Brief No. 3772, Nov. 13, 2012. Available: https://www.heritage.org/defense/report/the-alarming-trend-cybersecurity-breaches-and-failures-the-us-government-continues
11. M. Mosca, "Cybersecurity in an era with quantum computers: Will we be ready?" *IEEE Security & Privacy*, vol. 16, no. 5, pp. 38–41, 2018. Available: https://doi.org/10.1109/MSP.2018.3761723
12. L. Pokorny, "National cybersecurity governance and post-quantum cryptographic migration readiness: A qualitative policy discourse analysis of federal institutional preparedness," Zenodo, 2026. Available: https://zenodo.org/records/20369175
13. National Institute of Standards and Technology, "Post quantum cryptography standardization." Available: https://csrc.nist.gov/projects/post-quantum-cryptography/post-quantum-cryptography-standardization
14. M. Ajtai, "Generating hard instances of lattice problems," in *Proceedings of the Twenty-Eighth Annual ACM Symposium on Theory of Computing*, Philadelphia, PA, USA, 1996, pp. 99–108. Available: https://doi.org/10.1145/237814.237838
15. D. Micciancio and O. Regev, "Lattice-based cryptography," in *Post-Quantum Cryptography*, D. J. Bernstein, J. Buchmann, and E. Dahmen, Eds. Berlin, Germany: Springer, 2009, pp. 147–191. Available: https://doi.org/10.1007/978-3-540-88702-7_5
16. C. Peikert, "A decade of lattice cryptography," *Foundations and Trends in Theoretical Computer Science*, vol. 10, no. 4, pp. 283–424, 2016. Available: https://doi.org/10.1561/0400000074

17. O. Regev, "On lattices, learning with errors, random linear codes, and cryptography," *Journal of the ACM*, vol. 56, no. 6, Art. no. 34, pp. 1–40, 2009. Available: https://doi.org/10.1145/1568318.1568324
18. D. J. Bernstein, A. Hülsing, S. Kölbl, R. Niederhagen, J. Rijneveld, and P. Schwabe, "The SPHINCS+ signature framework," in *Proceedings of the 2019 ACM SIGSAC Conference on Computer and Communications Security*, London, UK, 2019, pp. 2129–2146. Available: https://doi.org/10.1145/3319535.3363229
19. D. J. Bernstein, T. Lange, and C. Peters, "Attacking and defending the McEliece cryptosystem," in *Post-Quantum Cryptography*, Berlin, Germany: Springer, 2008, pp. 31–46. Available: https://doi.org/10.1007/978-3-540-88403-3_3
20. J. Ding and D. Schmidt, "Rainbow, a new multivariable polynomial signature scheme," in *Applied Cryptography and Network Security*, New York, NY, USA, 2005, pp. 164–175. Available: https://doi.org/10.1007/11496137_12
21. R. Avanzi, J. Bos, L. Ducas, E. Kiltz, T. Lepoint, V. Lyubashevsky, J. M. Schanck, P. Schwabe, G. Seiler and D. Stehlé, "CRYSTALS-Kyber algorithm specifications and supporting documentation," NIST Post-Quantum Cryptography Standardization, Round 3 Submission, 2021. Available: https://pq-crystals.org/kyber/data/kyber-specification-round3-20210804.pdf
22. L. Ducas, E. Kiltz, T. Lepoint, V. Lyubashevsky, P. Schwabe, G. Seiler and D. Stehlé, "CRYSTALS-Dilithium: A lattice-based digital signature scheme," *IACR Transactions on Cryptographic Hardware and Embedded Systems*, vol. 2018, no. 1, pp. 238–268, 2018. Available: https://doi.org/10.13154/tches.v2018.i1.238-268

Part 4

Application and allied technologies for unmanned aerial vehicles

Chapter 11

Global awareness and adoption of UAVs in agriculture

A case study of urban India

Ipseeta Satpathy, B. Chandra Mohan Patnaik, Vishal Jain, and Anish Patnaik

11.1 INTRODUCTION

India is largely a nation whose economy is based on agriculture, but the nation is also lagging on the adoption of modern technologies that would help to improve agricultural output. Contrary to that, developed countries have started incorporating unmanned aerial vehicles (UAVs) into their precision farming systems, where photogrammetry and remote-sensing techniques are implemented (Zhang and Kovacs 2012). UAVs can also be used to speed up the agricultural activities and reduce the labor burden of farmers. Typically, these UAVs have sensors and cameras to keep an eye on crops, and they have sprayers to apply pesticides on them documented many UAV types that have seen service in both the military and commercial sectors. Finding out how UAVs may be used in various agricultural tasks including crop monitoring, height estimation, pesticide spraying, and soil and field surveys is an essential part of any comprehensive scientific study of UAVs in precision agriculture. Weight, flying range, cargo capacity, design, and cost are major elements that determine the efficiency of these UAVs (Anthony et al. 2014).

Drones were at times viewed as an expensive device, but their potential in the sphere of agriculture was not thoroughly studied, and it may become a barrier to the development of this industry. Drones are supplied with unique software allowing them to fly autonomously and have flight plans, which can be accompanied by GPS to specify a variety of parameters including speed, altitude, region of interest, geo-fencing, and failure measures (Huang et al. 2009). The preferred use of drones over conventional aircraft has a number of reasons which include; high spatial resolution, high turnover operating capabilities, low operating costs and it is easy to use. The described features are especially beneficial in precision agriculture because vast fields should be covered and processed effectively. Aerial vehicle feasibility has been greatly enhanced by the miniaturization of sensors and cameras, especially in the realms of infrared and sonar technology (Espinoza et al. 2017).

DOI: 10.1201/9781003598350-15

The Japanese were the first to implement the successful Unmanned Aerial Systems (UAS) technology in the 1980s to spray fertilizers on farms and then extended its application in crop dusting in the 1990s. There were 1,220 Yamaha unmanned helicopters in operation in Japan by 2001. Currently, over 40% of Japan's rice paddy is sprayed annually using over 2,000 Yamaha RMAX unmanned helicopters, which covers 2.5 million acres of land. On the other hand, the United States is far behind Japan in implementing UAV technology in agriculture, and the supporters of the technology have to overcome many privacy and legal issues that make it hard to introduce these technologies to the community (Park et al. 2017). The use of UAVs in agriculture has been on the rise recently, but the industry still faces a number of technological obstacles. The detection and quantification of plant stress is part of the many applications of UAVs that have received considerable attention, mostly because of the major benefits that early detection of the agent may offer to agricultural activity. As a result, there has been a great number of generated data, as well as a variety of suggested approaches, which makes it hard to comprehend what progress has already been achieved and what the key issues are that still remain. A thorough examination of the applications of UAVs in agriculture for the purpose of tracking and evaluating plant stressors such drought, diseases, nutritional deficits, pests, and weeds is planned for the current study (Du et al. 2018).

The pests, nutrient deficiencies, diseases, water stress, and the overall health of crops are also major components of precision agriculture that require monitoring. This monitoring has conventionally been based on aerial or ground survey; nevertheless, the methods are often constrained by the costs of operation and access to the areas (Dang et Al. 2018). While images produced by small planes are often more efficient, timely, and of high resolution, they are nonetheless rather costly per acre. Conversely, small UAVs or UAS may continuously collect data on a time- and space-centimeter scale.

11.1.1 Perks of using UAVs for farming

Because UAV technology allows for the real-time collecting and analysis of data, precision agriculture has been greatly improved by its debut. The high-resolution imaging and multispectral sensors enable farmers to identify diseases and manage crop health, and assess nutrient deficiencies to make sound and timely decisions. The UAVs are important in the optimization of resource distribution process, whereby fertilizers, pesticides, and water are distributed with accuracy, thus reducing wastage and environmental damages.

The first benefit of UAVs is that it can potentially reduce costs. The conventional farming systems normally imply that huge labor and machineries are used, which could increase the cost of operation. UAVs can be used to minimize this cost by automating some of the processes such as scouting of the field, irrigation, and crop analysis. The adoption of labor-intensive production methods helps reduce production costs, which in turn enhances the profitability of farming activities.

The use of UAVs to spray pesticides has many benefits, some of which are great in the view of efficiency. The traditional methods spray pesticides in excess, which is hazardous to the environment and the well-being of humans. On the other hand, using UAVs equipped with cutting-edge accuracy spraying will guarantee the homogeneity of pesticides distribution, reduce the risks of over-use of pesticides, and avoid the risks of contamination. Moreover, UAVs can work in challenging terrain and can act even in the case of unfavorable weather, thereby contributing to the overall positive performance of pest management practices.

The quantification of soil health is a highly significant point of sustainable farming practices, and the use of UAVs is a state-of-the-art approach to gauging soil conditions with precision. By using thermal imaging and multispectral sensors, UAVs may provide vital information on soil moisture, erosion, and nutrients. This information is important for the conservation of soils and for enhancing land management systems. Additionally, one can use UAVs to identify soil degradation before it becomes too late that some actions can be taken to prevent extreme devastation.

UAVs' use in the agriculture industry is known to not only increase production but also provide sustainable production. Reducing the use of agricultural chemicals while improving soil health and management practices is essential for promoting environmentally sustainable agriculture. The impacts of UAV use on the agricultural sector will continue to grow as the technology keeps improving, which will transform the agricultural sector to the next level.

All in all, UAVs bring an enormous advantage to the contemporary agriculture and make it more efficient, cheaper, and sustainable. UAVs play a crucial role in modernizing traditional farming by enabling precision agriculture, reducing production costs, optimizing pesticide application, and supporting improved soil health. The innovation of the UAV technology gives a radiant future to the global agriculture through food security and environmental protection.

11.1.2 Comparison between UAV-assisted and traditional farming techniques

The introduction of technology applications, such as UAVs, has been a game-changer in the agriculture sector, significantly altering the way the industry operates. The relative productivity of the UAV-based farming shows critical variations in efficiency, cost, precision, and environmental sustainability when compared with conventional farming.

11.1.3 Efficiency and productivity

The customary ways of farming are based mainly on human labor and old machinery, which in most of the occasions have proven to be time and labor consuming. UAV-assisted agriculture, on the other hand, employs drones in several activities, including crop patrol, pesticide spraying, and seeding, and therefore, time and energy costs are significantly lowered. Drones are able to travel long distances within a significantly limited period of time and, therefore, enhance productivity.

11.2 COST-EFFECTIVENESS

Even though the cost of entry into the UAV technology may be high in the first instance, the long-run economic gains will definitely surpass the capital investment. The traditional agricultural practices demand high human power and machinery, like tractors, that consume fuel, thereby making the operation more expensive. UAVs are able to effectively reduce costs of these practices by making sure there is increased resource utilization, wastage reduction, and greater reliance on huge workforces.

11.2.1 Accuracy and the use of data

The manual control usually involved in traditional agricultural systems can introduce subjectivity and inaccuracy. There are sophisticated sensors and imaging instrument in UAVs, and this gives accurate information about the condition of crops and soil and the presence of pests. Such reliance on information assists in implementing a specific system of farming to enhance the quality of crop production at the cost of resource usage.

11.3 ENVIRONMENTAL IMPACT

The traditional farming methods typically presuppose the prevalence of pesticides and fertilizers that cause degradation of soil and potential pollution of water. UAVs help improve the precision of these chemicals' application and minimizes their usage, thereby contributing to the elimination of environmental degradation. Moreover, another useful advantage of drones is the promotion of sustainable agricultural practices through saving water and reducing carbon emissions.

The UAV technology in agriculture has proved by far more efficient, economical, precise, and sustainable in comparison with the traditional agricultural activities. The major obstacle to the UAV use is the initial cost. However, benefits of the UAV technology emerge in the long run, and hence it can be viewed as a reasonable solution to the present-day farming and can contribute to increases in productivity rate and environmental sustainability.

11.4 NEED FOR THE CURRENT STUDY

Reviewing the research on UAVs within the agricultural sector is essential in improving efficiency, productivity, and sustainability. UAVs are able to employ aerial imaging to acquire real-time data to facilitate precision agriculture, crop monitoring, pest detection, and yield prediction. They are significant in agriculture as they reduce the cost of labor and optimize resource use while decreasing the negative effects on the environment caused by the usage of certain herbicides and fertilizers. Drones equipped with thermal and multispectral sensors may quickly diagnose crop diseases and assess water stress. With the help of AI and the Internet of Things, decision-making is even more refined. As worries about global warming and food insecurity are growing, the UAV technology is pivotal in evolving contemporary agriculture into a more intelligent, data-centric industry.

11.5 RESEARCH QUESTIONS

What is the existing awareness level of various benefits of UAV among farmers?

What is the current level of awareness among farmers on various challenges of using UAVs?

11.6 OBJECTIVES OF THIS STUDY

To know if farmers are aware of the many advantages of UAV.

To examine the awareness level of the different issues relating to UAV among farmers.

To make a contribution to the existing literature.

To analyze the problems or difficulties associated with UAV.

To study the challenges in India's UAV implementation, such as limited regulatory boundaries, safety concerns, and technical challenges.

11.6.1 Regulatory limitations

The laws are often behind the technological advancement. The drones must comply with the Drone Rules 2022 set by the Directorate General of Civil Aviation (DGCA). The No Permission–No Takeoff (NPNT) policy ensures that drones take off with proper authorization.

11.6.2 Safety issues

Concerns such as breaches of privacy, security risks, and misuse are feared. Moreover, there are risks of collision and crash. The perception of people also plays a major role especially in military operations or spy missions.

11.6.3 Technical challenges

- Drones have a limited range of flight time, cargo capacities, and communication capacities.
- Drones are sensitive to different weather conditions.
- Drones are prone to technical failures.
- Drones have limited battery capacity, which can be affected by temperature, humidity, altitude, and speed.

11.6.4 Other challenges

Airspace management is a challenging issue.

- Drones have challenges of integration in congested airspaces.
- There is a challenge of operating costs.
- There is a knowledge gap on the drone technology.

Connectivity problems are also an added complication to drone operations.
Attempts made to overcome UAV-associated problems are as follows:

Counter-Drone Technology: Invest and develop systems to detect and track possible threats. It includes anti-drone system development and incorporation with the Indian Air Force Air Defense Integrated Architecture (IAF ADIA).

Drone Database: Create and create a detailed database of drones imported or sold in India. This type of database would help security agencies to trace the activity of drones and overcome the threat of malicious use.

Drones Air Space Management Cell (DASMC): Develop a national airspace management in the ADIA to ensure national airspace protection. The DASMC would collaborate with the existing ADIA systems and the Digisky platform to increase coordination.

Quality Control: Implement quality control measures that would identify and correct mistakes at the initial stages. It must be done by cross-referring results to ground control points.

Drone Training: Offer subsidies on the cost of training and development to help develop a strong drone ecosystem.

Government-to-Government Transactions: Encourage technology transfer and formation of strategic alliance between governments.

Inter-Ministerial Committee: Form an inter-ministerial committee that mandates to monitor the policy of innovation and technology development and matters pertaining to reciprocity.

11.6.5 Present trends of UAV in India

The usage of fixed-wing drones is on the rise. This kind of drone can quickly survey a broad area, which is a major selling point. They are also comparatively easy to maintain and thus are beneficial to large-scale agricultural businesses:

Use of Thermal Images: The thermal imaging technology is a useful technology of quantifying soil moisture levels, which allows farmers to determine areas that need irrigation. Such an ability encourages better management of water and can increase the productivity of crops on the whole.

Precision Application of Drones: The drones are used to apply fertilizers, pesticides, and other agricultural chemicals to targeted areas. This precision reduces the quantity of water and chemicals applied, and this assists in increasing the crop yield.

Application of Drones in Data Collection: A variety of agricultural characteristics, including crop health and soil moisture, can be studied using drones, providing significant information to farmers when making informed decision regarding crop management.

11.7 REVIEW OF LITERATURE

11.7.1 Advantages of UAVs

According to many studies (Gago et al., 2015), sensing technologies need to improve their time-space resolution. Using precision farming methods gets an upgrade with this new invention. Moreover, it is useful for crop classification and monitoring (Inoue, 2020; Kalischuk et al., 2019; Maimaitijiang et al., 2017). It is also simple to generate fertilizer efficiency (Guan et al., 2019). Drought monitoring, biomass estimate, and yield prediction are among additional areas where the technologies have been used (Fawcett et al., 2020; Inoue, 2020). Hazard mitigation and wildlife and forest protection are other two areas they contribute to (Panday, Pratihast, et al., 2020). Water stress can be measured using these technologies (Inoue, 2020; Jinbo et al., 2019), and pests, weeds, and illnesses can be detected using them (Gašparović et al., 2020; Inoue, 2020).

11.7.2 Challenges of UAVs

UAVs have a number of issues, such as involvement of pilots, average quality of captured images, and moderate cost of implementation. Other areas of concern include stability, maneuverability, and reliability, as well as the concerns regarding standardization and engine power. Moreover, UAVs are also limited by small power sources, which means that the battery life is short and the duration of flights is also limited. They are prone to crashes and computer attacks, and have limited storage bandwidth. Handling of big data is normally constrained by poor data-processing capacity. Many studies have shown that agricultural drones are not widely used because of a lack of knowledge and regulation in the field (Bacco et al., 2018; Dawaliby et al., 2020; Hardin and Hardin, 2010; Hardin and Jensen, 2011; Lagkas et al., 2018).

Satellites can provide a continuous coverage at a regular time but are constrained by spectral resolution and susceptible to viewing problems, including cloud cover. They also have issues of unavailability and low data transfer rates, which result in the expensive spatial data collection process because of orientation and vignetting effects. Furthermore, the latency of providing data to end users is frequently high (Aboutalebi et al., 2019; Cen et al., 2019; Chen et al., 2019).

The use of aircrafts comes with high standards of adoption costs, cumbersome establishment requirements, and high maintenance costs. Provision of dependable aircrafts is also a challenge and further aggravated by the geometry of the images taken. Collection of data does not occur normally, and the vulnerability to lack of flexibility can cause grave accidents. Changes in sensor data under the influence of vibrations and georeferencing difficulties also make the use of aircraft in the data collection more complicated (Armstrong et al., 2011; Atkinson et al., 2018; Barbedo and Koenigkan, 2018).

11.7.2.1 Research gaps

In the literature analysis, it was noted that little research was done in the study areas associated with UAVs in regard to advantages and obstacles. In particular, the present study considers a range of factors related to the advantages and challenges associated with UAV adoption in agriculture.

11.7.3 Rationality of the study/need of the study

The topicality of the current research is due to the comprehensive discussion that exists globally in terms of the benefits and challenges related to making use of UAVs. Considering these discussions, this research is also going to evaluate the degree of awareness among farmers in the identified regions, and as a result, it will give information on what is happening on the ground.

11.8 RESEARCH DESIGN

In this research, both primary and secondary data will be used. Data were collected from chosen farmers of some rural and urban areas and students with agribusiness background. The study variables were mostly obtained through discussions in core groups with some additional insights provided by reviewing the available literature. A total of eight core group discussions were held. In the very beginning, more than 40 attributes were established, but with the progress of expanding the discussions with different groups, this number was narrowed down to 28. The last set contains 15 attributes about the advantages of UAVs and 13 attributes about the problems related to their implementation. A weighted mean method was used to assess several sub-variables throughout the selection phase. All aspects of perception were evaluated using a five-point Rensis Likert-type scale. The following weighting scheme was used in this scoring system: 4 for responses showing total agreement, 3 for responses showing agreement, 2 for neutral, 1 for disagreement, and 0 for responses showing total disagreement. A mean score of 2 was obtained by adding together all of the given weights (4 + 3 + 2 + 1 + 0) and then dividing by 5. In order to get the right samples for this investigation, we're using snowball and cluster sampling. The attributes identified through the literature review and focus group discussions were used to develop the study variables and the survey instrument.

11.9 SCOPE OF THIS STUDY

Participation in this research was limited to agribusiness majors and farmers from the capital region of Odisha, whether they lived in rural or urban regions. The students were included keeping in mind that they are the future of the nation for the overall development. The urban areas are in Bhubaneswar and Cuttack, and rural areas are in the outskirts of Bhubaneswar and Cuttack. The students included were from these two cities.

11.10 SAMPLE SIZE DETERMINATION

Research by Rummel and Schwab suggests that a ratio of 1:4 to 1:10 is optimal for sample sizes. For the purpose of determining the sample size, this guideline recommends that researchers utilize a minimum of 4 items and a maximum of 10 items. This implies that we would need at least 112 things and no more than 280 items for our research, which has 28 items. Following this protocol, we removed the outlier and compiled 182 observations. According to Rummel and Schwab's suggestions in their study, the current sample size of 182 is suitable (Table 11.1).

11.10.1 Method of survey and demographics of participants

The current research utilized snowball and cluster sampling. The demographic areas included in the Cuttack region are Badamba, Athagarh, Banki, Narasinghpur, Tigiria, Mahanga, Nischintakoili, Salepur, Cuttack Sadar, and Baranga. In Bhubaneswar, the sampling was done not only in the city but also in its peripheral areas including Chandaka, Bholo, Kantabada, Mendhasala, and Chhatabar (Table 11.2).

Table 11.1 Sampling frame

Category	*Gender*	*Frequency*
Rural farmers	Male (RFM)	34
	Female (RFF)	16
Urban farmers	Male (UFM)	43
	Female (UFF)	24
Agribusiness students	Male (ASM)	36
	Female (ASF)	29
Total		182

Source: Primary data.

RFM—Rural Farmers Male, RFF—Rural Farmers Female, UFM—Urban Farmers Male, UFF—Urban Farmers Female, ASM—Agribusiness Students Male, ASF—Agribusiness Students Female.

Table 11.2 Analysis of data

Attributes	*Code*	*RFM*	*RFF*	*UFM*	*UFF*	*ASM*	*ASF*
Awareness of benefits of UAVs							
Time conservation	AB1	1.09	1.18	3.48	3.46	3.42	3.59
Enhanced efficiency	AB2	1.18	0.87	3.37	3.25	2.94	3.93
Minimal operational expenses	AB3	0.97	1.16	3.12	3.58	3.33	3.72
Precise application of resources	AB4	1	1.56	3.19	3.42	3.05	3.31
Aids in yield estimation	AB5	1	1	3.09	3.63	3.17	3.76
Increased security and accessibility	AB6	1	1.13	3.06	3.38	3.25	3.76
Crop oversight and administration	AB7	1	0.88	2.93	3.29	3.06	3.86
Eco-friendly practices	AB8	0.91	1.38	3.23	3.46	3.08	3.59
Decrease in waste	AB9	0.91	1.16	3.12	3.46	3.11	3.55
Water conservation	AB10	0.85	1.31	3.19	3.17	2.92	3.45
Reduced maintenance requirements	AB11	0.88	1	3.06	3.29	3.14	3.86
Assists in the acquisition and collection of sensor data	AB12	1	1.13	3.44	3.29	2.92	3.48
Greater scalability	AB13	0.85	1.38	3.37	3.5	2.97	3.59
Promptness	AB14	1	1.25	3.28	3.38	2.92	3.59
Addresses the issue of labor shortages	AB15	0.91	0.81	3.60	3.58	3	3.90
Awareness of challenges associated with UAVs							
Significant upfront investment	AC1	1.06	0.75	1.4	1.17	2.83	3.38
Insufficient expertise and training	AC2	0.91	1.16	1	1.25	2.89	3.28
Anxiety regarding employment stability	AC3	0.94	1.56	1.16	1.21	3.25	3.38

(Continued)

Table 11.2 (Continued) Analysis of data

Attributes	*Code*	*RFM*	*RFF*	*UFM*	*UFF*	*ASM*	*ASF*
Dependence on weather conditions	AC4	0.82	1.06	1.4	1.17	3.03	3.34
Internet access and infrastructure limitations	AC5	1.06	1.13	1.05	1.21	2.81	3.45
Increased regulatory requirements and compliance	AC6	0.94	0.87	1.05	1.25	2.94	3.38
Technical sophistication	AC7	1.03	1.25	1.16	1.42	2.83	3.38
Restricted flight duration and distance	AC8	0.85	1.06	1.16	1.08	3.14	3.48
Enhanced compliance obligations	AC9	1.06	1.31	1.23	1.29	2.78	3.31
Difficulties in data management and analysis	AC10	1.06	1.06	1.12	1.25	3.17	3.62
Concerns regarding confidentiality and security	AC11	1.05	1.13	1.07	1.25	3	3.48
Adaptability and attitudinal challenges	AC12	0.94	1.31	1.14	1.29	3.17	3.34
Maintenance and support services	AC13	1.06	1.19	1.12	1.17	3.47	3.48

Source: Annexure A, B, C, D, E, and F.

11.11 RESULTS AND DISCUSSIONS

11.11.1 With reference to Table 11.2

11.11.1.1 Interpretation of the data on awareness of UAV benefits

The average score of awareness is 2 which is used as a criterion to determine the awareness that a certain group has about the benefits of the UAVs.

1. Rural Farmers (Male and Female):
   - RFM and RFF scored mean awareness below 2 on all benefits.
   - This shows that rural farmers lack awareness of UAV benefits, which implies that special awareness efforts should be conducted in rural communities.
2. Urban Farmers (Male and Female):
   - UFM and UFF are always scored above 3, which implies that people are highly aware of the benefits of UAV.
   - UFF is slightly better than UFM, especially in a low operation cost (AB3) and estimation of yield (AB5).
3. Agribusiness Students (Male and Female):
   - ASM and ASF exhibit the most awareness levels, and ASF dominates in most of the categories (e.g., AB2, AB3, AB5).
   - Their high awareness demonstrates that they have been exposed more to the use of UAVs through education.

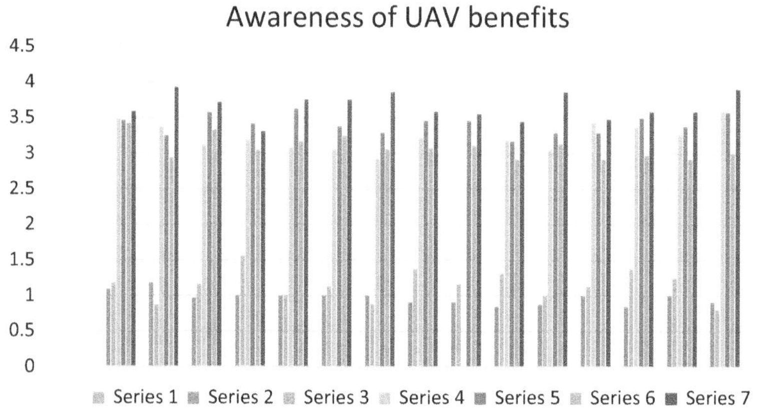

11.12 INTERPRETATION OF AWARENESS OF UAV CHALLENGES

The average level of awareness is 2 that will be used as a measure of the level of awareness among various groups.

11.12.1 Rural farmers (female and male—RFM and RFF)

- Both RFM and RFF are much less than 2 in their awareness scores in all the challenges.
- It means that the rural farmers are very unaware of the challenges that UAVs bring, such as initial investments, legal issues, and technical challenges.
- RFF shows slightly higher awareness than RFM in some areas (e.g., employment stability (AC3) and adaptability challenges (AC12)), but overall, awareness remains low.

11.12.2 Urban farmers (male and female—UFM and UFF)

- Urban farmers as well are under 2 for awareness, meaning that they are not aware of the challenges of UAVs.
- UFM and UFF show slight variations in individual aspects but remain largely unaware of concerns like data management, security, and maintenance issues.
- This implies that even urban farmers, albeit with greater levels of technology exposure, might not be well versed with the UAV challenges.

11.12.3 Agribusiness students (male and female—ASM and ASF)

The agribusiness students (ASM and ASF) are the most aware students, and they score above 2 in all respective categories.

ASF has the highest level of awareness, even out of ASM, especially on data management (AC10) and compliance (AC6) as well as confidentiality issues (AC11).

It implies that education helps greatly to raise awareness about UAV challenges, which proves the capabilities of academic institutions to train future agricultural workers.

11.12.4 Recommendations to the study findings

An awareness campaign should be established in the urban areas to clarify the benefits and problems associated with UAVs. This might take the kind of a series of seminars, theatrical acts, and conferences. Moreover, the fact that UAV implementation requires a significant amount of money is an urgent need to support farmers. Farmer training sessions are also essential regularly, whereby farmers ought to be sensitized on various schemes and their respective benefits of using UAV technologies.

11.12.5 Additional value of the present research

The study will help the stakeholders and policy makers to analyze the support system provided to farmers whether in rural or urban areas. The study

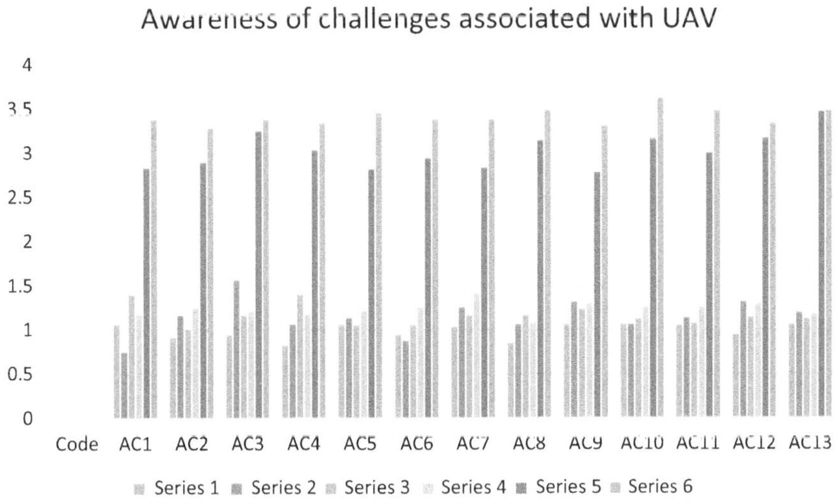

will also aim at sensitizing farmers on the various benefits of UAVs, and also enable an appreciation of the numerous challenges that are associated with their operation.

11.13 SCOPE FOR FUTURE STUDY

It is possible to replicate these research in other parts of India or even with a different group of farmers in a more rural setting.

11.14 LIMITATIONS OF THIS STUDY

Only the metropolitan regions and outskirts of Bhubaneswar and Cuttack in Odisha were included for this research. The variables were narrowed down to a list of variables chosen after carefully considering the available literature and deliberating on the same in a core group. Moreover, only 182 participants were used as the sample.

11.15 CONCLUDING OBSERVATIONS

According to the research, the rural farmers need to be provided with certain educational programs to increase their knowledge in the area of UAVs. Urban farmers demonstrate a high degree of awareness, which can indicate the willingness to use these technologies. The awareness of UAVs among agribusiness students is greatest, which highlights the impact of education on the awareness level of students. Both urban and rural farmers lack sufficient knowledge about the potential issues in UAV adoption to make well-informed judgments on this technology. The heightened awareness among agribusiness students is likely to be attributable to their academic training. Therefore, it is essential to implement awareness campaigns aimed at educating farmers about the potential obstacles to UAV adoption, particularly in relation to costs, regulatory frameworks, and technical specifications.

REFERENCES

Aboutalebi, M., Torres-Rua, A.F., Kustas, W.P., Nieto, H., Coopmans, C., McKee, M., 2019. Assessment of different methods for shadow detection in high-resolution optical imagery and evaluation of shadow impact on calculation of NDVI, and evapotranspiration. *Irrig. Sci.* 37(3), 407–429. https://doi.org/10.1007/s00271-018-0613-9.

Anthony, D., Elbaum, S., Lorenz, A., Detweiler, C., 2014. On crop height estimation with UAVs. *IEEE/RSJ International Conference on Intelligent Robots and Systems (IROS 2014)*, 14–18 September 2014, Chicago, IL, 4805–4812.

Armstrong, I., Pirrone-Brusse, M., Smith, A., Jadud, M., 2011. The flying gator: Towards aerial robotics in occam-π. *Commun. Process Architect.* 2011, 329–340. https://doi.org/10.3233/978-1-60750-774-1-329.

Atkinson, J. A., Jackson, R. J., Bentley, A. R., Ober, E., Wells, D. M., 2018. Field phenotyping for the future. In J. A. Roberts (ed.), *Annual Plant Reviews Online* (pp. 719–736). John Wiley & Sons, Ltd. https://doi.org/10.1002/9781119312994.apr0651.

Bacco, M., Berton, A., Ferro, E., Gennaro, C., Gotta, A., Matteoli, S., Paonessa, F., Ruggeri, M., Virone, G., Zanella, A., 2018. Smart farming: Opportunities, challenges and technology enablers. *2018 IoT Vertical and Topical Summit on Agriculture - Tuscany (IOT Tuscany)*, Tuscany, 1–6. https://doi.org/10.1109/IOT-TUSCANY.2018.8373043.

Barbedo, J. G. A., Koenigkan, L. V., 2018. Perspectives on the use of unmanned aerial systems to monitor cattle. *Outlook Agric.* 47(3), 214–222. https://doi.org/10.1177/0030727018781876.

Cen, H., Wan, L., Zhu, J., Li, Y., Li, X., Zhu, Y., Weng, H., Wu, W., Yin, W., Xu, C., Bao, Y., Feng, L., Shou, J., He, Y., 2019. Dynamic monitoring of biomass of rice under different nitrogen treatments using a lightweight UAV with dual image-frame snapshot cameras. *Plant Methods* 15(1), 32. https://doi.org/10.1186/s13007-019-0418-8.

Chen, A., Orlov-Levin, V., Meron, M., 2019. Applying high-resolution visible-channel aerial imaging of crop canopy to precision irrigation management. *Agric. Water Manag.* 216, 196–205. https://doi.org/10.1016/j.agwat.2019.02.017.

Dang, L. M., Hassan, S. I., Suhyeon, I., Sangaiah, A. K., Mehmood, I., Rho, S. et al., 2018. UAV based wilt detection system via convolutional neural networks. *Sustain. Comput. Inform. Syst.* 28, 100250.

Dawaliby, S., Aberkane, A., Bradai, A., 2020. Blockchain-based IoT platform for autonomous drone operations management. In: *Proceedings of the 2nd ACM MobiCom Workshop on Drone Assisted Wireless Communications for 5G and Beyond*, London, pp. 31–36. https://doi.org/10.1145/3414045.3415939.

Du, W., Xu, T., Yu, F., Chen, C., 2018. Measurement of nitrogen content in rice by inversion of hyperspectral reflectance data from an unmanned aerial vehicle. *Ciênc. Rural.* 48(6), 1–10.

Espinoza, C. Z., Khot, L. R., Sankaran, S., Jacoby, P. W. 2017. High resolution multispectral and thermal remote sensing-based water stress assessment in subsurface irrigated grapevines. *Remote Sens.* 9, 961.

Fawcett, D., Panigada, C., Tagliabue, G., Boschetti, M., Celesti, M., Evdokimov, A., Biriukova, K., Colombo, R., Miglietta, F., Rascher, U., Anderson, K., 2020. Multi-scale evaluation of drone-based multispectral surface reflectance and vegetation indices in operational conditions. *Remote Sens.* 12(3), 514.

Gago, J., Douthe, C., Coopman, R. E., Gallego, P. P., Ribas-Carbo, M., Flexas, J., Escalona, J., Medrano, H., 2015. UAVs challenge to assess water stress for sustainable agriculture. *Agric. Water Manag.* 153, 9–19. https://doi.org/10.1016/j.agwat.2015.01.020.

Gašparović, M., Zrinjski, M., Barković, Đ., Radočaj, D., 2020. An automatic method for weed mapping in oat fields based on UAV imagery. *Comput. Electron. Agric.* 173, 105385.

Guan, S., Fukami, K., Matsunaka, H., Okami, M., Tanaka, R., Nakano, H., Sakai, T., Nakano, K., Ohdan, H., Takahashi, K., 2019. Assessing correlation of high-resolution NDVI with fertilizer application level and yield of rice and wheat crops using small UAVs. *Remote Sens.* 11(2), 112.

Hardin, P. J., Hardin, T. J., 2010. Small-scale remotely piloted vehicles in environmental research. *Geogr. Compass* 4(9), 1297–1311. https://doi.org/10.1111/j.1749-8198.2010.00381.x.

Hardin, P. J., Jensen, R. R., 2011. Small-scale unmanned aerial vehicles in environmental remote sensing: Challenges and opportunities. *GIScí. Remote Sens.* 48(1), 99–111. https://doi.org/10.2747/1548-1603.48.1.99.

Huang, Y., Reddy, K. N., Fletcher, R. S., Pennington, D., 2018. UAV low-altitude remote sensing for precision weed management. *Weed Technol.* 32, 2–6.

Inoue, Y., 2020. Satellite- and drone-based remote sensing of crops and soils for smart farming–A review. *Soil Sci. Plant Nutr.* 66(6), 798–810. https://doi.org/10.1080/00380768.2020.1738899.

Jinbo, C., Xiangliang, C., Han-Chi, F., Lam, A., 2019. Agricultural product monitoring system supported by cloud computing. *Cluster Comput.* 22(4), 8929–8938.

Park, S., Ryu, D., Fuentes, S., Chung, H., Hernández-Montes, E., O'Connell, M., 2017. Adaptive estimation of crop water stress in nectarine and peach orchards using high-resolution imagery from an unmanned aerial vehicle (UAV). *Remote Sens.* 9, 828.

Zhang, C., Kovacs, J. M. 2012. The application of small unmanned aerial systems for precision agriculture: A review. *Precis. Agric.* 13(6), 693–712.

Annexure A

Awareness of Rural Farmers Male (RFM)-34

	CA	*A*	*N*	*DA*	*CDA*		
Attributes	*4*	*3*	*2*	*1*	*0*	*Weight*	*Mean*
Awareness of benefits of UAVs							
Time conservation	2	3	3	14	12	37	1.09
Enhanced efficiency	3	3	3	13	12	40	1.18
Minimal operational expenses	2	2	2	15	13	33	0.97
Precise application of resources	1	2	2	20	9	34	1
Aids in yield estimation	2	2	1	18	11	34	1
Increased security and accessibility	2	2	2	16	12	34	1
Crop oversight and administration	3	1	1	17	12	34	1
Eco-friendly practices	2	2	1	15	14	31	0.91
Decrease in waste	2	2	2	13	15	31	0.91
Water conservation	2	1	1	16	14	29	0.85
Reduced maintenance requirements	2	2	2	14	14	30	0.88
Assists in the acquisition and collection of sensor data	2	2	1	16	13	34	1
Greater scalability	2	2	1	13	16	29	0.85
Promptness	2	2	2	16	12	34	1
Addresses the issue of labor shortages	2	2	2	13	15	31	0.91
Awareness of challenges associated with UAVs							
Significant upfront investment	2	2	2	18	10	36	1–06
Insufficient expertise and training	1	2	2	17	12	31	0.91
Anxiety regarding employment stability	2	2	1	16	13	32	0.94
Dependence on weather conditions	2	2	2	20	8	28	0.82

(Continued)

(Continued)

	CA	A	N	DA	CDA		
Attributes	*4*	*3*	*2*	*1*	*0*	*Weight*	*Mean*
Awareness of challenges associated with UAVs							
Internet access and infrastructure limitations	3	1	1	19	10	36	1.06
Increased regulatory requirements and compliance	2	2	1	16	13	32	0.94
Technical sophistication	2	2	2	17	11	35	1.03
Restricted flight duration and distance	2	1	1	16	14	29	0.85
Enhanced compliance obligations	2	2	2	18	10	36	1.06
Difficulties in data management and analysis	2	2	1	20	9	36	1.06
Concerns regarding confidentiality and security	2	2	1	19	10	35	1.05
Adaptability and attitudinal challenges	2	2	2	14	14	32	0.94
Maintenance and support services	2	2	2	18	10	36	1.06

Source: Primary data.

Annexure B

Awareness of Rural Farmers Female (RFF)-16

	CA	A	N	DA	CDA		
Attributes	*4*	*3*	*2*	*1*	*0*	*Weight*	*Mean*
Awareness of benefits of UAVs							
Time conservation	1	1	1	10	3	19	1.18
Enhanced efficiency	0	0	0	14	2	14	0.87
Minimal operational expenses	1	0	1	13	1	19	1.16
Precise application of resources	2	1	1	12	0	25	1.56
Aids in yield estimation	0	1	1	11	3	16	1
Increased security and accessibility	1	0	1	12	2	18	1.13
Crop oversight and administration	0	0	0	14	2	14	0.88
Eco-friendly practices	1	1	1	13	0	22	1.38
Decrease in waste	0	1	1	14	0	19	1.16
Water conservation	1	1	1	12	1	21	1.31
Reduced maintenance requirements	0	1	0	13	2	16	1
Assists in the acquisition and collection of sensor data	0	1	1	13	1	18	1.13
Greater scalability	1	1	1	13	0	22	1.38
Promptness	1	0	1	14	0	20	1.25
Addresses the issue of labor shortages	0	0	0	13	3	13	0.81
Awareness of challenges associated with UAVs							
Significant upfront investment	0	0	0	12	4	12	0.75
Insufficient expertise and training	1	0	1	13	1	19	1.16
Anxiety regarding employment stability	2	1	1	12	0	25	1.56
Dependence on weather conditions	0	1	1	12	2	17	1.06
Internet access and infrastructure limitations	1	0	1	12	2	18	1.13

(Continued)

(*Continued*)

	CA	A	N	DA	CDA		
Attributes	*4*	*3*	*2*	*1*	*0*	*Weight*	*Mean*
Awareness of challenges associated with UAVs							
Increased regulatory requirements and compliance	0	0	0	14	2	14	0.87
Technical sophistication	1	1	1	11	2	20	1.25
Restricted flight duration and distance	0	1	1	12	2	17	1.06
Enhanced compliance obligations	1	1	1	12	1	21	1.31
Difficulties in data management and analysis	0	1	0	14	1	17	1.06
Concerns regarding confidentiality and security	0	1	1	13	1	18	1.13
Adaptability and attitudinal challenges	1	1	1	12	1	21	1.31
Maintenance and support services	1	0	1	13	1	19	1.19

Source: Primary data.

Annexure C

Awareness of Urban Farmers Male (UFM)-43

	CA	*A*	*N*	*DA*	*CDA*		
Attributes	*4*	*3*	*2*	*1*	*0*	*Weight*	*Mean*
Awareness of benefits of UAVs							
Time conservation	33	4	2	2	2	150	3.48
Enhanced efficiency	31	4	3	3	2	145	3.37
Minimal operational expenses	29	3	2	5	4	134	3.12
Precise application of resources	28	5	3	4	3	137	3.19
Aids in yield estimation	27	4	4	5	3	133	3.09
Increased security and accessibility	28	3	3	6	3	136	3.06
Crop oversight and administration	26	3	3	7	4	126	2.93
Eco-friendly practices	30	3	2	4	4	139	3.23
Decrease in waste	29	2	2	6	4	134	3. 12
Water conservation	27	6	4	3	3	137	3.19
Reduced maintenance requirements	28	4	3	6	2	136	3.06
Assists in the acquisition and collection of sensor data	32	4	3	2	2	148	3.44
Greater scalability	31	4	3	3	2	145	3.37
Promptness	29	5	3	4	2	141	3.28
Addresses the issue of labor shortages	35	3	2	2	1	155	3.60
Awareness of challenges associated with UAVs							
Significant upfront investment	2	2	2	31	6	49	1.4
Insufficient expertise and training	1	2	2	29	9	43	1
Anxiety regarding employment stability	2	2	1	34	4	50	1.16
Dependence on weather conditions	2	2	2	33	3	49	1.4

(Continued)

(*Continued*)

Attributes	CA 4	A 3	N 2	DA 1	CDA 0	Weight	Mean
Awareness of challenges associated with UAVs							
Internet access and infrastructure limitations	3	1	1	28	10	45	1.05
Increased regulatory requirements and compliance	2	2	1	29	9	45	1.05
Technical sophistication	2	2	2	32	5	50	1.16
Restricted flight duration and distance	2	1	1	37	3	50	1.16
Enhanced compliance obligations	2	2	2	35	2	53	1.23
Difficulties in data management and analysis	2	2	1	32	6	48	1.12
Concerns regarding confidentiality and security	2	2	1	30	8	46	1.07
Adaptability and attitudinal challenges	2	2	2	31	6	49	1.14
Maintenance and support services	2	2	2	30	7	48	1.12

Source: Primary data.

Annexure D

Awareness of Urban Farmers Female (UFF)-24

	CA	A	N	DA	CDA		
Attributes	*4*	*3*	*2*	*1*	*0*	*Weight*	*Mean*
Awareness of benefits of UAVs							
Time conservation	18	2	1	3	0	83	3.46
Enhanced efficiency	17	2	2	2	1	78	3.25
Minimal operational expenses	19	2	1	2	0	86	3.58
Precise application of resources	18	2	1	2	1	82	3.42
Aids in yield estimation	20	1	1	2	0	87	3.63
Increased security and accessibility	17	2	2	3	0	81	3.38
Crop oversight and administration	16	3	2	2	1	79	3.29
Eco-friendly practices	17	2	1	2	2	83	3.46
Decrease in waste	18	2	2	2	0	83	3.46
Water conservation	17	2	1	3	0	76	3.17
Reduced maintenance requirements	16	2	2	2	2	79	3.29
Assists in the acquisition and collection of sensor data	16	3	2	2	1	79	3.29
Greater scalability	18	2	2	2	0	84	3.5
Promptness	17	3	1	2	1	81	3.38
Addresses the issue of labor shortages	19	2	1	2	0	86	3.58
Awareness of challenges associated with UAVs							
Significant upfront investment	1	1	2	17	3	28	1.17
Insufficient expertise and training	1	2	2	16	3	30	1.25
Anxiety regarding employment stability	1	2	2	15	4	29	1.21
Dependence on weather conditions	1	1	2	17	3	28	1.17

(Continued)

(*Continued*)

	CA	*A*	*N*	*DA*	*CDA*		
Attributes	*4*	*3*	*2*	*1*	*0*	*Weight*	*Mean*
Awareness of challenges associated with UAVs							
Internet access and infrastructure limitations	2	1	1	16	4	29	1.21
Increased regulatory requirements and compliance	1	2	1	18	2	30	1.25
Technical sophistication	2	1	2	19	0	34	1.42
Restricted flight duration and distance	1	1	1	17	4	26	1.08
Enhanced compliance obligations	2	1	2	16	3	31	1.29
Difficulties in data management and analysis	2	1	1	17	3	30	1.25
Concerns regarding confidentiality and security	1	2	1	18	2	30	1.25
Adaptability and attitudinal challenges	2	1	2	18	1	31	1.29
Maintenance and support services	1	1	2	17	3	28	1.17

Source: Primary data.

Annexure E

Awareness of Agribusiness Students Male (ASM)-36

Attributes	CA 4	A 3	N 2	DA 1	CDA 0	Weight	Mean
Awareness of benefits of UAVs							
Time conservation	24	3	2	4	3	123	3.42
Enhanced efficiency	22	2	3	6	3	106	2.94
Minimal operational expenses	26	3	2	3	2	120	3.33
Precise application of resources	24	2	2	4	4	110	3.05
Aids in yield estimation	25	2	2	4	3	114	3.17
Increased security and accessibility	25	3	1	4	3	117	3.25
Crop oversight and administration	23	2	2	5	4	107	3.06
Eco-friendly practices	24	2	2	5	3	111	3.08
Decrease in waste	24	2	3	4	3	112	3.11
Water conservation	22	3	1	6	4	105	2.92
Reduced maintenance requirements	23	4	2	5	2	113	3.14
Assists in the acquisition and collection of sensor data	22	2	2	7	3	105	2.92
Greater scalability	23	2	2	5	4	107	2.97
Promptness	21	3	3	6	3	105	2.92
Addresses the issue of labor shortages	22	4	2	4	4	108	3
Awareness of challenges associated with UAVs							
Significant upfront investment	21	2	3	6	4	102	2.83
Insufficient expertise and training	22	3	1	5	5	104	2.89
Anxiety regarding employment stability	24	4	2	3	3	117	3.25
Dependence on weather conditions	23	3	2	4	4	109	3.03

(Continued)

(Continued)

Attributes	*CA* *4*	*A* *3*	*N* *2*	*DA* *1*	*CDA* *0*	*Weight*	*Mean*
Awareness of challenges associated with UAVs							
Internet access and infrastructure limitations	21	2	2	7	4	101	2.81
Increased regulatory requirements and compliance	20	5	3	5	3	106	2.94
Technical sophistication	21	2	2	8	3	102	2.83
Restricted flight duration and distance	22	4	3	4	3	113	3.14
Enhanced compliance obligations	21	2	2	6	5	100	2.78
Difficulties in data management and analysis	24	3	2	5	2	114	3.17
Concerns regarding confidentiality and security	22	3	4	3	4	108	3
Adaptability and attitudinal challenges	24	3	2	5	2	114	3.17
Maintenance and support services	20	5	3	4	4	125	3.47

Source: Primary data.

Annexure F

Awareness of Agribusiness Students Female (ASF)-29

	CA	A	N	DA	CDA		
Attributes	*4*	*3*	*2*	*1*	*0*	*Weight*	*Mean*
Awareness of benefits of UAVs							
Time conservation	23	3	1	1	1	104	3.59
Enhanced efficiency	27	2	0	0	0	114	3.93
Minimal operational expenses	25	2	1	0	1	108	3.72
Precise application of resources	22	1	2	1	1	96	3.31
Aids in yield estimation	27	0	0	1	1	109	3.76
Increased security and accessibility	26	1	1	0	1	109	3.76
Crop oversight and administration	25	4	0	0	0	112	3.86
Eco-friendly practices	23	3	1	1	1	104	3.59
Decrease in waste	21	6	0	1	1	103	3.55
Water conservation	20	6	1	1	1	101	3.45
Reduced maintenance requirements	19	7	2	1	0	112	3.86
Assists in the acquisition and collection of sensor data	22	2	3	1	1	101	3.48
Greater scalability	20	5	2	1	1	100	3.59
Promptness	21	6	1	0	1	104	3.59
Addresses the issue of labor shortages	26	3	0	0	0	113	3.90
Awareness of challenges associated with UAVs							
Significant upfront investment	20	5	1	1	2	98	3.38
Insufficient expertise and training	19	5	1	2	2	95	3.28
Anxiety regarding employment stability	22	2	1	2	2	98	3.38
Dependence on weather conditions	19	6	1	1	2	97	3.34

(Continued)

(Continued)

Attributes	CA 4	A 3	N 2	DA 1	CDA 0	Weight	Mean
Awareness of challenges associated with UAVs							
Internet access and infrastructure limitations	20	5	2	1	1	100	3.45
Increased regulatory requirements and compliance	19	6	1	2	1	98	3.38
Technical sophistication	21	3	2	1	2	98	3.38
Restricted flight duration and distance	20	6	1	1	1	101	3.48
Enhanced compliance obligations	19	5	2	1	2	96	3.31
Difficulties in data management and analysis	25	0	2	1	1	105	3.62
Concerns regarding confidentiality and security	22	3	1	2	1	101	3.48
Adaptability and attitudinal challenges	20	4	2	1	2	97	3.34
Maintenance and support services	23	2	1	1	2	101	3.48

Source: Primary data.

Chapter 12

Sensor-driven UAV technologies for disaster management

A comprehensive framework for real-time response and recovery

Bhavya, Arpit Mishra, Ujjwal Singh Katiyar, and Rahul Johari

12.1 INTRODUCTION

In an uncertain world characterized by climatic uncertainty, rapid urbanization, and rising geopolitical complexity, disasters – natural or human-induced – have become more intense and uncertain. From flash floods in cities that emerge with little warning to wildfires across forest ecosystems and industrial explosion in densely populated regions, the variety and scale of modern emergencies have outgrown the capability of conventional response systems.

These multifaceted crises demand a paradigm shift in disaster management – from reactive, late-gestation interventions to autonomous, data-driven, and real-time response systems that are able to provide services under unstructured and poor infrastructure conditions. Because of their ability to swiftly deploy, fly in the air, change sensors, and use on-board processing power, unmanned aerial vehicles (UAVs) have emerged as one of the most innovative tools of the twenty-first century. UAVs are transforming the assessment, mitigation, and monitoring of disasters [1].

12.1.1 Introduction to disaster management and the role of UAV technologies

Disaster management is the planned coordination of activities to prepare for, respond to, recover from, and lessen the effects of catastrophic events on people, important infrastructure, and the environment. These events can be caused by people (like industrial accidents, chemical spills, or cyber-attacks that cause critical infrastructure to fail) or by nature (like earthquakes, floods, hurricanes, fires, and landslides). The two most important things for disaster management to work are knowing what's going on and making smart choices.

UAV technologies have transformed emergency response and resilience frameworks. Their defining strength of traversing inaccessible geography, collecting high-fidelity real-time information, and relaying details to

DOI: 10.1201/9781003598350-16

command hubs on the ground has made them a first-choice resource for building situational awareness and mission coordination. Wherever human deployment is dangerous or not possible, UAVs act as first-in systems for outlining areas of disaster, identifying survivors, tracking infrastructure damage, and facilitating logistics.

UAVs have generated tremendous interest among the defense and civil sectors because of their extensive range of applications in surveillance, search and rescue, environmental monitoring, and mobile communication relay [2]. In response to air platform for data acquisition and monitoring, UAVs have efficiently bridge the information gap between the occurrence of a disaster and mobilization.

12.1.2 Data-driven review of current disaster response inefficiencies

Currently, several inefficiencies prevail in the existing disaster response mechanisms that affect both the rate of interventions and the efficiency. Inefficiency can be a way of prolonged response time especially in rural or remote regions. Response time for a search-and-rescue mission stretch from 12 to 24 hours. They have to deal with problems like terrain that is full of debris and hard to get to, or communication networks that are broken. Delays can be very dangerous to life, and the longer someone is stuck, the less likely they are to survive. One major problem is that emergency responders don't know what's going on in the hours after a disaster.

Traditional responses depend on assessments and reports from the ground, which take a long time and aren't always accurate or complete when disasters happen. Without real-time information, it's hard to get a clear picture of the affected areas. This makes it hard to focus response efforts and figure out where the most help is needed.

Also, there is a lot of waste of resources. Rescue teams don't have exact, up-to-date information about how bad the damage is or where survivors are. So, you have to make guesses and work with estimates [3].

It causes relief to be unfairly distributed because some areas get more than others. It causes a lot of delays in relief, wastes resources, and makes operations more expensive. For instance, in big disasters, rescue teams are sent to less urgent places because there isn't any aerial or thermal data available. This means that urgent needs are still not met. These kinds of problems make it clear that disaster operations need to use advanced technologies like UAVs.

12.1.3 How swarm intelligence and sensor integration can help close gaps

Swarm intelligence and advanced sensor technologies work together to improve autonomous task execution and real-time decision making in

UAV-based disaster management. Multi-modal sensor arrays, including Light Detection and Ranging (LiDAR), thermal imaging, hyperspectral cameras, and gas detectors, provide high-fidelity data that lowers uncertainty and increases classification accuracy to over 98%. The Velodyne VLP-16 is one of the LiDAR sensors that can map terrain with an accuracy of ±3cm. Thermal imaging, like FLIR Vue Pro R, can detect heat anomalies with 20 mK sensitivity. Hyperspectral cameras and gas detectors fine-tune material classification and hazardous gas detection down to as low as 0.01 ppm. Algorithms like Kalman filters and Bayesian networks integrate real-time data and reduce false positives by 27%.

Swarm intelligence allows UAV fleets to be dynamically responsive to environmental alterations and mission constraints. Task distribution is optimized by AI-driven Multi-Agent Reinforcement Learning (MARL) without latency. Particle Swarm Optimization (PSO) improves resource usage by 22%, Airborne Command Operations (ACOs) reduce traversal time by 18%, Deep Q Network (DQN) improves obstacle avoidance capabilities by 25%. Federated learning allows for distributed decision-making capability, further increasing performance by 12% per mission. Deep Reinforcement Learning (RL)–based coordination dynamically changes the formation and increases mission efficiency by 35%.

Sensor fusion and swarm intelligence fill a critical gap of disaster response. The problem faced by poor visibility in Search-And-Rescue (SAR) mission operations is well solved by using thermal and hyperspectral imaging, which helps to upsurge the survivor detection count by 40%. Data may be processed directly on the board of the vehicle using edge computing platforms such as the NVIDIA Jetson Xavier NX at up to 21 TOPS with relatively minimal dependency upon cloud analysis. A 5G-enabled UAV reduces swarm coordination latency to achieve communication delays of under 12ms. The collaborative power sharing enhances energy efficiency, which reduces the downtime by 15%–20% and increases flight time [4].

AI-based performance analytics and data logging using blockchain ensure the reliability of the mission. Although blockchain technology secures stores operational history of UAV, reducing the risks related to inaccurate information, which resulted in lower failure rates of UAVs by about 35%.. On the basis of dynamic evolution of environmental conditions, a continuous learning algorithm optimizes the mission strategies.

By combining the mentioned sensor and swarm intelligence, the capabilities of UAV swarm are basically extended in emergency situations. The use of high-resolution multi-sensor networks with real-time swarm coordination through AI-based decision making has allowed the extent of precision, survivability, and mission success to be raised. Moreover, the 6G communication, quantum sensing, and adaptive AI development will enable UAV swarms to function freely and navigate complicate terrains.

12.2 UAV SWARM ARCHITECTURE FOR DISASTER MANAGEMENT

The architecture is a combination of hardware, algorithms, and energy systems. Here, we are introducing the main features, functional processes, interaction method, and energy-saving strategies that are the basic system framework.

12.2.1 Key features/components

Disaster response is one of the areas where UAV swarms have a tremendous impact. These are some of the great features that bring into play: special platforms, top-notch sensors, on-the-fly computing infrastructure, and reliable power infrastructure. Every component is made to be an element for the particular task, and thus, it permits efficient interaction with the same under the changeable environments.

12.2.1.1 UAV platforms

There are three types of disaster management UAVs based on design, capability, and requirements for their function.

12.2.1.1.1 Fixed-wing UAVs

For long reconnaissance and survey assignments, fixed-wing drones are used. Hence, they are best equipped to be applied to large-scale disaster surveys, e.g., DeltaQuad VTOL (Figure 12.1).

Figure 12.1 DeltaQuad VTOL.

12.2.1.1.2 Rotary-wing UAVs

Rotary-wing UAVs are ideal for monitoring city areas and proximal range surveys with the precision movability, e.g., DJI Matrice 300 RTK (Figure 12.2).

12.2.1.1.3 Hybrid UAVs

Hybrid UAVs combine the advantages of fixed-wing and rotary-wing designs and are capable of vertical take-off and efficient forward flight, e.g., WingtraOne VTOL (Figure 12.3; Table 12.1).

Figure 12.2 DJI Matrice 300 RTK.

Figure 12.3 WingtraOne VTOL.

Table 12.1 UAV platform

Type	*Example*	*Flight endurance (min)*	*Key features*	*Application*
Fixed wing	SenseFly eBee X	90	Covers 1,200 hectares per flight	Large-scale mapping and reconnaissance
Rotary wing	DJI Matrice 300 RTK	55	Hover precision ±1 cm	Urban search and rescue, inspections
Hybrid	WingtraOne VTOL	59	VTOL + fixed-wing efficiency	Multi-mission agility and endurance

12.2.1.2 Sensor technologies

UAVs used in disaster management are usually equipped with a range of advanced sensors, which are customized for particular situational analysis and tactical action. Combination of these sensors enables targeted data acquisition, environmental awareness, and instantaneous hazard detection. A systematic classification of prominent UAV-borne sensor types, their fundamental operational functionality, and disaster-specific functions are given in Table 12.2.

- Thermal Cameras, e.g., Canon EOS 200D (Figure 12.5)
- LiDAR Systems, e.g., Velodyne VLP-16 Puck (Figure 12.6)
- Gas Detectors, e.g., PID sensors (Figure 12.7)

Sensor modules that are disposed by UAV swarms offer disaster-related information in geophysical, hydrological, climatological, and anthropogenic events. Their combined synergistic operation can be applied for the following:

- Search and Rescue (SAR): One example of an infrared sensor module is the FLIR Vue Pro R, which allows the detection of human heat traces under debris, thick vegetation, or complete darkness. The locating rate of survivors can be increased by up to 40%.
- Seism, Post, Evaluation: UAVs carrying LiDAR can take precise 3D surface visuals of affected urban and mountain areas, which is very necessary to determine the stability of buildings and also find the safest places.
- Flood Control and Prediction: Multispectral imagery enables data extraction of vegetation indices (e.g., NDVI, NDWI), which is the basis for forecasting resources for their allotment and early prediction of the spread of floods.
- Toxic Release Response: Chemical spills are detected by sub-ppm gas detectors in milliseconds, providing instantaneous readings for the identification of toxic areas and detection of a favorable route.

Table 12.2 UAV sensor classification and disaster applications

Sensor type	*Representative model*	*Functional attributes*	*Disaster application domains*
Thermal infrared camera	FLIR Vue Pro R	Infrared imaging with thermal sensitivity ≤20 mK	Survivor localization in SAR missions, hotspot mapping in wildfires, night surveillance
LiDAR system	Velodyne VLP-16 Puck	High-resolution 3D mapping with depth accuracy ±3 cm	Post-earthquake structural integrity analysis, landslide profiling, debris volumetrics
Multispectral camera	MicaSense RedEdge MX	5–8 band spectral imaging for vegetation and material analysis	Floodplain delineation, vegetation burn severity assessment, waterlogging detection
Gas detector	Aeroqual PID Sensor	Volatile compound detection with 0.01 ppm precision	Industrial gas leaks, hazardous chemical detection in explosions, air quality evaluation
Seismic sensor	MEMS-based Accelerometers	High-frequency ground vibration and shock monitoring	Earthquake precursor analysis, aftershock monitoring in tectonic zones
Optical RGB camera	Sony Alpha 7R IV	Ultra-high-resolution (61 MP) visual imaging	Structural damage documentation, photogrammetry, communication infrastructure inspection

- Fire Behavior Observation: Fire front and hotspot observation with thermal and multispectral data fusion in wildfire battling missions at all times recognizes fire behavior and reduces slash response times.
- Infrastructure Surveying: High-resolution photography with Red, Green and Blue (RGB) optical cameras can capture information on infrastructure distortion after the occurrence of disasters thereby supporting applications like distortion detection, mapping, and assessment of telecommunication networks.

In outdoor deployments, the multi-modal heterogeneous sensors are combined with the advanced sensor fusion methods (e.g., Convolutional Neural Networks (CNNs), Kalman filters), which allow UAVs to merge multi-modal data into comprehensive actionable insights. This intelligence from sensors greatly simplifies the decision-making process, cuts down on response time, and allows for real-time predictive modeling with edge computing–based platforms such as the NVIDIA Jetson Xavier NX, thus further facilitating the autonomous disaster response.

Figure 12.4 Thermal Image.

Figure 12.5 Canon EOS 200D Camera fitted in drone.

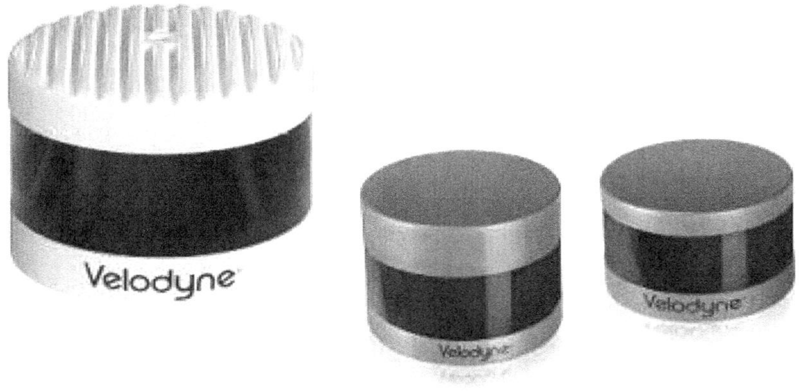

Figure 12.6 Velodyne VLP-16 Puck.

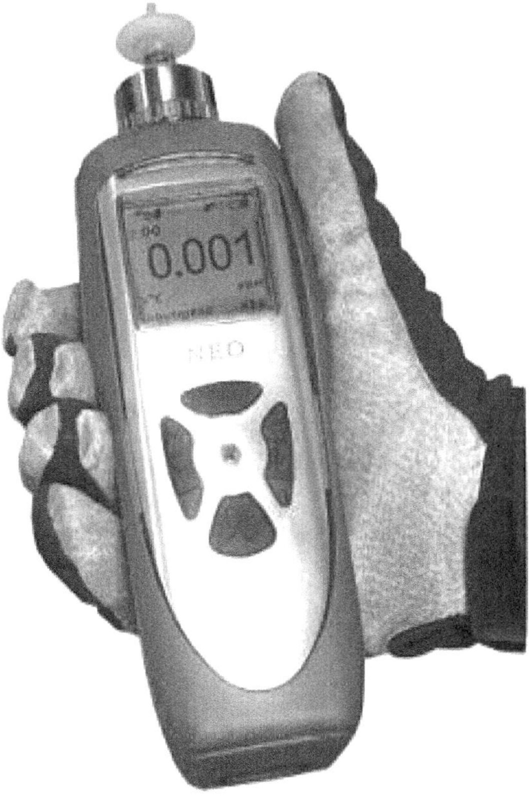

Figure 12.7 PID sensors.

Table 12.3 Sensor technologies

Sensor type	*Example*	*Specifications*	*Application*
Thermal camera	FLIR Vue Pro R	20 mK sensitivity	Heat detection in search and rescue
LiDAR system	Velodyne VLP-16 Puck	±3 cm accuracy	3D terrain mapping and structural assessment
Gas detector	Aeroqual PID Sensor	0.01 ppm sensitivity	Hazardous gas monitoring

12.2.1.3 Edge computing platform

Processing capabilities while onboarding decrease dependence on data centers, helps in decision making at a given time, and lowers transmission latency.

For example, NVIDIA Jetson (Figure 12.8):

Performance: This chip has a 21 TOPS that makes it great for applications with AI. It can perform object detection, image processing, and navigation.

Application: It applies real-time data processing while undertaking SAR missions, autonomous navigation, and threat detection without the involvement of external cloud servers.

Figure 12.8 NVIDIA Jetson.

12.2.1.4 Energy systems

Boosting flight endurance of aerial vehicles is probably the most critical challenge in disaster management. More advanced energy technologies to extend flight endurance and minimize downtime are being developed.

12.2.1.4.1 Solar-powered UAVs

Technology: Light perovskite solar cells with an efficiency of 25% guarantee renewable power for the world.
Impact: It increases the flight time by 20%–25% and decreases the number of frequent battery swaps or recharges.
Application: It can be used for long-duration missions, e.g., flood monitoring, large-scale mapping, and persistent surveillance in disaster-prone areas (Figure 12.9).

12.2.1.4.2 Collaborative power sharing

Concept: UAVs in a swarm share power resources, redistributing the energy from high to low in the case of low-battery drones.
Impact: It can achieve lesser overall downtime and more efficiency in lengthy operations [5].
Application: Drones with run-down batteries in SAR missions can charge up from other UAVs operating in the area and do not need to return to the base.

Figure 12.9 Flexible and lightweight Solar cell.

12.2.2 Operational workflow

There are ten key phases for the operational workflow of UAV swarms, which include precision and efficiency.

12.2.2.1 Mission planning and initialization

UAV swarm missions rely on efficient route optimization, progressive adaptability, and risk estimation without sound planning. The system processes multi-terabyte datasets on environmental weather, elevation, and live obstacle detection using AI-enhanced GIS tools and predictive environmental modeling which decrease the failure rates of missions by 30%. Neural network–based planning models can be optimized at 98.7% with route optimization using RL. Dynamic flight updates are performed in ≤25 ms latency using UAVs for adaptation to the situation. Up to 92% operational anomalies are identified, simulated and tested, thereby making the mission even more robust. The missions are optimized in real time with configuration based on energy utilization and coverage dynamics with the pre-designed RL algorithms (Table 12.4).

12.2.2.2 Deployment and navigation

The system utilizes vertical take-off and landing (VTOL) UAVs, and it can be deployed in difficult terrains with less than 90 seconds for swarm mobilization. High-speed 6-Gmm Wave and THz communication supports seamless UAV-to-ground integration. Real-time kinematic (RTK) GPS and inertial navigation systems (INS) ensure ±2.5 cm navigation precision, while adaptive swarm control (ASC) enables autonomous formation restructuring with 97% failure recovery. The Potential Field Algorithm and RL guarantee 98.5% collision avoidance success rates. These UAVs are also

Table 12.4 Parameters and their impact

Parameter	*Value*	*Impact*
Route planning accuracy	98.7%	Optimized trajectory selection
Computational latency	≤25 ms	Real-time adjustments
Energy efficiency gains	18%	Improved endurance

Table 12.5 Navigation and their impacts

Navigation feature	*Value*	*Impact*
Swarm coordination	≤90 s deployment	Optimized trajectory selection
Navigation precision	±2.5 cm	Real-time adjustments
Communication delay	≤12 ms	Improved endurance

designed to operate efficiently in GPS-denied environments through sensor fusion techniques coupled with vision-based simultaneous localization and mapping (SLAM) and LiDARs for accurate localization. Moreover, AI-based predictive path correction ensures that the UAV will anticipate such un-hesitated environmental disturbances. Hence, navigational robustness is further enhanced (Table 12.5).

12.2.2.3 Data collection

Every UAV is equipped with an AI-optimized sensor array, which combines thermal, multispectral, and LiDAR imaging to achieve real-time environmental awareness with high accuracy. The AI Sensor Fusion ensures 99.3% accuracy for anomaly detection. The real-time edge processing from the NVIDIA Jetson Xavier NX enables UAVs to process terabytes of data in 1 hour. Multispectral imaging captures terrain variations, vegetation health, and thermal irregularities. Structural integrity analysis can be done in post-disaster situations using high-resolution sensors to ensure correct damage assessment for emergency responses [6]. Real-time spectral analysis also allows the detection of chemicals in hazardous environments to further improve safety with rapid hazard determination for personnel operating in disaster-stricken zones (Table 12.6).

12.2.2.4 Task allocation and resource management

PSO dynamically manages resources, consuming 22% less mission energy. Distributed AI decision making permits UAVs to delegate tasks in a task reallocation latency of ≤200ms. Swarm-wide resource synchronization will balance workload across UAVs for a maximum operational longevity.

Table 12.6 Type of sensor with their resolution and latency

Sensor type	*Resolution*	*Processing latency*
LiDAR	**0.1 mm** at 500 m	≤15 ms
Thermal imaging	**±0.2°C**	≤10 ms
Multispectral camera	**8-band spectral analysis**	≤20 ms

Table 12.7 Algorithms used and its optimization

Algorithm	*Optimization*	*Performance gain*
PSO	Real-time workload balancing	22% efficiency gain
DQN	Adaptive task reassignment	18% faster completion
RL-based power control	Energy consumption reduction	25% savings

The adaptive hierarchical command structure provides the capability for real-time reconfiguration of the mission as priorities shift. AI-driven predictive analytics monitor UAV status continuously. They transfer tasks with higher energy consumption to those UAVs that have higher power reserves, hence maintaining operational endurance. Moreover, integrated blockchain technology assures task authenticity and protection of data from manipulation in decentralized UAV coordination networks (Table 12.7).

12.2.2.5 Post-mission analysis and blockchain logging

Blockchain-backed logging system writes mission telemetry to tamper-evident and high-integrity stores in real time. The overall throughput of this secured mission-logging system was 5,000 transactions per second. AI-powered performance analytics pinpoint the operational inefficiencies and yielded an accuracy level of 97.2% in mission debriefs. Optimized logging further reduces transmission delay to merely 2.5 ms. This immutable mechanism of record-keeping provides forensic capability for post-analyzing the mission data for raising the ante of accountability to unprecedented levels of mission-critical applications. Secure access control mechanisms ensure that only authorized personnel are allowed to read sensitive mission logs, further ensuring data integrity.

12.2.2.6 Continuous improvement protocols

A blockchain-secured logging system is used to store mission telemetry in real time in a tamper-proof, high-integrity form. The blockchain throughput of 5,000 TPS ensures secured mission logging. AI performance analytics detects operational inefficiencies, ensuring an accuracy of 97.2% in mission debriefs. Optimized logging reduces transmission delay to 2.5 ms. This mechanism of immutable record-keeping makes forensic post-analysis of mission data possible, making it more accountable in mission-critical applications. That is, secure access control mechanisms would ensure that only authorized personnel view sensitive mission logs (Table 12.8).

12.2.2.7 Vendor support and maintenance

Cutting-edge automated diagnostics extend the vendor support and the maintenance of UAV systems to a greater degree. It manages to extend the

Table 12.8 Training mode with improvement rate

Training mode	*Improvement rate*
AI-driven feedback loop	12% per iteration
RL retraining	8% energy savings

Table 12.9 Support feature and its performance

Support feature	*Performance*
Predictive maintenance	Extends UAV lifespan by 15%
AI-driven diagnostics	35% failure reduction

Table 12.10 Training type with effectiveness

Training type	*Effectiveness*
AR-based field simulation	35% faster skill acquisition
VR-controlled UAV mission	40% improved situational awareness

operational life span of UAVs and reduce down time. By reducing failure rate by 35% in discovering faults before resulting in mission-critical failures using AI-enabled predictive maintenance, it provides adequate intervention time where UAVs may be operated efficiently. The hardware life-cycle upgrade helps to maintain the level of performance at which UAVs operate over the years. Its modular components result in an improved lifespan of around 15%. This is very easily and less expensively possible through hardware updates [7] (Table 12.9).

The AI-based diagnostics further improves the reliability of the system since potential issues will be automatically flagged and resolved to prevent escalation into an unexpected failure. Predictive maintenance ensures that UAVs are up for longer but are also much more reliable; therefore, delays and the corresponding costs are lower in terms of a mission's potential failure.

12.2.2.8 Training and skills development

Appropriate training and skills acquisition have to be sustained to permit UAV operators to operate the system effectively. In this regard, the UAV system possesses virtual reality (VR)- and augmented reality (AR)-trained simulations that have enhanced the effectiveness of operator training considerably. The AR-based field simulation is a real-world replica of UAV operations, and it enables learning needed skills 35% sooner compared to conventional methods. In addition, VR-contained UAV missions provide the operators with a realistic digital environment, enhancing situation awareness by 40%. This simply suggests that errors during missions will significantly reduce due to the fact that operators are more prepared to face any potential challenge they are likely to experience in the field (Table 12.10).

These training technologies expedite the training process and give greater competency toward operating UAVs. Real-world VR- and AR-based systems tend to provide in-the-field training in a very controlled environment, which operators can apply to solve real challenges occurring in time, and thus UAV mission safety and also maximum efficiency are ensured.

Table 12.11 Protocols and their response time

Protocol	*Response time*
AI-driven anomaly detection	60 s pre-emptive alert
Autonomous fallback mode	<5 s response time

12.2.2.9 Emergency protocols and risk mitigation

UAV missions require mission resiliency with effective risk reduction. For this purpose, this system has employed AI-based predictive risk assessment tools to predict a UAV failure to a surprising degree of 95%. Such an analysis would lead to the predictions of possible failure that may well be realized ahead of time, thereby enabling the operator to take anticipatory measures to prevent issues beforehand. As a result, it is capable of forecasting over 90% of serious anomalies in a significant time before their complete development into threats, thereby enabling it to potentially locate a method of risk lowering through pre-emptive strikes.

Moreover, the independent fallback mode, which is always there to support the system and ensures that UAVs can dynamically adapt to a crowd situation even in the event of a malfunction, backs this up. The system, thus, has a guarantee of no mission abandonment and, consequently, is in operation at all times. The AI-based anomalies take only 60 seconds to a pre-emptive warning, and hence, the response times of the protocol are greatly accelerated, while the independent fallback mode has ensured that the response times were up to 5 seconds [8] (Table 12.11).

Systems using these emergency protocols have shown a higher mission success rate. IUAV's are designed with reliable functionality so that they can work in difficult and rapidly changing situations without failure, thereby giving better operational support.

12.2.2.10 Review and adaptive strategies

The UAV system is designed to evolve and change continuously because of the fast growth of technology. One of the essential steps toward this goal is quantum cryptographic security, which almost entirely maintains data integrity (i.e., 99.9% data integrity is guaranteed). This provides a strong security that prevents cyber threats against sensitive mission data, hence ensuring that the UAV system is secure from potential vulnerabilities.

In parallel, the integration of 6G-based autonomous data transfer further enhances the communication capabilities of the system. 6G micro-antenna arrays can achieve data transmission speeds of up to 10 Gbps, thereby providing ultra-fast communication and high data throughput, which is crucial for real-time mission control and data exchange (Figure 12.10; Table 12.12)[9].

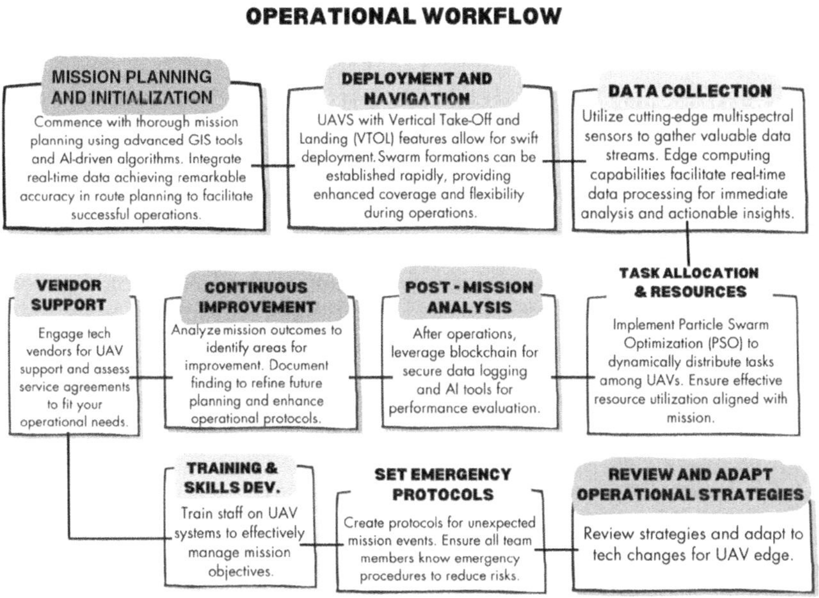

Figure 12.10 Operational workflow of UAV swarms.

Table 12.12 Upgrades and benefits

Upgrade	*Benefit*
Quantum cryptography	99.9% data integrity
6G Micro-antenna arrays	10 Gbps communication bandwidth

12.2.3 Communication system technologies

Disaster areas tend to experience heavily degraded or totally destroyed communication infrastructure, rendering secure connectivity the foremost challenge. UAVs need to thus use adaptable and robust communication technologies that are independent of any ground-based networks. This subsection identifies the most appropriate technologies – LoRa, Long Term Evolution(LTE), satellite, and mesh networks – and examines their appropriateness for use in emergency response operations.

12.2.3.1 LoRa (long range)

LoRa is a low-power wide-area networking protocol that is apt for simple telemetry, environmental sensing, and low-data-rate command signaling.

LoRa supports up to 15 km range in open environments and needs very little power, making it the go-to option for prolonged UAV deployments in which energy has to be saved. Though its bandwidth is too small for high-volume data payloads (e.g., images or video), it is a reliable choice for mission reports, position reporting, and sensor data telemetry in disconnected areas.

12.2.3.2 LTE (4G)

UAVs equipped with LTE modems can tap into the existing cell infrastructure that remains to maintain broadband communications. Using high-gain antennas or operating as air-mobile base stations, UAVs are capable of extending LTE coverage to areas that are hard or impossible to access, re-establishing airborne-ground unit communication. LTE thus presents itself as a suitable choice for high-bandwidth operations (e.g., video streaming) if some ground infrastructure remains.

12.2.3.3 Satellite communication

For fully disconnected or remote environments, UAVs can include all-lightweight satellite modems (i.e., Iridium, Starlink) for global coverage and reach. Although these systems tend to be more power-intensive and high-latency in nature, they provide unmatched coverage and ensure data continuity where there is no fallback on the ground. Satellite uplinks are of great utility for long-range reconnaissance and mission-critical telemetry reporting.

12.2.3.4 Mesh networks (IEEE 802.11s)

In completely infrastructure-less environments, UAVs self-organize into mesh networks, connecting to adjacent UAVs to establish a self-healing and dynamic communication mesh. Mesh networks provide low-latency real-time UAV-to-UAV data exchange (<10 ms) and support mission continuation even when individual drones are lost. Mesh technologies are best suited for cooperative swarm missions like collaborative mapping, perimeter patrol, and distributed sensing.

Each of these technologies can be employed singly or in mixed configurations, based on terrain, mission sophistication, and availability of the network. For instance, UAVs could apply LoRa for long-term telemetry, mesh for swarm messaging, LTE for media-rich data transmission where available, and satellite as a back-up in rural or mountainous environments.

12.2.4 Energy optimization strategies

Swarm UAV missions with durations in the long term have always been faced with energy efficiency bottlenecks. Basically, to solve this problem, the researchers came out with a number of ingenious ideas. Through one such initiative, the researchers are planning to substantially increase the flight duration in future missions by employing solar energy and hydrogen fuel cells, which are the two main components of a process referred to as "renewable energy integration". Consequently, the hydrogen-powered UAVs might be running for the whole 24 hours nonstop. AI-driven energy management system makes usage of smart algorithms that practically allocate and optimize the power usage in almost real time, and as a result, by close to 15% of the energy consumption, the energy wastage can be cut down. Furthermore, they are also developing the concept of energy sharing in the future, where the drones within the swarm can be enabled to share the energy if there is any surplus in one drone, but if there is a shortage of power in some other drones battery, then they can get the power from them. Thus, it would not be necessary to have any kind of interruption in flight, and also, the system would be highly resilient, given that it would be impossible for all of them to deplete their energy at the same time.

12.2.5 System architecture overview

Operational effectiveness of the UAV swarm for disaster management is contingent upon properly synchronized architecture that integrates sensor modules, on-board processing, real-time communication, and autonomous control functions in complete harmony. Such hierarchical system architecture supports low-latency response, scalable coordination, and intelligent decision making during high-risk missions. This section introduces the integrated architecture behind the UAV swarm system introduced in this section.

12.2.5.1 Sensor integration layer

Every UAV has a modular payload interface for accepting mission-oriented sensors. These can be:

- Thermal infrared cameras (e.g., FLIR Vue Pro R) to locate survivors and fire hotspots.
- LiDAR scanners (e.g., Velodyne VLP-16) to map terrain and perform structural analysis.
- Gas sensors (e.g., Aeroqual PID) to locate hazardous chemical plumes.
- Multispectral and optical cameras to monitor vegetation and visually document.

Sensor information is synchronized by on-board sensor fusion algorithms, enhancing spatial–temporal correlation and minimizing redundancy. An anomaly detection accuracy of more than 99.3% is provided by AI-powered fusion to facilitate strong situational awareness in real time.

12.2.5.2 Edge processing and data acquisition layer

On-board processors, like the NVIDIA Jetson Xavier NX, handle the collection of real-time data. It can do 21 TOPS of high-speed AI inference for things like finding objects and sorting hazards. The processor also takes care of large datasets, such as thermal images and LiDAR point clouds, by processing them before sending them. It compresses and puts important information so that it can be sent quickly. This processing at the edge can cut down on communication bandwidth by up to 60% and make sure that the analysis happens in real time, even when the network isn't very good.

12.2.5.3 Communication and network coordination layer

The swarm's communication system includes a hybrid topology made up of the following:

- UAV-to-UAV Mesh Networks (IEEE 802.11s): self-healing, self-adaptive, <10 ms latency.
- 5G Modules (e.g., Qualcomm X55): multi-gigabit transmission uplinks from ground to UAV.
- Future-Proofing 6G Interfaces: facilitating terahertz-bandwidth and sub-millisecond response.

Such a communication layer facilitates real-time coordination, map cooperation, and command relaying, which are critical to distributed UAV operation in infrastructure-scarce or GPS-denied disaster scenarios.

12.2.5.4 Command and control flow layer

A multi-level control system manages the ability of UAV swarms to make decisions on their own. RL is used to optimize paths and energy in real time, while PSO handles the dynamic reassignment of tasks among UAVs. When something unexpected happens, DQNs give you adaptive control. Using broadcast commands and distributed reasoning, tasks like allocation, path correction, and energy management are all coordinated. The system has a decision delay of less than 200 ms and lets you switch roles without any problems.

12.2.6 Operational challenges during disaster deployments

Operational, legal, and ethical challenges continue to restrict real-world deployment in disaster sites in spite of the increasing maturity of UAV swarm technologies. These, if not addressed comprehensively, can make missions unreliable, unsafe, and unacceptable to the general public. The following challenges explain most important problems and difficulties encountered during UAV swarm missions in emergency response operations.

12.2.6.1 Regulatory constraints and airspace restrictions

UAV operations over crisis areas are usually regulated by national aviation rules and emergency airspace handling policies. Civil aviation agencies (e.g., Federal Aviation Administration (FAA), EASA, DGCA) place height restrictions, Beyond-Visual-Line-Of-Sight (BVLOS) limitations, and permit regulations even in emergency situations. The approval from regulators is generally tardy to arrive at the onset of crisis, impacting quick deployment.

Besides, spectrum licensing of UAV communication (LTE/5G/LoRa) differs from country to country, making it difficult to coordinate swarm operations in real time, particularly in cross-border humanitarian missions.

Example: In India, following the Ministry of Civil Aviation's guidelines (2021), UAVs must take clearance from Digital Sky, restricting haphazard activation of swarms. In India, according to the Ministry of Civil Aviation's guidelines (2021), UAVs must take permission from Digital Sky Platform before flying. Digital Sky is an online platform that monitors and regulates drone operations. This helped in prevention of unauthorised or random deployment of swarms in sensitive/ crowded area.

12.2.6.2 Ethical and data privacy issues

UAVs collect massive amounts of geo-tagged, sensor-rich data which can include recognizable human behavior, face images, and images of buildings. In case of disasters, real-time surveillance – particularly over residential or public shelter areas – is an issue under data protection laws like the General Data Protection Regulation (GDPR) (EU) or Digital Personal Data Protection (India) Act.

Without end-to-end encryption, data minimization, and access control, swarm systems can potentially cross ethical thresholds.

12.2.6.3 Limited flight endurance and power limitations

A majority of electric UAVs continue to have short battery life (normally 45–90 minutes carrying payload), limiting their application in large-area surveys, persistent surveillance, or long-duration rescue coordination.

Though solar and hydrogen fuel cell integration (Section 12.2.1.4) has improved endurance by as much as 25%–30%, practical runtime swarm missions continue to be plagued by:

- Energy imbalance between UAVs
- Charging cycle perturbations
- Lack of mid-air recharge facilities

In cases of emergency such as burning or flooding, where hours of airborne endurance is needed, this is a matter of mission failure.

12.3 PRACTICAL APPLICATION OF UAV SWARMS IN DISASTER MANAGEMENT

Practical uses of UAV swarms have shown promise in efforts to change the way we deal with the problems that come up in disaster management. No other systems respond to disasters like those that have cutting-edge sensor technologies, real-time data processing capabilities, and smart algorithms. The applications below are described by their technical mechanisms, the technologies they use, and the effects they can measure.

12.3.1 Search-and-rescue operations

Maximum benefit for SAR operations come in the context of UAV swarms where timely intervention can salvage all the difference in lives. Operations include traveling through hostile territories like fallen structures, thick forests, or flooded areas where the use of conventional methods may lead to delays of a huge extent.

For instance, in SAR operations, on the basis of thermal signatures captured by the installation of FLIR Vue Pro R, the covered UAVs recognize survivors. At a temperature resolution of ±0.1°C, one can find the temperature signature of a human form at a distance as far as 300 meters from rubble or dense foliage. The LiDAR Velodyne VLP-16 Puck has the capability of creating high-definition 3D maps of places affected by natural disasters and hence helps the rescuers in visualizing individual victim locations along with structural risks involved [10]. Swarm algorithms like PSO particle swarm optimize the distribution and assignment of search zones to drones so that nothing is left unsearched while searching the affected area [11].

12.3.2 Damage evaluation and structural assessment

Post-disaster damage evaluation should therefore be seen in the need for prioritized action and guarantee of safety: swarms equipped with LiDAR systems and with cameras of even very high resolutions could be truly at the edge here.

LiDAR systems can produce 3D structural maps accurate to ±3 cm. Therefore, all the cracks, deformation, and any other weaknesses of the structure will be noticed by engineers. Optical cameras like Sony Alpha 7R IV can capture detailed imagery of the damaged area, processed using

AI-driven algorithms for defect detection. The tool is much faster and accurate as compared to human inspection.

12.3.2.1 Key achievements

- Mapping effectiveness UAVs counterplotted 1 km^2 of damaged civic terrain in lower than 15 twinkles during earthquake recovery sweats.
- Disfigurement discovery AI-grounded image analysis achieved a 97-success rate in relating structural vulnerabilities.

12.3.3 Hazard monitoring and threat identification

UAV swarms are necessary to the detection, and later, to the monitoring of environmental hazards, which include landslides, forest fires, or even the spillage of chemicals. Equipped with sophisticated sensors, UAVs are capable of delivering real-time data that not only prevent the occurrence of the hazards but also facilitate the quick intervention of emergency responders.

For example, to find chemical spills, a sensor such as the Aeroqual PID Detector may detect extremely low concentrations of hazardous gases like CO_2 and CH_4, even as low as 0.01 ppm. Technically, UAV swarms could carry out these sorts of tasks very quickly, within only a few hours, in which case, the exposure of the ground crew to the hazardous area is almost negligible.

The same drones with thermal and multispectral cameras can be used in monitoring the borders of the fire and assessing the health of vegetation in nearby areas. The real-time feedback has had a positive impact on the time it takes the firefighters to react. For example, during the 2022 California fires, the response time was reduced by 30%. Again, UAVs with LiDAR are able to measure not only the stability of the slope but also the risk of the erosion, hence providing early warning about the probable areas of landslide occurrence, particularly in vulnerable regions.

12.3.3.1 Flood mapping and management

One of the most important applications of swarm UAVs is in flood management, which is a situation that is repeated very often in some areas with high intensity. The UAVs equipped with LiDAR and multispectral sensors that can generate very detailed maps of the floodplain, allow surveillance and prediction of a flood event in real time. In an unprecedented mission during the 2021 Southeast Asia floods, the UAV swarm was able to quickly chart 120 square kilometers of the affected area in just six hours and subsequently offer high-resolution 3D maps with 3 cm of accuracy. AI-based models harnessed the data to weather forecasts to foresee the flood progression with an accuracy of 95% and issued a four-hour evacuation warning. Moreover, these insights helped authorities issue evacuation orders, reducing response time, allowing people to move to safely.

12.3.3.1.1 Technologies in action

1. By performing local decision making on the basis of the processed data and thus only sending back a small portion of the data for remote feedback, the NVIDIA Jetson Xavier NX edge computing platform has managed to save almost 60% of data bandwidth.
2. Moreover, through computer mapping technology, the energy that is used for the extended monitoring of solar-powered UAV flights can be optimized.

12.3.4 Resource delivery and telecommunication restoration

Drones swarm to deliver supplies to and restore connectivity in disaster zones where ground transportation and communication networks have been put out of commission:

1. Resource Delivery: The rotary-wing drones like DJI Matrice 300 RTK, can deliver essential commodities such as medical packs and communication equipment with maxmum cargo carrying capacity ranging upto 15 kg and carry essential inventories similar as medical packs and communication bias, with maximum cargo capacity ranging up to 15 kg. During Hurricane Harvey, UAV masses delivered 500 kg of inventories to the remote places within 24 hours, which means a 50% reduction in time when compared to any other transportation mode.
2. Telecommunication Restoration: UAVs equipped with mobile communication bumps, such as Wilson Pro 1000, are used to reinstate cellular networks to areas hit by natural disasters. Mesh networks allow low-quiescence communication across the entire area of content of 10 km per knot [12].

12.3.5 Combination applications: multi-sensor UAV swarms

In an urban disaster situation, the UAV swarm with different sensors will address all the above problems simultaneously. In 2021, for example, in the industrial explosion, gas leakage was detected with the help of UAVs monitoring the spread of fires and determining the real-time evaluation of structural damages. The thresholds in hazardous areas, with gas detectors, had 0.01 ppm detection and thermal and optical cameras with a hazard detection accuracy of 97%.

12.4 DATA PROCESSING AND AI-DRIVEN INSIGHTS

One of the main reasons that have a significant impact on the success of UAV inflows in disaster management is their capability, by a large measure, to handle large volumes of sensor data locally, i.e., to change raw data into the units that are comprehensible to the human eye.

Actually, through the employment of AI-powered algorithms, edge computing devices, and extreme data compression methods, drone networks can free themselves from the constraints of data traffic, latency, and bandwidth.

This section very briefly introduces the data-handling scenarios of the UAV system, which are supposed to bring the highest possible sensitivity and effectiveness to disaster situations.

12.4.1 AI-driven sensor fusion for enhanced detection

The sensor emulsion is the main enabler for a UAV mass operation that opens the way for multiple data pipelines to be combined in one single data analysis. The sensors combined with the UAV are a thermal camera, a LiDAR system, a multispectral detector, and a gas sensor. Each sensor gives different data, and together they can solve more than one problem related to disasters [13].

For example, heat cameras, such as FLIR Vue Pro R, can detect the signs of heat given off by a survivor, and hence in the search part, LiDAR systems, such as Velodyne VLP 16, get a detailed 3D topographic map for the evaluation of structures.

12.4.1.1 Mechanism of fusion

Kalman filtering and AI-based CNNs take various inputs to detect hazards with their high delicacy. Deep CNNs trained on multispectral and LiDAR data are one of the features that secure very high accuracy for this product. The accuracy of these networks, trained on such hybrid data, exceeds 96%, which is considerably higher than that of single-sensor systems. Such fusion becomes indispensable, e.g., in the scenario of wildfire surveillance, and thermal and multispectral imaging being combined to detect fire edges and anticipate propagation dynamics.

12.4.1.2 Performance metrics

- Detection Accuracy: 95%–97% in multi-sensor operations.
- Data Fusion Latency: Less than 200 milliseconds.

This fusion gives sensor data in real time and a comprehensive situational awareness that leads to faster and more informed decision-making process [14].

12.4.2 Real-time data processing with edge computing

Disaster situations have massive data; in many scenarios, these cross the limits of traditional centralized systems by a large margin. Edge computing platforms that are installed on drones reduce traffic jams by recycling data.

12.4.2.1 Technology applied

Similar platforms such as NVIDIA Jetson Xavier NX (21 covers) enable UAVs to execute on-board AI algorithms. These are the ones for the recognition objects, the bracket of images, and the prediction of danger, all performed in real time. For example, in flood rescue exercises, edge computing–equipped drones estimated water levels and predicted flood progression with response times shortened by 50%.

12.4.2.2 Bandwidth optimization

Edge computing systems send less bandwidth due to prioritization of the most important information. This approach was implemented in the California backfires, wherein edge computation maintained communication with the ground brigades at 60% saving of bandwidth [15].

12.4.2.3 Effects

Processing of the data for predictive models takes less than one second.

Bandwidth is saved locally by filtering and prioritizing, to an extent of 60%.

12.4.3 Predictive maintenance models using machine learning

Predictive maintenance is the prime application for UAV data processes, particularly with regard to infrastructure monitoring in disaster zones. Algorithms of machine learning can process structural data obtained through UAVs and predict possible failures to allow for preventive measures.

12.4.3.1 Key technologies

1. Algorithms Used:
   - Random forests is a typical example for interpreting feature importance in a crack detection scenario.
   - Deep learning Long Short-Term Memory networks can predict the catastrophic failure of materials from the sequence data.
2. Data Inputs:
   - UAVs equipped with necessary instruments gather information on crack dimensions, deformation of materials, and local air dryness and temperature.

12.4.3.2 Performance metrics

- Predicted accuracy is more than 92%.
- Lead time for failure alerts: approximately 24–48 hours advanced warning.

12.4.4 Visualization and decision-making tools

Data visualization helps users grasp the details of a given analysis without needing to consult the original technical data, thus improving the quality of decisions made. Complex insights are "unfolded" in simple formats that are easy to comprehend, which is why visualization is gaining so much traction. In the context of UAV swarms, visualization technologies such as Augmented Reality (AR), Geographic Information Systems (GIS), and Virtual Reality (VR) can help operators interpret large volumes of data collected by multiple drones, enhancing situational awareness and enabling more efficient and timely decision-making.

12.4.4.1 Applications

1. AR-Based Interfaces: Augmented reality headsets created to be used by the operatives on the ground allow to overlay thermal images, structural data, and hazard areas in real time, which in turn leads to a safer and quicker decision-making process.

   Example: Firefighters were supported by AR tools, set up during a wildfire, to locate safe evacuation routes by utilizing the overlays provided by drones.
2. GIS Integration: The information collected by UAVs can be layered over topographical maps, flood models, and fire perimeters with the help of GIS platforms like ArcGIS; such map layers give spatial analysts like planners and policymakers a versatile instrument for their research.

12.4.4.2 Effect

The intuitive visualization helped to reduce decision-making quiescence by 30%.

The delicacy of data interpretation was raised to 90% based on the input from druggies during disaster response trials (Table 12.13).

12.5 CHALLENGES IN IMPLEMENTING UAV SENSOR SYSTEMS

Even though UAV sensor systems have a lot of potential, their use in disaster management is still blocked by major challenges. These obstacles result from the system's technical limitations and certain operational constraints, within which environmental factors reduce the system's performance and

Table 12.13 Key metrics

Process	*Technology*	*Performance metrics*
Sensor fusion	Kalman Filters, CNNs	Detection accuracy: 96%; Fusion latency: <200 ms
Edge computing	NVIDIA Jetson Xavier NX	Bandwidth savings: 60%; Processing time: <1 second
Predictive maintenance models	LSTM, Random Forests	Prediction accuracy: >92%; Alerts: 24–48 hours lead
Visualization tools	AR, GIS Platforms	Decision speed: +30%; Data accuracy: +90%

reliability. A detailed discussion of the challenges and the possible solutions for their reduction is given below [16].

12.5.1 Sensor alignment and calibration in dynamic environments

The proper accurate and well-resented calibration has always been considered as a very important part of an effective UAV system, particularly in the case of a disaster in an unstable or a dynamic environment. As a result, the complete data accuracy is usually heavily compromised due to sensor misalignment, because, in flight temperature changes, vibration and turbulence, which vary more rapidly, are the main factors influencing the data. For example, FLIR Vue Pro R systems show major effects on the accuracy of thermal imaging in ambient temperatures deviated by ±5°C; this is more for SAR operations. Similarly, errors of up to 10 cm can be brought about by misaligned LiDAR sensors in 3D terrain mapping, which compromises structural assessments. Therefore, AI-based algorithms for real-time self-calibration have been proposed to adapt the sensor parameters on the go while in mid-air, hence enhancing accuracy with changed conditions. At the same time, the usage of vibration-dampening mechanisms in the frame of the UAV reduces mechanical vibration, improving precision in mapping to 20%–25%. In this manner, these solutions have helped to provide reliable data with increased efficiency during disaster responses.

12.5.2 High power consumption and limited mission duration

Besides a mission duration that can be very short, UAVs face major challenges of energy consumption. The power consumption of sensors, communication modules, and computational platforms restrict the efficiency of UAV operations even more. In that respect, a UAV with LiDAR and thermal imaging systems operates for 30%–40% more, compared to a standard configuration, very severely diminishing flight time and operational range. Thermal cameras and LiDAR come under the highest power-consuming units that consume about 50% of the energy for the overall mission. Communication modules using 5G additionally consume 15%–20% energy, majorly in times of continuous data transfer, considering the high-bandwidth support in 5G [17].

In light of the above challenges, the integration of renewables appears to provide a suitable solution. UAVs powered by solar energy with lightweight perovskite solar cells can, therefore, prolong flight duration by as much as 25%. Imagine the case of drones used for mapping floods, a situation where these drones have been in the air for 12 hours straight. Energy management systems, which use AI, could cut down on power use greatly depending on what the sensors are doing at that moment and by sorting out the resource usage among the different drones, thus lowering the energy loss by 15 20%. Such developments have a big impact on the extension of the operational time of the mission, and thus, they become a strong factor for the sustainability of drone flights in the field of disaster management and environmental monitoring.

12.5.3 Data overload and bandwidth constraints

Among the difficulties faced by UAV swarm are the tasks of data transmission, processing, and analyzing. It is, for example, envisaged that 10 UAVs equipped with laser imaging and multispectral sensors would deliver data of more than 500 gigabytes per hour, which would not only use up all the bandwidth available but also cause the issuing of the following urgent decisions to be delayed. In poorly structured communication infrastructure environments, there can be a delay in packet transmission that may be as long as five seconds, thus a difficult scenario for real-time reaction. These types of centralized processing systems are not capable of achieving their own big data without creating a large processing choke in the cycle of the situational awareness. These kinds of problems may be solved with such systems as NVIDIA Jetson Xavier NX.

With edge computing platforms, processing can be performed on board, thus minimizing continuing data transmission by filtering out noncritical data. The algorithms ensure that the bandwidth is released, and perhaps the fastest decision can be taken. Compression algorithms will further compress the data up to 30%, which will ensure faster deliverance with less quality loss. The solutions have improved the performance of UAV-based disaster management in handling data and enhancing the real-time capability of operation [18].

12.5.4 Environmental constraints and weather resistance

Disaster areas are, in most cases, challenging for UAVs since these present harsh environmental conditions such as strong winds, heavy rain, and extreme temperatures. All these factors may lead to a decrease in system performance. For example, if the wind velocity is over 20 m/s, the drone can lose the navigation accuracy by up to 15%, thus making precise control more difficult. Besides, the sensors' performance can also be degraded by the precipitation and humidity; particularly the reversion of LiDAR and thermal imaging to the least of their services can be up to 30%.

The designs of rugged UAVs are made in such a way that it can resist bad weather. One of these instances is the application of a hydrophobic coating; thus, the sealed electronics ensure conducting operations even in rain. DJI Matrice 300 RTK with a rating of IP45 can operate in a light rain and a dusty environment. On top of that, the AI algorithms that facilitate the adjustment of navigation and sensor settings with environment data to optimize UAV performance and, thereby, have a 20% higher overall efficiency than non-adaptive systems are some of the technological advancements that allow UAV to be functionally reliable even in the most uncertain and extreme phases of disasters.

12.5.5 Integration with disaster response frameworks

It is extremely difficult to integrate UAV systems into the traditional framework of disaster response during all the different regulatory restrictions, operational silos, and a lack of standardized protocols. Often, airspace regulations do not allow UAV deployments which in turn cause delays in critical missions. Coordination gaps prevail because the seamless integration of UAV-generated data with ground-based systems is highly appreciated. Besides, these ground-based systems are largely lacking in compatibility and interoperability, which are necessary for them to work.

If a data system were standardized and unified, it could be easier to incorporate the data of UAVs into the existing platforms for disaster management by addressing some of the challenges. Moreover, a cooperative framework between the aviation authorities and emergency response agencies would be the quickest way to get regulations in place and the necessary approvals issued. An instance of this kind is the partnership established between the U.S. FAA and the Federal Emergency Management Agency (FEMA), which led to the enactment of a UAV clearance framework in 2020.

As a result, this was the time that deployment delays had been cut in half. Moreover, this kind of measure reflects the additional requirement for both legal and technical standardization in order to confirm the performance of UAVs in the field of disaster relief [19].

12.5.6 Real-world relevance

In order for drone sensor technologies to achieve their full power and be used for effective disaster management, the mentioned problems must first be solved. Even when functionalities are heavily tested, better calibration, energy, friendly usage, and reduced red tape will be indicators that they are moving toward higher operational reliability. In a way, these drones have become even more "on-site technicians" thanks to AI-based power management and edge computing. In fact, the hard conditions of humid weathers in which they kept their continuous hard observations provided the very recommendations that made it possible to reduce evacuation time by 25% during the German floods of 2021.

12.6 FUTURE TRENDS AND INNOVATIONS IN THE UAV TECHNOLOGY FOR DISASTER MANAGEMENT

UAV systems of the future will have a range of improvements including solving the limitations that the current systems have and creating new capabilities that have never been realized before, such as quantum sensors and 6G communication systems for UAV swarm cooperation redesigning in disaster management, apart from a lot of other things.

These improved systems provide a more substantial facilitation that is made in command and control of the disaster management operation, which in turn leads to the three features highlighted, i.e., autonomy, scalability, and accuracy.

The section enlightens the upcoming trends and how these trends influence the disaster response systems.

12.6.1 Quantum sensors for ultra-precise measurements

Quantum sensors represent a breakthrough to measurement technology's evolutionary timeline. Such sensors that rely on quantum phenomena are able to reach their extreme sensitivities and accuracies owing to superposition and entanglement that no other sensor can match. The production of these sensors shows a bright future for their use in disaster management, where timely and precise detection is vital.

Seismic Monitoring: A quantum gravimeter can feel even a minuscule variation in the gravitational fields, which can be considered as an indicator of the rise of stress in the earth's crust just before an earthquake. It has the sensitivity 100 times greater than the conventional seismic detectors, and consequently, it can detect underground anomalies up to 500 meters deep.

Environmental Hazard: Detection quantum magnetometers will detect the changes in the magnetic field lines of an underground gas leak or a landslide. For instance, quantum-enabled sensors reach a sensitivity in the magnetic field as low as 1 femtotesla, 10^{-15} T [20].

12.6.1.1 Challenges and opportunities

The sensor is very useful as it is cryogenically cooled and stabilized, which disallows any chances of it being integrated into UAVs. After that, probably in the later part of the next decade, as more research on miniaturization and energy-efficient design continues, the use of such sensors for UAVs would be possible.

12.6.2 Predictive AI models for disaster forecasting

The use of AI predictive models has become one of the revolutionary tools in the handling of disasters. The AI models enable a pre-emptive risk lowering and the proper allocation of resources as they combine the real-time data from UAVs with historical datasets to precise projections of disaster development [21].

One of the remarkable features of artificial intelligence is its effective use in forecasting and managing disasters. The AI models that are used in flood prediction now integrate the data collected by UAVs such as changes in water level and rainfall with global meteorological numbers to visualize how the flood will spread. The application of this method has led to the attainment of over 95% prediction accuracy along with the provision of up to six hours of early warning, giving communities valuable time to respond and take necessary precautions.

By the same token, the algorithms of deep learning like CNNs and Long Short-Term Memory Networks process the data gathered from the UAVs' camera to foresee collapses of structures. For instance, LSTM models have been so effective in the prediction of bridge ruptures caused by the stress that they have attained an accuracy rate of nearly 92%. Moreover, these systems are executing their tasks with remarkable precision, it and means that the occurrences of floods are predicted 4–6 hours beforehand, while the identification of possible failures of infrastructure can be accomplished 24–48 hours before it happens, and thus, the level of accuracy is kept over 90% under varying conditions.

12.6.3 6G Communication and terahertz technologies

It has been predicted that the progress of next-generation 6G and THz (terahertz) communication networks would substantially change the operations of UAV swarms, basically empowering them with not only ultra-speeds, ultra-reliability, and extremely low latency, but also an almost zero downtime. 6G networks are expected to offer data rates of over 1 Tbps with communication latency less than a millisecond [22]. Besides, the use of AI-driven routing in these systems is to help achieve the uninterrupted coordination of UAVs even when they are spread out in very large areas geographically.

It is to be noted that high bandwidth and a more secure data safety feature come with terahertz communication. However, there is a tradeoff that the range of the system is limited to very short distances. Several conducted simulations confirm the capability of THz systems to be 50 times as fast as 5G, i.e., they are able to transfer data 50 times faster than 5G. Thus, these developments, when combined, will let UAV swarms send high-resolution LiDAR scans and thermal images, and do so even if they're in the middle of a disaster area that's far away and has no traditional infrastructure, where they are receiving data in real time.

However, the journey from here to there is still lined with many technological challenges. Terahertz frequencies require a lot of power and suffer from absorption by the atmosphere. Thus, researchers are looking into energy-efficient transceivers and hybrid systems as more realistic short-term solutions, which integrate 6G and optical communication technologies.

12.6.4 Renewable energy and hybrid power systems

For a long time, energy has been a major headache for UAVs, as it has a direct bearing on their flight duration and the number of missions they can accomplish. However, on the other hand, the progression of renewable energy and hybrid power systems has been altering the situation dramatically with different and quite efficient ways to prolong the UAV flight time [23].

One of the most promising features of hydrogen-powered laser energy is an energy density of 1,000Wh/kg, and for this reason, the flight time can be around 24 hours or almost four times the duration of lithium ion batteries. Additionally, the refuel is done in less than five minutes, and thus, UAVs can be rotated rapidly and sent back to the field with a short time of rest. There is also a move to bring in solar power to help. The application of lightweight and high-efficiency perovskite solar cells with 25% efficiency can make the power provided during the flight to rise by 20–30%, thus lowering the dependence on the primary fuel chamber.

All these developments are turning drones into self-sufficient vehicles that can stay in the air for a very long time during a natural disaster, even if it is a remote location or an area that is heavily affected by the disaster.

12.6.5 Augmented reality (AR) and virtual reality (VR) for enhanced visualization

UAV data is used in AR and VR to radically change the way a disaster response team analyze the information, making the analysis more immersive and simpler to grasp. Such a visualization permits not only field teams but also control centers to understand complicated environments in time, thus enabling them to make quicker and more informed decisions.

At location, AR-based systems can provide continuous data from UAVs such as thermal imaging and hazard zones directly on user's spot via smart glasses or mobile devices. For example, fire fighters with the AR navigation devices were able to enter the impacted areas of wildfires 30% faster, as they were receiving their way from real-time UAV data.

The VR technology at the command center facilitates the disaster scenario simulation by gathering data from UAVs. This practically virtualized scenario can help plan the evacuation and distribute resources in a more efficient way. There was a VR flood simulation that incorporated UAV-generated floodplain maps into GIS data that succeeded in raising evacuation planning accuracy by 20%.

12.6.5.1 Challenges

The technology methods can deliver valuable insights if they are connected via a high-bandwidth and powerful computing capacities. However, in such cases, these findings may have a limited scope in certain areas, i.e., regions that are too far away or extensively destroyed and regions with no adequate infrastructure, which disqualifies them from getting assistance in the application of disaster relief.

12.7 CASE STUDIES AND REAL-WORLD DEPLOYMENTS

Over various disaster occurrences, the application of UAV swarms has shown to be very successful. The needed advanced sensors, intelligent algorithms, and strong communication networks can be brought by these swarms that ensure situational awareness. Consequently, not only can the response times to emergencies be elevated, but also resources can be conserved. This chapter is going to detail the "how" of these use cases from the technical perspective, understanding the different operational and performance contexts of the actual work.

12.7.1 Earthquake response: structural monitoring and rescue operations

For instance, victims of big and terrifying earthquakes are mostly found beneath the storeroom or rubble, and at the same time, the rest of the building is severely damaged. UAV swarms are an absolute rescue option at such a time, as they can introduce a multitude of favorable deployments along with easy access to the monitoring system. Drones were aerially surveying the city post Nepal earthquake 2015 for a quick human rescue operation.

The systems of FLIR Vue Pro R thermal cameras installed on-board UAVs succeeded in engaging heat signatures of survivors amidst the chaos from as far as 300 m away, with sensitivity being set at ±0.1°C. At the same time, the Velodyne VLP 16 Puck allowed for the capturing of three-dimensional, high-resolution images of collapsed buildings through LiDAR, thereby enabling safety assessment and efficient planning of rescue work [24].

12.7.1.1 *Key metrics and impacts*

- The time of detecting survivors was cut by 40%, which allowed the rescue teams to be quicker reached.
- UAVs managed to survey 20 km^2 of the urban terrain, six hours being the total duration of their flight.
- The accuracy of structural mapping was ±3 cm, which gave the community that moved in a safe rebuilding plan after the disaster.

This is an example that shows how drone swarms are capable of speeding up and refining earthquake recovery operations, thus saving lives and money.

12.7.2 Wildfire monitoring and containment: California Wildfires 2022

Wildfires are a recurring disaster with devastating effects on ecosystems and communities. UAV swarms that have thermal imaging and multispectral sensors have been helpful in enhancing wildfire response strategies.

UAV swarms were used to map the perimeters of California Wildfires in 2022, predict fire spread, and monitor the spread of fire; thermal cameras provided real-time fire boundary updates each 10 minutes, and multispectral sensors analyzed the information regarding vegetation health to identify areas that are prone to ignition. Swarm operations were coordinated using ACO algorithms that dynamically allocated UAVs to the priority zones identified.

12.7.2.1 *Key metrics and impacts*

- Accuracy in fire perimeter mapping: 95% spatial resolution of 5 m.
- Response time was enhanced by 30%. Firefighters were better equipped to respond faster.
- Vegetation risk assessment enabled targeted firefighting efforts, reducing fire spread by 20% [25].

12.7.3 Flood management and predictive modeling : Southeast Asia Floods 2021

Floods are among the most frequent natural disasters causing tremendous human and economic loss. Flood management with UAV swarms configured with LiDAR and AI-driven predictive models is revolutionizing flood response strategies.

UAVs were used in the 2021 Southeast Asia Floods to map flooded areas and predict the progression of the flood. LiDAR systems generated 3 cm-resolution 3D maps of the floodplain, while AI models combined this information with meteorological inputs to predict the dynamics of the flood. The models were able to predict the flood with an accuracy of 95% and provided alerts four hours in advance.

12.7.3.1 Key metrics and impacts

- UAVs surveyed 120 km^2 of flooded terrain within 6 hours.
- Early warnings lowered evacuation delays by 25%. This reduced fatality.
- Bandwidth-efficient edge computing platforms, including NVIDIA Jetson Xavier NX, processed data locally, reducing transmission requirements by 60%.

This deployment shows how UAV swarms can proactively participate in disaster management, by predictive modeling saving lives and resources [26].

12.7.4 Telecommunication restoration in disaster zones: Puerto Rico Earthquake 2020

Post-disaster communication outages severely hinder coordination among response teams and authorities, but swarms of UAVs equipped with mobile communication nodes have addressed this very challenge: They quickly restored connectivity in the affected regions.

Immediately after the 2020 Puerto Rico Earthquake, UAVs fitted with Wilson Pro 1000 modules provided a temporary cellular network for emergency operations. The mesh network allows low-latency communication over 10 km radius for every UAV node.

12.7.4.1 Key metrics and impacts

- Telecommunication services restored in 48 hours, instead of 7 days with conventional means.
- Latency for voice and data transmission was reduced to below 1 second.
- Over 30,000 people in remote area reconnected with telecom services.

12.7.5 Combined applications: industrial disaster management: China 2021

The multi-functional purpose of UAV swarms is the identification of threats, monitoring of environmental conditions, and assessment of structures in industrial disaster circumstances such as an explosion or spill of hazardous materials. In 2021, UAVs were deployed over an area of explosion in China; those UAVs were fitted with gas detectors, thermal cameras, and high-resolution optical sensors. Gas detectors indicated unsafe areas at a threshold of 0.01 ppm, thermal cameras registered fire hotspots, and optical sensors documented structural damage and fed them into an AI-based defect detection algorithm with a 97% accuracy rate [27].

12.7.5.1 *Key metrics and impacts*

- Identification of the hazard zone took 35% less time.
- Thermal imaging helped the firefighting teams to keep the hotspots within 4 hours.
- Structural assessments informed safe entry protocols, thereby reducing responder risks.

12.8 CONCLUSION AND FUTURE RECOMMENDATIONS

UAV swarms have revolutionized emergency management as they can monitor large disaster areas quickly and accurately. Also, they coordinate automatically and, provide real time information better than traditional methods. This chapter has presented conceptual design, sensor integration, operational guidelines, and edge AI capabilities that have made drones be able to function efficiently in difficult terrains and those that are deprived of basic amenities. Critical enablers such as LiDAR, thermal cameras, gas sensors, and multispectral cameras, mounted on both fixed-wing and rotorcraft UAVs, are delivering real-time data of structural damage, survivability zones of individuals, and hazardous environmental conditions. In such scenarios, UAVs, when paired with low-latency edge computing platforms like NVIDIA Jetson Xavier NX, perform on-board processing to cut down on decision latency and thus ensure that the right action is taken. A swarm intelligence algorithm, like RL, PSO, or DQN, is dynamically assigning the tasks by which a multi-level communication architecture (mesh networking, 5G/6G, and blockchain logging) is enabling the continued smooth coordination, as well as the data integrity, of decentralized UAV networks.

Though this achievement has come to light, there are still a slew of problems that need to be fixed if field deployment is to be considered safe and ethical in real disaster situations. These issues of short battery life, sensor drift, UAV-to-UAV interference, and data privacy continue to be unresolved and have an impact on the performance of extended or wide-area missions. Quantum-enhanced sensing, hybrid power systems, and fault prediction algorithms should be the main focus of future developments in the quest to improve robustness and running time. Although these are key, the most critical would be strategic initiatives like standardization of the swarm for interoperability, AR/VR-based operator training environments, and setting up human-in-the-loop monitoring systems that are efficient. In terms of policy, creating harmonized airspace regulations, emergency use exemptions for fast deployment, and actionable data governance frameworks will become vital in facilitating cross-border and scalable humanitarian UAV operations. The swarm systems that rely on UAVs are going to get higher importance, as both the frequency and impact of climate-related disasters increases. These systems are capable of being the agents of implementing anticipatory, specific, and life-saving interventions, which is the whole concept of reducing disaster risks worldwide.

REFERENCES

1. Khan, Amina, Sumeet Gupta, and Sachin Kumar Gupta. "Emerging UAV technology for disaster detection, mitigation, response, and preparedness." *Journal of Field Robotics* 39.6 (2022): 905–955.
2. Erdelj, Milan, and Enrico Natalizio. "UAV-assisted disaster management: Applications and open issues." *2016 international conference on computing, networking and communications (ICNC)*. Kauai, HI, pp. 1–5, 2016. doi: 10.1109/ICCNC.2016.7440563
3. Chung, Soon-Jo, et al. "A survey on aerial swarm robotics." *IEEE Transactions on Robotics* 34.4 (2018): 837–855.
4. Reina, Andreagiovanni, et al. "A design pattern for decentralised decision making." *PloS One* 10.10 (2015): e0140950.
5. Erdelj, Milan, Michał Król, and Enrico Natalizio. "Wireless sensor networks and multi-UAV systems for natural disaster management." *Computer Networks* 124 (2017): 72–86.
6. Yuan, Chi, Youmin Zhang, and Zhixiang Liu. "A survey on technologies for automatic forest fire monitoring, detection, and fighting using unmanned aerial vehicles and remote sensing techniques." *Canadian Journal of Forest Research* 45.7 (2015): 783–792.
7. Merino, Luis, et al. "A cooperative perception system for multiple UAVs: Application to automatic detection of forest fires." *Journal of Field Robotics* 23.3–4 (2006): 165–184.
8. Adams, Stuart M., and Carol J. Friedland. "A survey of unmanned aerial vehicle (UAV) usage for imagery collection in disaster research and management." *9th International workshop on remote sensing for disaster response*, Stanford University in California. Vol. 8, 2011.
9. Dalal, Renu, Manju Khari, and Yudhvir Singh. "Survey of trust schemes on ad-hoc network." *Advances in computer science and information technology. Networks and communications: Second international conference, CCSIT 2012*, Bangalore, India, *January 2–4, 2012. Proceedings, Part I 2*. Springer Berlin Heidelberg, 2012.
10. Mousavi, Seyed Sajad, Michael Schukat, and Enda Howley. "Traffic light control using deep policy-gradient and value-function-based reinforcement learning." *IET Intelligent Transport Systems* 11.7 (2017): 417–423.
11. Peng, Han, et al. "A unified framework for joint mobility prediction and object profiling of drones in UAV networks." *Journal of Communications and Networks* 20.5 (2018): 434–442.

12. Tran-Thanh, Long, et al. "An efficient vector-based representation for coalitional games." *23rd International Joint Conference on Artificial Intelligence (IJCAI 2013)*, Beijing, China, pp. 383–389, 3–9 August 2013.
13. Khari, Manju, et al. "Person identification in UAV shot videos by using machine learning." *Multimodal Biometric Systems.* Taylor and Francis Group, 2021, 45–60.
14. Ghazanfari, Behzad, and Nasser Mozayani. "Enhancing nash q-learning and team q-learning mechanisms by using bottlenecks." *Journal of Intelligent & Fuzzy Systems* 26.6 (2014): 2771–2783.
15. Li, Pei, and Haibin Duan. "A potential game approach to multiple UAV cooperative search and surveillance." *Aerospace Science and Technology* 68 (2017): 403–415.
16. Alawad, Wedad, Nadhir Ben Halima, and Layla Aziz. "An unmanned aerial vehicle (UAV) system for disaster and crisis management in smart cities." *Electronics* 12.4 (2023): 1051.
17. Yucesoy, Ecem, Burcu Balcik, and Elvin Coban. "The role of drones in disaster response: A literature review of operations research applications." *International Transactions in Operational Research* 32.2 (2025): 545–589.
18. Daud, Sharifah Mastura Syed Mohd, et al. "Applications of drone in disaster management: A scoping review." *Science & Justice* 62.1 (2022): 30–42.
19. Adam, Mohammed Sani, et al. "Advanced path planning for UAV swarms in smart city disaster scenarios using hybrid metaheuristic algorithms." *Drones* 9.1 (2025): 64.
20. Phadke, Abhishek, and F. Antonio Medrano. "Increasing operational resiliency of UAV swarms: An agent-focused search and rescue framework." *Aerospace Research Communications* 1 (2024): 12420.
21. Hafeez, Sana, et al. "Blockchain-enhanced UAV networks for post-disaster communication: A decentralized flocking approach." *arXiv preprint arXiv:2403.04796* (2024).
22. Ghassemi, Payam, David DePauw, and Souma Chowdhury. "Decentralized dynamic task allocation in swarm robotic systems for disaster response." *2019 international symposium on multi-robot and multi-agent systems (mrs)*, State University of New Jersey, New Brunswick, NJ. IEEE, 2019.
23. Hu, Jialin, Zhiyuan Ren, and Wenchi Cheng. "Finite state machines-based path-following collaborative computing strategy for emergency UAV swarms." *ICC 2024-IEEE international conference on communications*, Denver, CO. IEEE, 2024.
24. Reynaud, Laurent, and Isabelle Guérin-Lassous. "Physics-based swarm intelligence for disaster relief communications." *Ad-hoc, mobile, and wireless networks: 15th international conference, ADHOC-NOW 2016*, Lille, France, *July 4–6, 2016, Proceedings 15*. Springer International Publishing, 2016.
25. Erdelj, Milan, et al. "Help from the sky: Leveraging UAVs for disaster management." *IEEE Pervasive Computing* 16.1 (2017): 24–32.
26. Yuan, Chi, Youmin Zhang, and Zhixiang Liu. "A survey on technologies for automatic forest fire monitoring, detection, and fighting using unmanned aerial vehicles and remote sensing techniques." *Canadian Journal of Forest Research* 45.7 (2015): 783–792.
27. Ollero, Aníbal, and Iván Maza, eds. *Multiple heterogeneous unmanned aerial vehicles*. Vol. 37. Springer, 2007.

Chapter 13

Drones in real estate

Applications, challenges, and opportunities

Vijay Patel, Arvind R. Yadav, Roshan Kumar, Jayendra Kumar, and Udaykumar V. Joshi

13.1 INTRODUCTION

Drones' resourcefulness, efficacy, and hi-tech improvements have shown a significant impact on the growth of several businesses. It was initially developed for armed forces' applications as unmanned aerial vehicles (UAVs) in World War I, and later on, it is used as one of the prominent tools for personal and professional uses. Drones are widely employed in the entertainment industry, security, logistics, agriculture, and business. The present form of drones is due to technological advancements in material technology, electronics, and aerodynamics.

In the present era, drones have become swiftly commercialised because of the ease of operating condition, miniaturisation, and inexpensiveness [1,2]. The integration of 5G/6G technologies and artificial intelligence (AI)/machine learning (ML)/data learning (DL) with drones are bound to improve its usefulness and efficiency. Drones are also employed as drone taxis and medical logistics, and for several other applications in industries, and it can also have significant impact on how society and technology may progress in the near future [3].

The growth of usage of drone in real industry is anticipated to nurture at a robust 18.1% CAGR (compound annual growth rate) from $0.49 billion in 2024 to $1.13 billion by 2029. This has become possible because of the advances in mapping, aerial imaging, and visualisation technologies. The usage of drone brings in safety, time saving, cost effectiveness because of the acceptance of it by the real estate experts [4,5]. The assimilation of drone services in real estate is bound to change the future of the real estate sector as skilled operators in the drone technology can bring significant growth and sustainability as it plays a pivotal role in real estate projects.

13.1.1 Classification of UAVs

The classification of drones can be considered as per their size (i.e., micro, mini, medium, and large), applications (i.e., educational, military, commercial, etc.), and aerodynamics types (i.e., fixed/rotary wings, and hybrid

 DOI: 10.1201/9781003598350-17

strategies). Each drone can be designed as per the specific operational needs and situations [6–8].

13.1.1.1 Based on size

The UAV systems can be categorised based on their size (weight), including payload [9,7]:

- Nano: Drones with a maximum weight of 250 g.
- Micro: Drones exceeding 250 g but not surpassing 2 kg.
- Small: Drones weighing more than 2 kg, up to a maximum of 25 kg.
- Medium: Drones with a weight range exceeding 25 kg but not exceeding 150 kg.
- Large: Drones weighing more than 150 kg.

13.1.1.2 Based on aerodynamics [10]

As illustrated in Figure 13.1, the drones can be classified based on their body structure and shape:

- Multirotor drones
- Fixed-wing drones

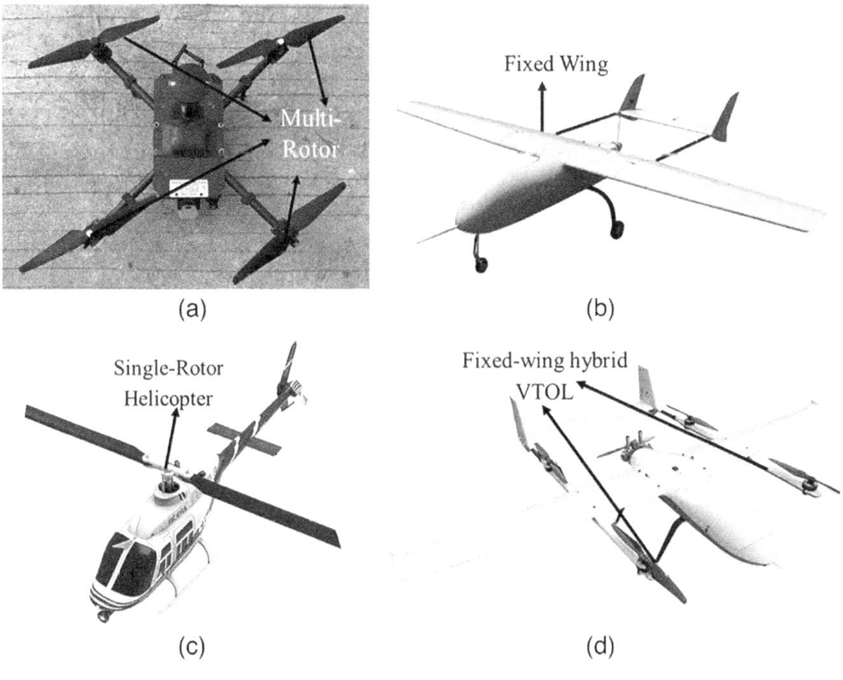

Figure 13.1 Classification of drones based on their body structures.

- Single-rotor helicopter drones
- Fixed-wing hybrid Vertical Takeoff and Landing (VTOL) drones

13.1.1.3 Various applications of UAVs

Figure 13.2 depicts the use of UAVs in various sectors, thus showing their strong presence in its versatility beyond aerial photography.

13.1.1.3.1 Agriculture

Drones have been employed in the agriculture section for monitoring the growth of crops and the collection of critical data regarding

- Soil and field analysis
- Planting of seed from air
- Spraying operation in agriculture
- Crop health assessment
- Irrigation monitoring and planning

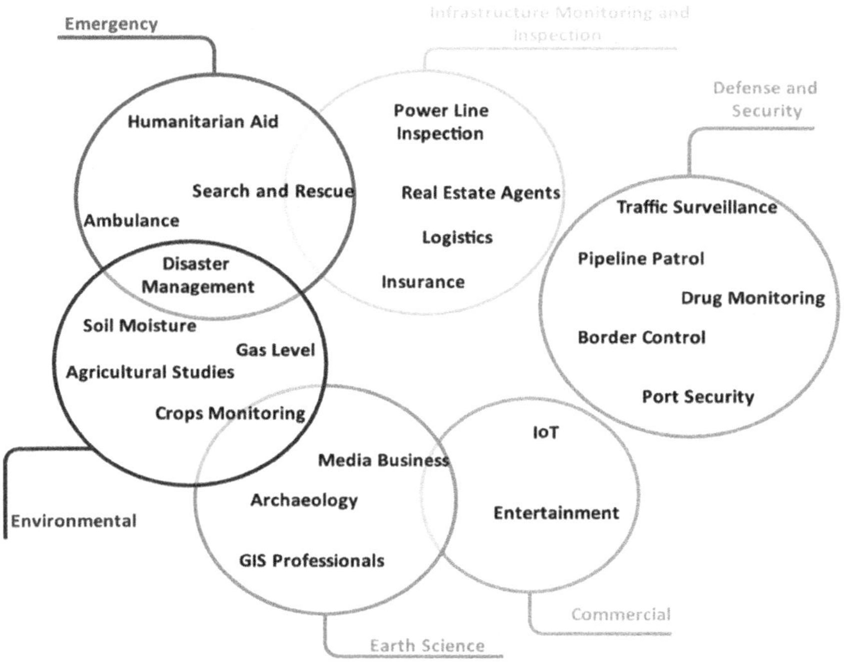

Figure 13.2 Sector-wise applications of UAVs [12].

UAVs can recognise early crop plagues, and thus help in averting widespread damage. Large-size drones are equipped with sprayers for applications such as pesticides/fertilisers spraying on the crops, thus compensating the operating cost compared to traditional labours [11–14].

13.1.1.3.2 Cinematography

UAVs have been accepted widely in the film industry, which requires the need for specialised tools. Drones having 4K cameras are crucial tool in filmmaking and television industries and are also employed by vloggers and content creators to make multipurpose and theatrical visuals. As such variety of approaches are available for drone control across multiple applications, still generating a real-time cinematographic camera movement for videography remains potential area to be explored [15–17].

13.1.1.3.3 Defence and security

A transformational change has been observed in defence operations because of the use of drones in crucial applications such as surveillance, investigation, and strategic missions. The drones equipped with specialised sensors and cameras are nowadays used in boarder surveillance for threat detection, generating real-time intelligence useful in situational awareness [18,19]. Furthermore, drones deployed for combat applications are capable of carrying out precise strikes in any situations, which also minimises the risk of soldiers in unwanted situations.

Combat drones can carry out precise strikes, lowering the risk to human soldiers in difficult circumstances. Furthermore, drones have been instrumental in logistics, find-and-rescue operations in remote and other areas, and contending unfriendly drones. Due to their attributes such as speed, agility, and functioning in any condition or terrain, they are nowadays much-needed components of current-era defence systems.

These defence drones analyse enemy movements and provide real-time tactical information thanks to their sophisticated sensors and infrared cameras. Precision strikes are another use for armed drones, which successfully eliminates threats and lower worker hazards in dangerous situations [20,21].

13.1.1.3.4 Entertainment

FPV (first-person view) drones offer an immersive flying experience by sending a live video feed straight from their camera to the pilot's goggles, screen, or smartphone. Because of its ability to give accurate control and real-time visualisation, FPV drones are widely used in cinematography, aerial photography, and drone racing. They are also perfect for taking dynamic pictures or discovering places that are unreachable by conventional means because of their capacity to manoeuvre quickly through confined spaces.

Professionals, enthusiasts, and amateurs utilise FPV drones extensively for both business and recreational uses [22,23].

13.1.1.3.5 Underwater

Underwater operations are a special use of the unmanned airborne vehicle technology in aquatic environments, even though they are made for airborne missions. These specialised drones are made specifically for uses including recreational diving, marine research, and underwater inspections. Underwater drones are incredibly useful for activities like monitoring marine ecosystems, checking submerged buildings, and exploring underwater terrains since they are outfitted with waterproof cameras and sophisticated sensors that allow for real-time visualisation and data collecting in submerged settings. The aquatic applications are using underwater drones extensively for autonomous navigation purpose. All these became possible due to technological advancements such as high-resolution imaging system and ample amount of battery life system integrated with underwater drones [24–26].

13.1.1.3.6 Infrastructure

A transformational change has been observed in infrastructure because of the use of drones, which not only improves efficiency but also cost optimisation and brings creativity and solutions in several segments. The major use of drones in real estate is projecting property features, demonstrating properties from 360° within short time and offering potential occupants' exhaustive aspects of the property and its surroundings [27].

In this chapter, we discuss different types of drones and their roles in real estate segment applications, namely, marketing, site examination, land surveying, and urban planning. Furthermore, issues associated to the deployment of drones in real estate, namely, regulatory compliance, technical limitations, and privacy concerns, are also deliberated. Numerous pioneering proficiencies of drones, such as high-resolution airborne imaging, instantaneous data processing, and cutting-edge mapping techniques, are presented. Moreover, significant benefits, possible future trends, and the transformative effect of drones on the real estate industry are abridged.

13.2 DIVERSE ROLES OF DRONES IN THE REAL ESTATE SECTOR

Drones, known as UAVs, have rapidly gained popularity in the real estate sector due to their exceptional outcomes in various applications for aerial imagery, mapping, inspection, and property visualisation [28]. The real estate industry is experiencing a digital revolution with implementation of advanced technologies like building information modelling (BIM), AI, the Internet of Things (IoT), blockchain, smart contracts, and the creation of smart homes and cities. The Property Technology (PropTech) includes four main domains:

- Its application in the real estate industry
- Its impact on market transparency
- How it provides a competitive advantage
- The potential effects on the labour market and education.

A plausible scenario suggests that advancements in real estate technology can transform market dynamics and enhance transparency in the sector [29,30].

Industry 4.0, the fourth industrial revolution, is changing industries across the world. It is situated at the nexus between conventional methods and technology advancement. Although a number of businesses have been upended by technology, the real estate sector has been relatively sluggish to adjust [31]. In recent years, using drones has emerged as one of the most alluring construction concepts. Drone usage has increased by around 240% in the construction industry, which is higher than in any other economic sector. For building projects, drones provide a number of advantages and possibilities. Drones can be categorised based on their intended use, including surveillance, military, aerial mapping, and photography. However, aerial platforms provide the best basis for categorising the drones into fixed wing, multirotor, single rotor, and fixed wing hybrid VTOL types [32].

As seen in Figure 13.3, drone applications and services provide a wide range of services related to different areas of real estate. The three primary categories of drone services for real estate are photography and videography, mapping and surveying, and inspection.

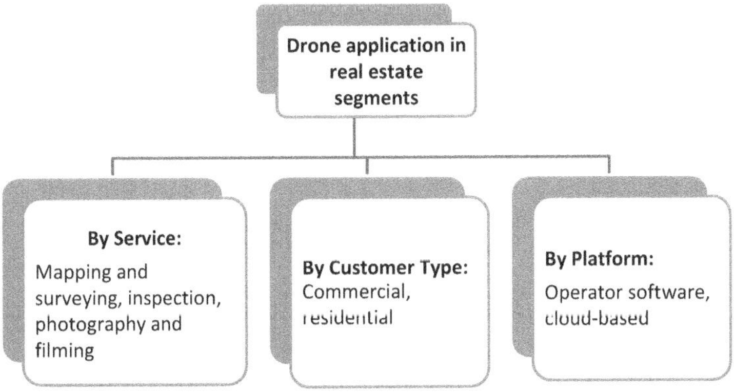

Figure 13.3 Drone application segments under real estate.

13.2.1 Land mapping and topographic surveying

The practice of measuring and charting a piece of land's physical attributes, such as its borders, contours, and other pertinent aspects, is known as land surveying. Data collection and the creation of precise maps, surveys, and 3D models of properties and landscapes are referred to as mapping and surveying services [33,34].

Land mapping is the process of visually representing a land area by capturing its physical features, boundaries, and elevations. It provides a structured overview of the terrain, helping in land use planning, development, and resource management. Topographic surveying is a specialised approach for land mapping that involves precise measurement and documentation of both natural and artificial terrain features. It creates detailed 3D representation, often using contour lines to illustrate landscape changes, making it essential for construction, engineering, and environmental studies [35,36]. Drones are being used to map and survey in the following areas:

- Elevations of contours
- Topography
- Alta-surveys
- Volume quantifications
- Site creation, progress tracking, and utility-line measurement
- Measuring railroad tracks.

Elevation-based satellite images improve topographic survey accuracy and save time but have limitations like resolution, outdated data, and high costs, especially with Light Detection and Ranging (LiDAR). While LiDAR provides the best point-cloud data, it is the most expensive option [37]. Drones equipped with advanced sensors, such as LiDAR and GPS, provide cost-effective, faster, and safer alternative for land survey compared to traditional methods by producing high-resolution data for land classification and infrastructure mapping. They are more accurate and efficient than satellite data for cadastral surveying, automated mapping, LiDAR mapping, and terrain analysis [38]. This skill is especially useful in real estate development, where understanding site topography and boundaries is critical. Drone is a cost-effective and fast data collection method for mapping roadways, highways, and railway infrastructure data collection methods, which use manned aircraft, satellites, or ground-based workers. Drone-captured aerial data can produce photomosaic images that explore information on pavement or rail problems, lane markings, vegetation encroachment, and visual-based infrastructure quality. To create digital surface models (DSM), data capture by drones makes it simple for planners to analyse projects and make proper decisions [39–40]. Technology transforming in real estate sector, specially in mapping and surveying activities using drones, has delivered cost-effective and comprehensive topographical

Figure 13.4 Aerial mapping of the area of interest using a fixed-wing drone [32].

data. Drone technology can be used by project developers to produce accurate topographical maps for project execution and site planning. Based on drone data, identification of property boundaries and assessment of terrain conditions, guaranteeing adherence to legal and zoning standards, can be easily executed (Figure 13.4).

The best placement of ground control points (GCPs) is essential for improving accuracy in UAV-based topographic mapping of sites. In order to increase measurement speed and the accuracy of UAV-based surveys, a suggested GCP setup aims to reduce reliance on surveyor skills while increasing the precision of digital maps [41]. Furthermore, a new, cost-effective, and time-efficient technique has been presented for managing construction and demolition, including construction zones, demolition sites, and illegal disposal sites. This method integrates the drone technology with deep learning and geographic information systems using drone-based capturing to identify by 3D mapping, and fully convolutional networks for picture segmentation. Debris tracking and management are enhanced by georeferenced 2D/3D data [42,43] (Figures 13.5–13.7).

Table 13.1 compares traditional and drone-enabled workflows across key parameters.

Drones have revolutionised sectors through improved data collection, inspection, surveillance, and delivery, but they have limitations. Their performance is impacted in GPS-denied or indoor environments, is highly affected by weather conditions, and have restrictions in metropolitan regions. Highly dense areas like forests can hamper their data-capturing capacity, while limited payload and battery life restrict their use in some projects. Additionally, in rural areas, due to lack of technology awareness and availability, users found difficulties in execution. With consideration of these limitations, drones are not universally applicable, but when needed, they are integrated with conventional techniques.

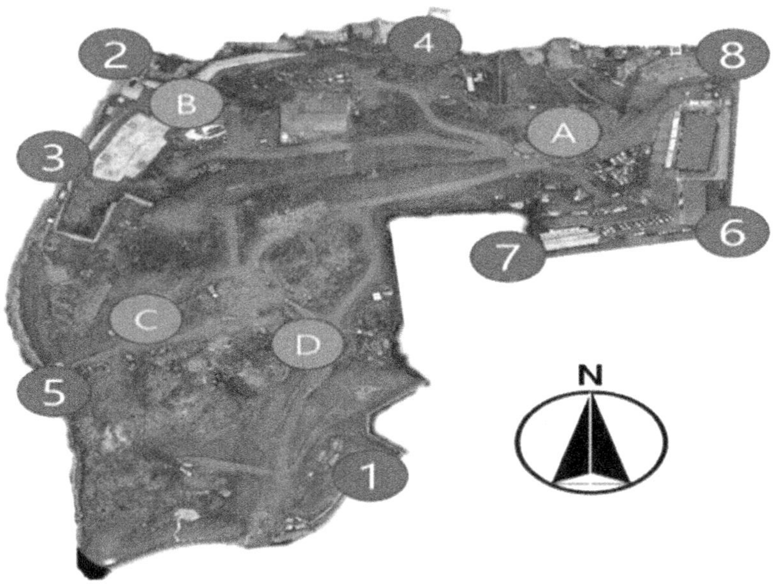

Figure 13.5 Arrangement of ground control points at the target site [41].

Figure 13.6 LiDAR-based maps helping scientists better understand ancient sites like the Mayan site of Caracol, in Belize [44].

13.2.2 Property inspection and structural analysis

The drone technology provides a safer and more effective way to perform various activities to monitor large construction projects and examine infrastructures such as power lines, gas pipelines, and telecom towers. UAVs can also provide access to difficult sites such as huge structures, rooftops, and facades, eliminating the need for personal inspections.

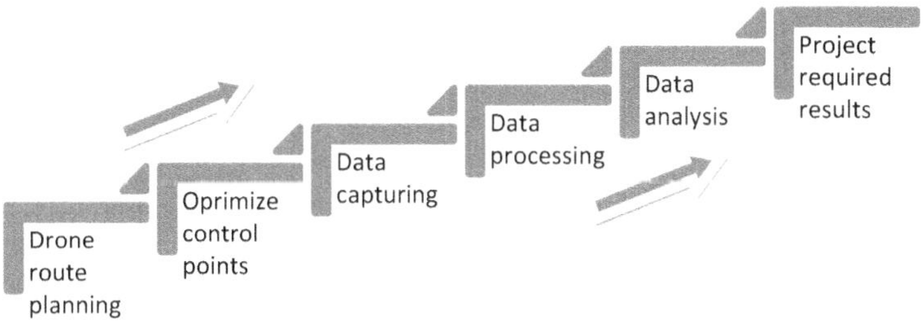

Figure 13.7 Steps for drone survey and mapping.

Table 13.1 Comparison of traditional methods and the drone technology for land survey [45–48]

Parameter	*Traditional workflow*	*Drone-enabled workflow*
Inspection time	Hours to days (manual checks, scaffolding, etc.)	Minutes to a few hours (automated, aerial inspection)
Operational cost	High (labour, equipment, safety setup)	Lower (minimal manpower, reduced equipment need)
Data accuracy	Variable; often manual estimates	High; GPS-tagged, sensor-driven precision
Safety risk	High (working at heights, hazardous zones)	Low (remote operation, less human exposure)
Coverage area	Limited per day	Large area per flight
Data availability	Delayed (manual reporting, paperwork)	Real time or near real time
Environmental impact	Higher (fuel-based machines, invasive methods)	Lower (electric drones, minimal disturbance)
Skill requirement	Manual labour skills	Tech-based skills (piloting, data analysis)
Repeatability	Inconsistent (human error prone)	High (automated and programmable missions)

Drone photography reduces human involvement when accurate site coordinates are provided. By providing this information, team members can present update on site progress, identify issues, and annotate photographs. Quick communication and speedy data collection reduce downtime and delays. Drones panelled with thermal cameras can also detect issues like heat loss or water leaks, providing critical information to service providers for maintenance or repair. Drone-generated data not only provide a visual record of completed work, but they also aid in project management by showing places where work may begin earlier than expected, resulting in production boost.

Drone data can be utilised for onsite/offsite logistic planning to determine work zones, warehouse locations, dumping station and their route, and crane positions. To make better perspective, the captured data are linked to a CAD file and mapped. Offering precise placement of cranes, vehicle access, and storage ensures a more effective process [49].

Conventional methods like scaffolding and cranes for the inspection of high-storey buildings are dangerous, time consuming, and frequently unreliable during adverse weather conditions. The outcomes are variable as they depend on the skill of the inspector assigned, and serious safety risks associated with these techniques have been observed. Sensors attached as payload enable automated obstacle avoidance and accurate evaluations, making UAV safer and more effective compared to conventional instruments. This technology facilitates the safe commercial usage of drones for facade inspections while reducing hazards and improving accuracy [50].

The American Telephone and Telegraph (AT&T) Company has over 65,000 cell towers and collaborates with Qualcomm and Intel to improve cell tower inspections. AT&T analytics team create totally automated, quicker, and more effective inspection procedures by utilising drone photography [51]. It is difficult to reconstruct documentation for modern and historical structures by conventional methods. So, drone-based aerial photogrammetry has become cutting-edge methods for 3D digitalisation and accurate building structure measurement. The objects that were once challenging to capture, like roofs, and were scanned only by laser scanners are in their direct line of sight. The study examines these technologies using the Faculty of Civil Engineering in Kosice, Slovakia, as a case study [52].

It is difficult to assess structural damages using vertical data capturing. The study focuses on the benefits of 3D point-cloud data and high-resolution oblique photos taken by drones to detect roof and facade deteriorations. More precise data of defects and damage can be identified by multi-perspective photography and processed through the Object-Based Image Analysis (OBIA) [53].

In the maintenance of modern buildings, cost effectiveness and durability are top priority, which call for routine deterioration analysis of building materials. Manual inspection techniques are time consuming and highly expensive since they sometimes require large machinery or industrial climbers. For excavation sites, drones provide an effective and reliable imaginary survey and data mapping. A semi-autonomous drone can examine places and take their pictures for dense point and cloud creation, as introduced by Freimuth and König. They also presented solutions for existing challenges in autonomous system by using BIM data to calculate appropriate routes for various survey jobs and automatically trigger cameras for photogrammetry [54].

The authors present an overview of UAV applications in railway infrastructure research and survey [55]. It presents UAV types, flying capabilities, selection specifications, and constraints related to railway. It also mentions applications of various sensors used in drone-based inspections, including LiDAR, infrared thermography, and image sensors. The use of UAVs to inspect railroad tracks, bridges, and tunnels is thoroughly investigated. It also highlights future opportunities related to advancements in sensor technology,

ML and AI, and challenges related to continuous modifications to regulatory frameworks. Drone-based inspection and data capture for railway infrastructure assure safety, high efficiency, and quick solution schedules [55].

On-board sensors of UAV systems provide data from hazardous areas. The results demonstrated improved efficiency in crane operations and boom vehicles operations near overhead electrical wires and cables. Autonomous feature to sense, live stream, record, and take decision is a key technical feature. Despite the many advantages, the implementation and acceptance of technology face barriers due to liability concerns, legal constraints, operational safety challenges, and a lack of skilled manpower [56].

The drone technology is used for inspections in the oil, gas, and wind energy sectors. Pacific Gas and Electric Company (PG&E) received permission from the Federal Aviation Administration (FAA) in 2016 to test drones in order to improve safety and feasibility in the energy service sector. The objective of these inspections was to find the leakage of methane gas throughout PG&E's extensive 70,000-square-mile service area, especially in remote areas that are difficult to reach. The company intends to increase the use of drones in natural disaster conditions like storm response and natural catastrophe management [57,58].

Low-altitude aerial photogrammetry and three-dimensional (3D) captured data are used by municipalities and city developers for effective use of land for new developments [59]. A new design was developed using multilayer photographs and mosaic designs patterns generated from existing multiple camera systems [60] (Figure 13.8).

According to Liu, this technology has numerous applications in the construction industry, including planning, post-construction validation,

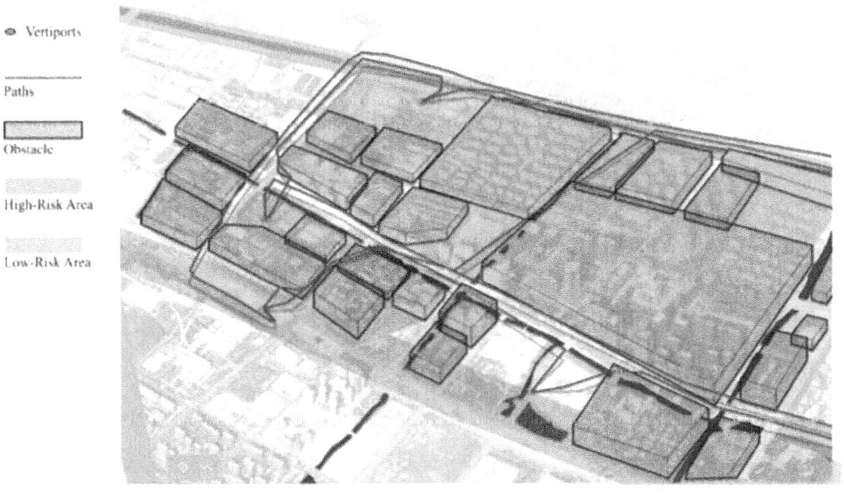

Figure 13.8 Land planning layout using a drone [61].

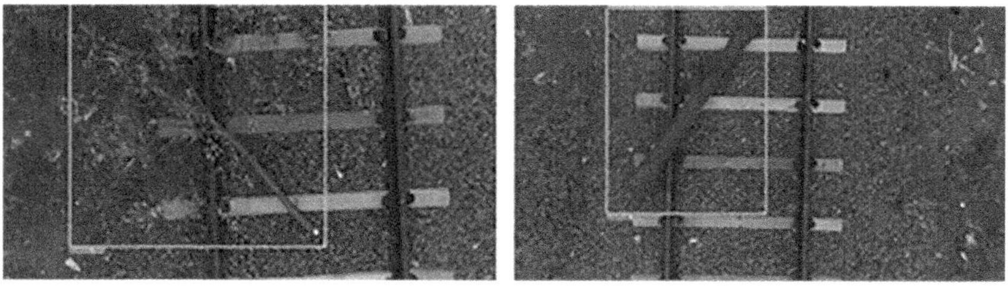

Figure 13.9 Inspection of railway track by a drone [55].

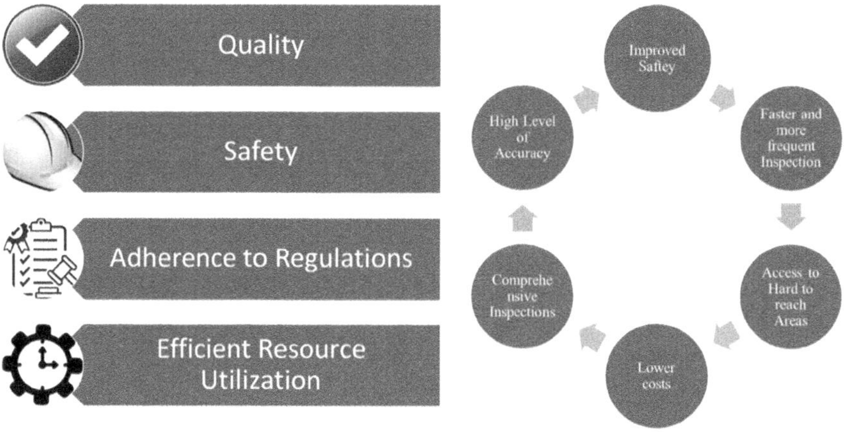

Figure 13.10 (a) Goal of construction monitoring and (b) benefits of drones in infrastructure inspection.

mapping, monitoring, detailed examination, and marketing [62]. Yuhan et al. have presented a method using drone-based ortho-image pairs to estimate real-time construction site elevation. By capturing low- and high-altitude ortho-image pairs, the method extends the baseline and adjusts spatial resolution for accurate elevation determination. Authorities confirmed the effectiveness of this method in excavation and facility layout, which depend on precise elevation data [63].

Conventional train track assessments are expensive, time consuming, and hazardous. They disrupt rail services in addition to requiring continuous availability of manpower on site, which poses serious safety issues [55] (Figure 13.9).

The purpose is to oversee and evaluate various phases of a construction project to ensure that it progresses according to the plans, specifications, and timeline (Figures 13.10 and 13.11).

Figure 13.11 Monitoring of a multi-storey construction site.

13.2.3 Smart cities and sustainable development

The 21st century is witnessing rapid urbanisation, placing increasing pressure on existing infrastructure and available resources. This has led to the development of smart cities—urban environments that leverage technology to enhance efficiency, sustainability, and quality of life. Within this transformation, the drone technology has emerged as a versatile platform for data collection, analysis, and intervention [64].

Urban Planning and Development: Drones provide high-resolution surveys and 3D modelling capabilities, which are important for urban planning. The captured data are utilised to create digital twin cities, which are virtual versions of actual cities which include master planning, real-time monitoring, and analysis [65,66].

Infrastructure and Traffic Management: Drone applications are important for planning and controlling infrastructure and traffic flow. They also optimise road networks and reduce congestion by keeping live data streaming. The data captured by drones are also used for security and monitoring, which improves the intelligence of traffic control systems [67,68].

Sustainability and Environmental Monitoring: Drones used for environmental monitoring identify various zones based on the data captured by various sensors [64,67]. By continuous observation of crop growth, soil health, and water management, users can take decision in the initial phase and act accordingly [69].

Emergency Services and Disaster Response: During rescue operations in emergency or natural disaster conditions, drones can improve public safety and response times by delivering vital relief to locations that are inaccessible with traditional vehicles [67,70].

Integration with New Technology: Emerging technologies like AI/ML/IoT enhance capabilities of drones and explore more applications in smart cities [71]. Drones are a perfect tool for to real-time data transfer and operation [69]. With innovative solutions for traffic control, urban planning, environmental sustainability, and emergency services, the drone technology is revolutionising the field of smart cities [72].

13.2.4 Aerial photography and videography for real estate marketing

With the use of aerial photography and videography, properties are observed from above, displaying their layout, characteristics, and surrounds, which makes the drone technology an essential asset which is better than the conventional photography. Property visibility via aerial photography and videography makes it possible for potential purchasers to comprehend a property's dimensions, design, and accessibility to neighbouring facilities. The expansive estates or waterfront vistas would particularly take benefits of aerial photography, which highlights closeness to parks, lakes, or golf courses, providing prospective purchasers with a better understanding of the lifestyle the property provides [73,74]. Figure 13.12 depicts the standard procedure for drone-based real estate applications.

- Establish goals, choose a course, and evaluate the weather and rules while planning a flight.
- Make flight plan, capture photo and video, and make sure that the data are of high quality.
- Process data including mapping pictures, generating 3D models, and editing video.

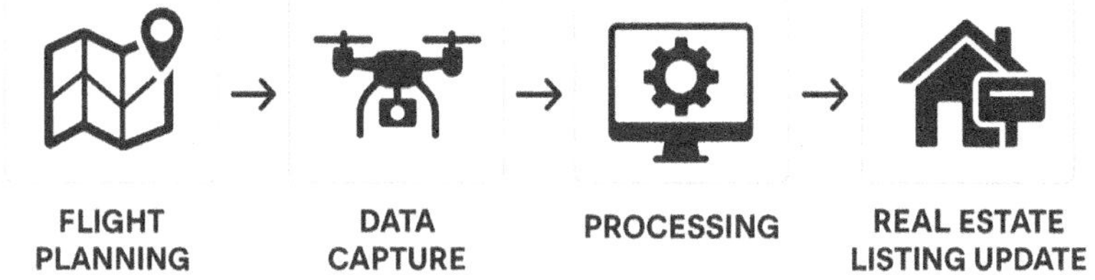

Figure 13.12 Typical drone-based workflow for real estate.

- Create visualisation with marketing-related visual materials such as flyovers, maps, and tours.
- Update real estate listings by adding images and incorporating them into advertisements and listings.

The unique features of a home like pools, gardens, or architectural elements that could be difficult to capture with the conventional photography are highlighted in aerial pictures. In a competitive market, consultants may set their houses apart from the competition and draw in more buyers by showcasing these attributes [75,76].

It has been demonstrated that using aerial photography and videography in real estate marketing greatly boosts interest and accelerates transactions. According to surveys, properties that use drone photography for marketing sell up to 68% faster than those using conventional methods. Using aerial photography, agents can generate a contemporary, polished look that draws in customers and makes an impact [77].

With drone's unique perspective, the data captured by drone improve property exposure, draw attention to standout characteristics, and boost buyer engagement by offering creative alternatives in real estate marketing [78,79].

13.3 TECHNIQUES AND TOOLS

13.3.1 Drone swarm technology

Drone swarm technology draws inspiration from nature by mimicking the behaviour of bird flocks or bee swarms, in which each unit operates independently while remaining synchronised through behavioural protocols and continuous communication. Swarms of drones can be deployed with coordinate patterns for the task like urban property assessments and construction site monitoring. Swarm has advantages over single drone mission like to provide simultaneous data collection, quicker coverage of complex terrains, and enhanced performance where high volumes of real-time information are required [80].

Major research areas in swarm technology are to develop algorithms that enable drones within the swarm to self-assign roles, avoid collisions, and adapt to changing obstacles such as pedestrians, vehicles, or temporary site changes. Moreover, researchers are working on synchronising swarm-generated data with BIM systems so that consultant may compare digital designs with actual site. Swarm data is used via simulator before being used in real-world [81,82].

However, deployment of large number of drone in swarms poses significant challenges, such as airspace coordination, constant communication between drones, GPS signal disruptions, and adhering data privacy and regulatory requirements. A successful operation is directly dependent on advanced AI and ML systems that can handle real-time coordination, sensor data and its response through autonomous decision taking capabilities.

13.3.2 Geospatial and mapping tools and techniques

Drones equipped with advanced sensors and GPS systems have been developed into useful, affordable instruments for producing high-resolution data about both infrastructures and natural areas. Various disciplines, including land surveying, construction design, environmental evaluation, agriculture, disaster response, and urban planning, depend heavily on these geospatial data and techniques. Photogrammetry is one of the most used UAV-based mapping methods. By capturing many overlapping photographs from various angles, 2D orthomosaic photographs and 3D models are generated using various software [83,84].

Using sophisticated image-processing techniques, drone-captured data can be converted into dense point clouds and 3D mesh models, which are essential for architectural design, construction management, and urban planning. The precise integration of these complex digital data representations with BIM enables a smooth connection between digital construction operations and actual site conditions [85,48]. In areas with dense vegetation and where exceptional precision is essential, LiDAR provides a robust alternative. Drones equipped with LiDAR systems emit rapid laser pulses to measure distances between the aircraft and the terrain, enabling the creation of detailed Digital Elevation Models (DEMs) and Digital Terrain Models (DTMs). This highly precise and accurate technology is extensively used in forestry management, flood precasting, mineral exploration, and archaeology, where the delivered results have centimetre-level accuracy [86]. Drones typically use GPS correction techniques called Real-Time Kinematic (RTK) or Post-Processing Kinematic (PPK) to increase geospatial data accuracy. These techniques significantly reduce positional errors of typical GPS, with sub-centimetre accuracy, which is essential for activities like cadastral mapping, infrastructure design, and land boundary delineation. Drone surveys are more accurate when RTK makes real-time corrections in flight and PPK refines the data later [87,88]. Drones are most widely equipped with Red, Green, and Blue (RGB) cameras as imaging systems. These optical sensors capture high-resolution data for general-purpose applications, which makes them ideal for aerial photography, visual inspection, and media production. UAVs with a gimbal control system deliver clear and smooth footage in dynamic flight circumstances. By detecting changes in the infrared radiation (heat) from objects, thermal cameras detect and visualise temperature differences in the observed scene. Thermal cameras offer crucial information in low-visibility or low-light situations, where RGB sensors are used for enhanced situational context [13].

13.4 CHALLENGES IN IMPLEMENTING THE DRONE TECHNOLOGY

The incorporation of the drone technology into the real estate industry involves many difficulties. It is needed to be addressed in order to guarantee the integration of the drone technology in secure, effective, and sustainable ways. The drone technology currently has several advantages and also limitations. Limitations like image processing, operational complications, safety concerns, legal restrictions, privacy and security challenges, and technological restraints were addressed in several publications [89].

13.4.1 Workforce readiness

To switch from old ways of working to get the things done with drones needs many upgrades from technology to workforce. The workers need to be motivated and trained to use these tools appropriately. To get the best results from drones, there's a big need for skilled drone pilots. Just knowing how to fly a commercial drone isn't enough to work with them. Pilots should not only learn to operate remotes, but they should also know how to use navigation tools, handle emergencies, follow airspace rules, keep people safe, and gather data [90].

Qualified drone pilots are key for safe, legal, and useful flying activities. As drones are becoming more popular, a greater number of skilled pilots are the need of the hour. This fact equally applies to areas like real estate that handle tons of data too. Drones can create 3D models of buildings, take clear aerial photos, and provide virtual tours of locations with technological advancements. Real estate agents who use this data may do better than their competitors with such vital information too. The way to train workforce is to develop a pathway to help people learn new skills and improve the ones they have related to the drone technology. Planning the courses for drone pilots and workshops on digital skills for real estate agents could be included in training programmes [91].

13.4.2 Technical challenges

It is easy for drones to take good pictures, but the critical tasks like recording weather, light, and purposeful camera specifications need the technology at the next level. Bad images can deteriorate the performance of computer vision programmes employed for analysis. The ability of drones to view structures is limited to its camera capacity. So, drones need to fly multiple times to get all pictures of big projects, which slows down data gathering. To train computer vision models, properly annotated data are required. Manually labelling photos for infrastructure analysis is really time consuming and requires much work. To make the model work well, uniquely and properly labelled data are the mandatory requirement [92,93].

The drones may collide with obstacles while flying, potentially affecting data collection. So, handling drone flight to avoid obstacles is very important for getting data safely. Big buildings can make it hard for drones to reach some parts because taller and closely oriented buildings may be hurdle for proper image capture [94].

The next big deal is privacy and security. While taking drone shoots of infrastructures, sometimes sensitive spots may be covered, breaking the privacy of residents and other community. The weight-carrying capacity of drones is also a limitation. Added sensors and cameras may affect the free fly of drones [95].

13.4.3 Operational challenges

For drone inspections to work, GPS must be accurate to provide positioning and also its communication must be reliable. Drone-based inspection may face problems due to these issues [96]. The battery life and hence limited flight time are challenges for drones to fly for long time and gather larger data [97–99].

Mechanical vibrations of drones on flight sometimes limit the data accuracy and may further mess up data collection, which can make images and sensors less accurate. The development and successful implementation of advanced sensors and stabilisation systems can help solve this problem [100]. The communication stability plays a key role in the usage of drones, and the ground control station should have a stable communication link. Technical problems can stop real-time control and data transfer [101].

The lack of widely available and publicly accessible datasets with adequate annotations may impede the advancement of drone inspection technologies and solutions to the problems encountered [95].

13.4.4 Data processing challenges

Data collection is followed by data processing. It includes data labelling and annotation appropriately for properly training ML models. The development of robust models supporting computer vision for accurate, fast, and acceptable results is the demand and a good opportunity. Creating strong computer vision models working up to acceptable results is still tough.

Development of good algorithms and parallel processing are the key areas to explore with the applications of big infrastructure projects. The system development for real-time analysis will also be required to look into drone pictures for managing infrastructure with smart algorithms and strong processing capacity. The developmental changes and deviation in timely infrastructure changes can be observed and kept track on it.

13.4.5 Real-world challenges

The presence of natural and unexpected uncertainties, calamities, and extreme weather conditions may hinder reliable detections. The damage detection of real-world infrastructure may deviate from the analytical capabilities of deep learning models, as these models requires consistent and extensive training datasets. Deep learning data models are less effective to handle data imbalances due to damaged data points in the infrastructure images recorded.

13.4.6 Privacy and surveillance concerns

A sensitive challenge to monitor properties is the people who are concerned. Drones working on monitoring the infrastructure of properties might infringe the rights of nearby property owners and tenants. However, drones can help protect property, following strict compliance with privacy regulations [102,103]. Using the drone technology in real estate, several problems can be addressed. Drones can be employed for property marketing, management, and security. New technologies can assist property security in the real estate industry. Drones with cameras can watch over properties for monitoring purpose.

13.5 POTENTIAL FUTURE PROSPECTS FOR THE DRONE TECHNOLOGY IN THE REAL ESTATE INDUSTRY

The development of big data tools, IoT, and AI is the big support to the UAV technology. These tools help UAVs to work on their own with very little human control and intervention. ML, the core component of AI, is playing a vital role in this development. It helps UAVs to adapt guided thinking and dataset-based actions. Novel deep learning methods are making UAVs better to study, analyse, and take corrective actions based on information data. For example, AT&T uses UAVs with deep learning to check towers and find faults in live HD video. Industrial Sky Works uses ML to look at 2D and 3D images and spot defects or unusual patterns [94].

Deep learning is the key algorithm to handle the raw data coming from UAV sensors. It finds, understands, and analyses patterns. Convolutional neural network (CNN) is a deep learning method that is very good in recognising and sorting pictures. It helps UAVs to follow patterns based on more accurate and curated data to take decisions, which makes its purpose more satisfactory [104]. Recent advancements allow UAVs to use advanced image analysis algorithms to capture images of buildings, bridges, and other structures with the help of cameras and sensors. This assists in checking structural damages and keeping an eye on the progress of work. The UAVs transmit the collected data and pictures to the ground control station

(GCS), enableing real-time structural health monitoring (SHM). The hat transforms and Hue, Saturation and Value (HSV) thresholding are the popular two image-processing methods that help find cracks in concrete surfaces faster and more accurately [105, 106,34].

The coming years should aim for research on improving the battery technology. Better batteries will allow UAVs to stay in the air for longer periods and travel over greater distances without the need for frequent charging or battery replacement [107,108]. This improvement is important for carrying out long-duration missions such as environmental monitoring, agricultural surveys, and infrastructure inspections. Further studies should also aim to create advanced inspection systems that can work automatically and deliver results in real time. These systems must be able to detect faults or irregularities with high accuracy while reducing the need for human supervision. The use of modern tools such as ultrasonic sensors, multispectral cameras, and improved image-processing methods can make this possible [109]. By combining these technologies, future UAV systems will be able to perform safer, faster, and more reliable inspections across various applications.

13.6 CONCLUSION

Employing the drone technology in the real estate segment has brought a noteworthy revolution through reforming property assessment, and marketing and monitoring approaches. This work presents different types of UAVs (ranging from nano, micro to larger, fixed-wing and rotary system drones) deployed based on precise functioning requirements. Drones are widely used in real estate applications such as land measurement, terrain modelling, structural examinations, and visually persuasive publicity content through real-time and aerial imagery. Incorporation of thermal imaging, photogrammetry, LiDAR, and GPS correction technologies like RTK and PPK has greatly enhanced the precision and usability of drone data. However, quite a few challenges persevere, namely, regulatory compliance, privacy concerns, considerable early investment, and, last but not the least, requirement of skilled drone operators. Technological growths in AI-driven analytics, immersive simulated tours, BIM integration, and imminent air mobility solutions recommend an energetic progression of drone use in real estate. To reveal the benefits of UAVs, synchronised efforts amid government, educational organisations, and industry players are essential. Establishing adaptive guidelines, enhancing specialised exercise, and nurturing incessant invention will be critical for fruitful and accountable implementation. Eventually, drones are redefining best practices and setting new scales in the real estate segments. For experts, embracing these developments will be vital to sustain keenness and bring into line with forthcoming industry trends.

REFERENCES

1. Papa, U. (2018). Introduction to unmanned aircraft systems (UAS). In: Kacprzyk, J. (ed) *Embedded Platforms for UAS Landing Path and Obstacle Detection. Studies in Systems, Decision and Control*, vol 136. pp. 1–11, Springer, Cham. https://doi.org/10.1007/978-3-319-73174-2_1.
2. Edulakanti, S. R., Ganguly, S. (2023). Review article: The emerging drone technology and the advancement of the Indian drone business industry. *The Journal of High Technology Management Research*, 34(2), 100464, ISSN 1047-8310. https://doi.org/10.1016/j.hitech.2023.100464.
3. Boccadoro, P., Striccoli, D., Grieco, L. A. (2021). An extensive survey on the Internet of Drones. *Ad Hoc Networks*, 122, 102600, ISSN 1570-8705. https://doi.org/10.1016/j.adhoc.2021.102600.
4. The Business Research Company. (2025). *Real estate drone services global market report 2025: By service (mapping and surveying, inspection, photography and filming), by platform (operator software, cloud-based), by customer type (commercial, residential) – Market size, trends, and global forecast 2025–2034.* The Business Research Company. https://www.thebusinessresearchcompany.com/report/real-estate-drone-services-global-market-report.
5. Nwaogu, J. M., Yang, Y., Chan, A. P.C., Chi, H.-L. (2023). Application of drones in the architecture, engineering, and construction (AEC) industry. *Automation in Construction*, 150, 104827, ISSN 0926-5805. https://doi.org/10.1016/j.autcon.2023.104827.
6. Gupta, S. G., Ghonge, M., Ghonge, M., Jawandhiya, P. M. (2013). Review of unmanned aircraft system (UAS). *International Journal of Advanced Research in Computer Engineering & Technology (IJARCET)*, 2(4). Available at SSRN: https://ssrn.com/abstract=3451039 or https://dx. doi.org/10.2139/ssrn.3451039
7. Hassanalian, M., Abdelkefi, A. (2017). Classifications, applications, and design challenges of drones: A review. *Progress in Aerospace Sciences*, 91, 99–131, ISSN 0376-0421. https://doi.org/10.1016/j.paerosci.2017.04.003.
8. Hossain, R. (2022). A short review of the drone technology. *International Journal of Mechatronics and Manufacturing Technology*, 7(2), 53–68.
9. Merkert, R., Bushell, J. (2020). Managing the drone revolution: A systematic literature review into the current use of airborne drones and future strategic directions for their effective control. *Journal of Air Transport Management*,89, 101929.
10. González-Jorge, H., Martínez-Sánchez, J., Bueno, M., Arias, A. P. (2017). Unmanned aerial systems for civil applications: A review. *Drones*, 1(1), 2. https://doi.org/10.3390/drones1010002.
11. Sabour, M. H., Jafary, P., Nematiyan, S. (2023). Applications and classifications of unmanned aerial vehicles: A literature review with focus on multi-rotors. *The Aeronautical Journal*, 127(1309), 466–490. https://doi.org/10.1017/aer.2022.75.
12. Rejeb, A., Abdollahi, A., Rejeb, K., Treiblmaier, H. (2022). Drones in agriculture: A review and bibliometric analysis. *Computers and Electronics in Agriculture*, 198, 107017, ISSN 0168-1699. https://doi.org/10.1016/j.compag.2022.107017.

13. Aasen, H., Burkart, A., Bolten, A., Bareth, G. (2015). Generating 3D hyperspectral information with lightweight UAV snapshot cameras for vegetation monitoring: From camera calibration to quality assurance. *ISPRS Journal of Photogrammetry and Remote Sensing*, 108, 245–259, ISSN 0924-2716. https://doi.org/10.1016/j.isprsjprs.2015.08.002.
14. Mahroof, K., Omar, A., Rana, N. P., Sivarajah, U., Weerakkody, V. (2021). Drone as a Service (DaaS) in promoting cleaner agricultural production and Circular Economy for ethical Sustainable Supply Chain development. *Journal of Cleaner Production*, 287, 125522, ISSN 0959-6526. https://doi.org/10.1016/j.jclepro.2020.125522.
15. Torres-González, A., Capitán, J., Cunha, R., Ollero, A., Mademlis, I. (2018). A multidrone approach for autonomous cinematography planning. In: Ollero, A., Sanfeliu, A., Montano, L., Lau, N., Cardeira, C. (eds) *ROBOT 2017: Third Iberian Robotics Conference. ROBOT 2017. Advances in Intelligent Systems and Computing*, vol. 693, pp. 337–349. Springer, Cham. https://doi.org/10.1007/978-3-319-70833-1_28
16. Mademlis, I., Torres-González, A., Capitán, J. et al. (2023). A multiple-UAV architecture for autonomous media production. *Multimedia Tools and Applications*, 82, 1905–1934. https://doi.org/10.1007/s11042-022-13319-8.
17. Pueyo, P., Dendarieta, J., Montijano, E., Murillo, A. C., & Schwager, M. (2024). Cinempc: A fully autonomous drone cinematography system incorporating zoom, focus, pose, and scene composition. *IEEE Transactions on Robotics*, 40, 1740–1757.
18. Park, S., Kim, H. T., Lee, S., Joo, H., Kim, H. (2021). Survey on anti-drone systems: Components, designs, and challenges. *IEEE Access*, 9, 42635–42659. https://doi.org/10.1109/ACCESS.2021.3065926.
19. Chiper, F.-L., Martian, A., Vladeanu, C., Marghescu, I., Craciunescu, R., Fratu, O. (2022). Drone detection and defense systems: Survey and a software-defined radio-based solution. *Sensors*, 22(4), 1453. https://doi.org/10.3390/s22041453
20. Garg, N., Roy, N. (2020). Enabling self-defense in small drones. *HotMobile'20: Proceedings of the 21st International Workshop on Mobile Computing Systems and Applications.* pp 15-20, Austin, TX. https://doi.org/10.1145/3376897.3377866
21. Guitton, M. J. (2021). Fighting the locusts: Implementing military countermeasures against drones and drone swarms. *Scandinavian Journal of Military Studies*, 4.1, 26–36.
22. Delmerico, J., Cieslewski, T., Rebecq, H., Faessler, M., Scaramuzza, D. (2019). Are we ready for autonomous drone racing? The UZH-FPV drone racing dataset. *2019 International Conference on Robotics and Automation (ICRA)*, Montreal, QC, pp. 6713–6719. https://doi.org/10.1109/ICRA.2019.8793887.
23. Alsoliman, A., Rigoni, G., Callegaro, D., Levorato, M., Pinotti, C. M., Conti, M. (2023). Intrusion detection framework for invasive FPV drones using video streaming characteristics. *ACM Transactions on Cyber-Physical Systems*, 7, 2, Article 12, 29 pages. https://doi.org/10.1145/3579999.
24. Bursuc, A., et al. (2024). Overview on sea drones evolution and their use in modern warfare. *Land Forces Academy Review*, 29(2), 195–209. https://doi.org/10.2478/raft-2024-0021.

25. Mondal, K., Banerjee, T., Panja, A. (2019). Autonomous underwater vehicles: recent developments and future prospects. *International Journal for Research in Applied Science & Engineering Technology (IJRASET)*, 7(XI), 215–222.
26. Roberts, G. N., Sutton, R. (2006). Advances in unmanned marine vehicles. *Institution of Engineering and Technology*, 1(12), 69, 28–44.
27. Shakhatreh, H. et al. (2019). Unmanned aerial vehicles (UAVs): A survey on civil applications and key research challenges. *IEEE Access*, 7, 48572–48634. https://doi.org/10.1109/ACCESS.2019.2909530.
28. Yahya, M. Y., Shun, W. P., Md Yassin, A., Omar, R. (2021). The challenges of drone applicationâ in the construction industry. *Journal of Technology Management and Business*, 8(1), 20–27.
29. Siniak, N., Kauko, T., Shavrov, S., Marina, N. (2020). The impact of proptech on real estate industry growth. *IOP Conference Series: Materials Science and Engineering*, 869(6), 062041. https://iopscience.iop.org/article/10.1088/1757-899X/869/6/062041/meta.
30. Naeem, N., Rana, I. A., Nasir, A. R. (2023). Digital real estate: A review of the technologies and tools transforming the industry and society. *Smart Construction and Sustainable Cities*, 1, 15. https://doi.org/10.1007/s44268-023-00016-0.
31. Newman, C., Edwards, D., Martek, I., Lai, J., Thwala, W. D., Rillie, I. (2021), Industry 4.0 deployment in the construction industry: A bibliometric literature review and UK-based case study. *Smart and Sustainable Built Environment*, 10(4), 557–580.
32. Matus, T., Mésároš, P. (2019). Utilizing drone technology in the civil engineering. *SSP - Journal of Civil Engineering*, 14(1), 27–37.
33. Dastgheibifard, S., Mahsa A. (2018). A review on potential applications of unmanned aerial vehicle for construction industry. *Sustainable Structures and Materials, An International Journal*, 1(2), 44–53. https://www.researchgate.net/publication/326266278_A_Review_on_Potential_Applications_of_Unmanned_Aerial_Vehicle_for_Construction_Industry#fullTextFileContent.
34. Dupont, Q., Chua, D., Tashrif, A., Abbott, E. (2017). Potential applications of UAV along the construction's value chain. *Procedia Engineering*, 182, 165–173. https://doi.org/10.1016/j.proeng.2017.03.155.
35. Collier, P. (2002). The impact on topographic mapping of developments in land and air survey: 1900–1939. *Cartography and Geographic Information Science*, 29(3), 155–174.
36. Usery, E. L., Varanka, D. E., Davis, L. R. (2018). Topographic mapping evolution: From field and photographically collected data to GIS production and linked open data. *The Cartographic Journal*, 55(4), 378–390.
37. Ahmed, R., Mahmud, K. H. (2022). Potentiality of high-resolution topographic survey using unmanned aerial vehicle in Bangladesh. *Remote Sensing Applications: Society and Environment*, 26, 100729, ISSN 2352-9385. https://doi.org/10.1016/j.rsase.2022.100729.
38. Du, M., Li, H., Roshanianfard, A. (2022). Design and experimental study on an innovative UAV-LiDAR topographic mapping system for precision land levelling. *Drones*, 6(12), 403.
39. McNeil, B., Snow, C. (2016). The truth about drones in mapping and surveying. *Skylogic Research*, 200, 6.

40. Liang, X., Si, F., Jiang, Y., Zhou, H., Zhan, K., Chang, Z., Qiu, X., Yang, D. S. (2021). First high-resolution tropospheric NO_2 observations from the Ultraviolet Visible Hyperspectral Imaging Spectrometer (UVHIS). *Atmospheric Measurement Techniques*, 14, 435–454. https://doi.org/10.5194/amt-14-435-2021.
41. Park, J. W., Yeom, D. J. (2021). Method for establishing ground control points to realize UAV-based precision digital maps of earthwork sites. *Journal of Asian Architecture and Building Engineering*, 21(1), 110–119. https://doi.org/10.1080/13467581.2020.1869023
42. Jiang, Y., Huang, Y., Liu, J., Li, D., Li, S., Nie, W., Chung, I.-H. (2022). Automatic volume calculation and mapping of construction and demolition debris using drones, deep learning, and GIS. *Drones*, 6(10), 279. https://doi.org/10.3390/drones6100279
43. Nixon, A. (2019). Best Drones For Agriculture 2019: The Ultimate Buyer's Guide. Retrieved August 3, 2019, from https://bestdroneforthejob.com/drone-buying-guides/agriculture-drone-buyers-guide/
44. Ornes, S. (2014). Mapping the lost megalopolis. *Proceedings of the National Academy of Sciences of the United States of America*, 111(43), 15283–15285. https://doi.org/10.1073/pnas.1417601111.
45. Gutiérrez, G., Searcy, M. T. (2016). Introduction to the UAV special edition. *The SAA Archaeological Record, Special Issue Drones in Archaeology*, 16(2), 6–9.
46. Matsimbe, J., Mdolo, W., Kapachika, C., Musonda, I., Dinka, M. (2022). Comparative utilization of drone technology vs. traditional methods in open pit stockpile volumetric computation: A case of njuli quarry, Malawi. *Frontiers in Built Environment*, 8, 1037487. https://doi.org/10.3389/fbuil.2022.1037487.
47. Fitzpatrick, B. (2015). *Unmanned aerial systems for surveying and mapping: Cost comparison of UAS versus traditional methods of data acquisition*, Master of Science. University of South Carolina, Carolina.
48. Nex, F., Remondino, F. (2014). UAV for 3D mapping applications: A review. *Applied Geomatics*, 6(1), 1–15. https://doi.org/10.1007/s12518-013-0120-x.
49. Adams, W., Brooks, T., Meneely, J., Millar, R. (2021). Drones in construction. In: *Proceedings of the CiTA BIM Gathering*, Dublin, September 21–23, 2021.
50. Adhikari, S. (2023). *Usage of drone for building facade inspection*. Master's Theses. 7. https://epublications.marquette.edu/theses_open/754
51. Gilbert, M. (2017). Drones & AI: The Next Phase of Automation. Accessed February 22, 2026. [Online]. Available: https://about.att.com/innovationblog/drones_automation
52. Mesároš, P., Tkáč, M., Mandičák, T. (2019). Effective connection of terrestrial laser scanner and unmanned aerial vehicle for as built measurement of buildings. *14th International Conference Organization, Technology and Management in Construction*, Zagreb, September 04–07, 2019, pp. 518–532.
53. Fernandez Galarreta, J., Kerle, N., Gerke, M. (2015). UAV-based urban structural damage assessment using object-based image analysis and semantic reasoning. *Natural Hazards and Earth System Sciences*, 15, 1087–1101. https://doi.org/10.5194/nhess-15-1087-2015.

54. Freimuth, H., König, M. (2015). Generation of waypoints for UAV-assisted progress monitoring and acceptance of construction work. In: *15th International Conference on Construction Applications of Virtual Reality (CONVR) 2015*, Alberta, CA, pp. 77–86.
55. Aela, P., Chi, H.-L., Fares, A., Zayed, T., Kim, M. (2024). UAV-based studies in railway infrastructure monitoring. *Automation in Construction*, 167, 105714, ISSN 0926-5805. https://doi.org/10.1016/j.autcon.2024.105714.
56. Gheisari, M., Esmaeili, B. (2019). Applications and requirements of unmanned aerial systems (UASs) for construction safety. *Safety Science*, 118, 230–240, ISSN 0925-7535. https://doi.org/10.1016/j.ssci.2019.05.015.
57. PG & E. (2016). Testing Safety Drones to Inspect Electric and Gas Infrastructure. Accessed: February 22, 2026. [Online]. https://investor.pgecorp.com/news-events/press-releases/press-release-details/2016/PGE-Testing-Safety-Drones-to-Inspect-Electric-and-Gas-Infrastructure/default.aspx.
58. Kim, S., Kim, S. (2018). Opportunities for construction site monitoring by adopting first personal view (FPV) of a drone. *Smart Structure Systems*, 21, 139–149.
59. Al-Rimawi, T., Nadler, M. (2025). Leveraging smart city technologies for enhanced real estate development: An integrative review. *Smart Cities*, 8(1), 10. https://doi.org/10.3390/smartcities8010010.
60. Xie, F., Lin, Z., Gui, D., Lin, H. (2012). Study on construction of 3d building based on UAV images. *ISPRS - International Archives of the Photogrammetry, Remote Sensing and Spatial Information Sciences*, XXXIX-B1, 469–473. https://isprs-archives.copernicus.org/articles/XXXIX-B1/469/2012/.
61. He, X., He, F., Li, L., Zhang, L., Xiao, G. (2022). A route network planning method for urban air delivery. *Transportation Research Part E: Logistics and Transportation Review*, 166, 102872. https://doi.org/10.1016/j.tre.2022.102872.
62. Liu, P., Chen, A., Huang, Y.-N., Han, J.-Y., Lai, J.-S., Kang, S.-C., Wu, T.-H. R., Wen, M.-C., Tsai, M.-H. (2014). A review of rotorcraft Unmanned Aerial Vehicle (UAV) developments and applications in civil engineering. *Smart Structures and Systems*, 13, 1065–1094. https://d1wqtxts1xzle7.cloudfront.net/92456579/sss.2014.13.6.106520221015-1-1mmxj2d-libre.pdf?1665802931=&response-content-disposition=inline%3B+filename%3DA_review_of_rotorcraft_Unmanned_Aerial_V.pdf&Expires=1779370379&Signature=L5u4bXbBTsTOBt~H0gjUfq~nOqbWNjpCBbyulAsXxiZS9ST0fv1ShUJ~NewqECMO45Y572XUsA4eiuMEGCpo9idMNUFpDmdH~1aPeCC8ZFa6I-uI7vAJPDLXmENwItt8WH1HqX2i8oqLECSZfaqK-HRl8J0Rli9eonAczG100Tts8y4YsmruhYPW0~ZcGVdVBHhAZjG-Kk1EdHR9~UpyOE3iTc9uVBnNUVg6VHD3Ke7utqfZbQfz8d~dCV0F7d3AEB4~qtEb1I9EmG3q7J8T1ZEcA8PnxwqQatd~W9TFDAQqWqRScpujwdm8BWD5prAgp2X5zGgmf5YuZGphq4HDJg__&Key-Pair-Id=APKAJLOHF5GGSLRBV4ZA.
63. Jiang, Y., Bai, Y. (2020). Determination of construction site elevations using drone technology. *Construction Research Congress 2020: Computer Applications*, Tempe, pp. 296–305. https://doi.org/10.1061/9780784482865.032.
64. Muhmad Kamarulzaman, A. M., Wan Mohd Jaafar, W. S., Mohd Said, M. N., Saad, S. N. M., Mohan, M. (2023). UAV implementations in urban planning and related sectors of rapidly developing nations: A review and future perspectives for Malaysia. *Remote Sensing*, 15(11), 2845. https://doi.org/10.3390/rs15112845

65. Yıldız, S., Kivrak, S., Arslan, G. (2021). Using drone technologies for construction project management: A narrative review. *Journal of Construction Engineering, Management & Innovation*, 4, 229–244. https://doi.org/10.31462/jcemi.2021.04229244.
66. Salem, T., Dragomir, M., Chatelet, E. (2024). Strategic integration of drone technology and digital twins for optimal construction project management. *Applied Sciences*, 14(11), 4787. https://doi.org/10.3390/app14114787.
67. Alkaabi, K., Fawair, E., Rhman, A. (2022). Drones applications for smart cities: Monitoring palm trees and street lights. *Open Geosciences*, 14(1), 1650–1666. https://doi.org/10.1515/geo-2022-0447.
68. Nextech. Drones and the Smart City. https://nextech.online/smart-city/
69. Katekar, V., Cheruku, J. K. (2022). The application of drone technology for sustainable agriculture in India. *Current Agricultural Research*, 10(3). https://doi.org/10.12944/CARJ.10.3.19.
70. Rathee, G., Kumar, A., Kerrache, C. A., Iqbal, R. (2022). A trust-based mechanism for drones in smart cities, 4, 255–264. *IET Digital Library*. https://digital-library.theiet.org/doi/full/10.1049/smc2.12039.
71. Santosh Kumar, R., Prakash, K., Suryanarayana, G. (2023). Drones Enable IoT Applications for Smart Cities. *Drone Technology: Future Trends and Practical Applications*, 207–241. Wiley Online Library. https://onlinelibrary.wiley.com/doi/abs/10.1002/9781394168002.ch9.
72. Mukati, N., Thakur, M. S., Deepak Goud, Er., Abhishek Chourasiya, Er. (2023). *Role of Drone Technology for Smart City Applications Using 5G*. IJRTI, 8(9), 342–346. https://ijrti.org/papers/IJRTI2309043.pdf.
73. Matterport. (2024). Real Estate Drone Photography: Benefits & Statistics. https://matterport.com/learn/real-estate-photography/drone. Accessed: February 22, 2026. [Online].
74. Parkers Properties. (2024). Elevating Real Estate Marketing with Drone Photography and Videography. https://www.parkersproperties.co.uk/guides/selling/elevating-real-estate-marketing-with-drone-photography-and-videography/3. Accessed: February 22, 2026. [Online].
75. Hommati. (n.d.). Benefits of Using Aerial Photography and Video in Real Estate Marketing. https://www.hommati.com/blog/benefits-of-using-aerial-photography-and-video-in-real-estate-marketing4. Accessed: February 22, 2026. [Online].
76. Luxury Presence. (n.d.). The Ultimate Guide to Real Estate Drone Photography in 2026. https://www.luxurypresence.com/blogs/level-up-your-listings-with-drones/7. Accessed: February 22, 2026. [Online].
77. Twist Tours. (n.d.). Aerial Photography is The Next Level of Real Estate Marketing. https://www.twisttours.com/aerial-photography-is-the-next-level-of-real-estate-marketing/8. Accessed: February 22, 2026. [Online].
78. Luxury Locations. (n.d.). Harnessing the Power of Aerial Photography & Videography in Real Estate Marketing. https://www.luxurylocations.com/news/article/144/harnessing-the-power-of-aerial-photography-videography-in-real-estate-marketing. Accessed: February 22, 2026 [Online].
79. Real Estate Photography. (n.d.). Benefits of Drone Real Estate Photography For Your Property. https://v1realestatephotography.com/benefits-of-drones-in-real-estate-photography/5. Accessed: February 22, 2026. [Online].

80. Kravchuk, S. O., Afanasieva, L. (2019). Formation of a wireless communication system based on a swarm of unmanned aerial vehicles. *Information and Telecommunication Sciences*, 1, 11–18, ISSN 2411-2976.
81. Saeed, R. A., Omri, M., Abdel-Khalek, S. et al. (2022). Optimal path planning for drones based on swarm intelligence algorithm. *Neural Computing & Applications*, 34, 10133–10155. https://doi.org/10.1007/s00521-022-06998-9.
82. Volk, R., Stengel, J., Schultmann, F. (2014). Building information modeling (BIM) for existing buildings—Literature review and future needs. *Automation in Construction*, 38, 109–127, ISSN 0926-5805.
83. Quamar, M. M., Al-Ramadan, B. M., Khan, K. A., Shafiullah, M., El-Ferik, S. (2023). Advancements and applications of drone-integrated geographic information system technology - A review. *Remote Sensing*, 15, 5039.
84. Budiharto, W., Irwansyah, E., Suroso, J. S., Chowanda, A., Ngarianto, H., Gunawan, A. A. (2021). Mapping and 3D modelling using quadrotor drone and GIS software. *Journal of Big Data*, 8(48), 1–12, ISSN 2196-1115. https://doi.org/10.1186/s40537-021-00436-8
85. Colomina, I., Molina, P. (2014). Unmanned aerial systems for photogrammetry and remote sensing: A review. *ISPRS Journal of Photogrammetry and Remote Sensing*, 92, 79–97.
86. Wallace, L., Lucieer, A., Watson, C., Turner, D. (2012). Development of a UAV–LiDAR system with application to forest inventory. *Remote Sensing*, 4(6), 1519–1543.
87. Song, X., Yang, G., Yang, C., Wang, J., Cui, B. (2017). Spatial variability analysis of within-field winter wheat nitrogen and grain quality using canopy fluorescence sensor measurements. *Remote Sensing*, 9(3), .237. https://doi.org/10.3390/rs9030237.
88. Turner, D., Lucieer, A., Watson, C. (2012). An automated technique for generating georectified mosaics from ultra-high resolution unmanned aerial vehicle (UAV) imagery, based on structure from motion (SfM) point clouds. *Remote Sensing*, 4(5), 1392–1410. https://doi.org/10.3390/rs4051392.
89. Sony, S., Dunphy, K., Sadhu, A., Capretz, M. (2021). A systematic review of convolutional neural network-based structural condition assessment techniques. *Engineering Structures*, 226, 111347.
90. Rakha, T., Gorodetsky, A. (2018). Review of Unmanned Aerial System (UAS) applications in the built environment: Towards automated building inspection procedures using drones. *Automation in Construction*, 93, 252–264.
91. Aerial Southeast. How Drone Photography Can Elevate Your Marketing Campaign. https://aerialsoutheast.com/how-drone-photography-can-elevate-your-marketing-campaign/. Accessed: April 12, 2026. [Online].
92. Deng, J., Singh, A., Zhou, Y., Lu, Y., Lee, V. C. S. (2022). Review on computer vision-based crack detection and quantification methodologies for civil structures. *Construction and Building Materials*, 356, 129238.
93. Spencer Jr, B. F., Hoskere, V., Narazaki, Y. (2019). Advances in computer vision-based civil infrastructure inspection and monitoring. *Engineering*, 5(2), 199–222.
94. Guan, S., Zhu, Z., Wang, G. (2022). A review on UAV-based remote sensing technologies for construction and civil applications. *Drones*, 6(5), 117.
95. Shakhatreh, H., Sawalmeh, A. H., Al-Fuqaha, A., Dou, Z., Almaita, E., Khalil, I., Guizani, M. (2019). Unmanned aerial vehicles (UAVs): A survey on civil applications and key research challenges. *IEEE Access*, 7, 48572–48634.

96. Ai, D., Jiang, G., Lam, S. K., He, P., Li, C. (2023). Computer vision framework for crack detection of civil infrastructure—A review. *Engineering Applications of Artificial Intelligence*, 117, 105478.
97. Sivakumar, M., Tyj, N. M. (2021). A literature survey of unmanned aerial vehicle usage for civil applications. *Journal of Aerospace Technology and Management*, 13, e4021.
98. Fan, J., Ala Saadeghvaziri, M. (2019). Applications of drones in infrastructures: Challenges and opportunities. *International Journal of Mechanical and Mechatronics Engineering*, 13(10), 649–655.
99. Outay, F., Mengash, H. A., Adnan, M. (2020). Applications of unmanned aerial vehicle (UAV) in road safety, traffic and highway infrastructure management: recent advances and challenges. *Transportation Research Part A: Policy And Practice*, 141, 116–129.
100. Gheisari, M., Irizarry, J., Walker, B. N. (2014). UAS4 SAFETY: The potential of unmanned aerial systems for construction safety applications. Construction Research Congress May 2014, Atlanta, GA, pp. 18011810.
101. Zhang, S., Bogus, S. M., Lippitt, C. D., Kamat, V., Lee, S. (2022). Implementing remote-sensing methodologies for construction research: An unoccupied airborne system perspective. *Journal of Construction Engineering and Management*, 148(9), 03122005.
102. Duque, L., Seo, J., Wacker, J. (2018). Synthesis of unmanned aerial vehicle applications for infrastructures. *Journal of Performance of Constructed Facilities*, 32(4), 04018046.
103. Beata Stępień-Załucka, D. (2022). Real estate video surveillance, and neighbourhood right to privacy – The potential area of normative impact from the perspective of the polish law. *Journal of Intelligent & Robotic Systems*, 106, 14.
104. York, D. D., Al-Bayati, A. J., Al-Shabbani, Z. Y. (2020). Potential applications of uav within the construction industry and the challenges limiting implementation. In Grau, D., Tang, P., El Asmar, M. (eds) *Construction Research Congress 2020: Project Management and Controls, Materials, and Contracts*, pp. 31–39. https://doi.org/10.1061/9780784482889.004
105. Caballero-Martin, D., Lopez-Guede, J. M., Estevez, J., Graña, M. (2024). Artificial intelligence applied to drone control: A state of the art. *Drones*, 8(7), 296. https://doi.org/10.3390/drones8070296.
106. Sankarasrinivasan, S., Balasubramanian, E., Karthik, K., Chandrasekar, U., Gupta, R. (2015). Health monitoring of civil structures with integrated UAV and image processing system. *Procedia Computer Science*, 54, 508515.
107. Vishwath, N. A., Yadav, A. R., Mehta, D., Belani, J., Chauhan, R. R. (2022). A guide to novice for proper selection of the components of drone for specific applications. *Materials Today: Proceedings*, 65, 3617–3622.
108. Vishwath, N. A., Yadav, A. R., Mehra, V., Patel, V., Patel, M. K. B., Sondagar, V., Sharma, A., Singh, G. (2025). Computational analysis and designing of flight controllers to improve the drone performance for a novice pilot. *International Journal on Interactive Design and Manufacturing (IJIDeM)*, 19(2), 1395–1408.
109. Hatoum, M. B., Nassereddine, H. (2022). The use of drones in the construction industry: Applications and implementation. *ISARC. Proceedings of the International Symposium on Automation and Robotics in Construction*, Vol. 39, Bogotá, Colombia, July 13–15, 2022. IAARC Publications.

Chapter 14

Application of UAVs in industrial manufacturing and transportation

Ujjwal Singh Katiyar, Arpit Mishra, Bhavya, and Rahul Johari

14.1 INTRODUCTION

The manufacturing and transport industries of today are undergoing a complete change due to the influence of global markets having more complex operations with a greater demand for sustainability. The two sectors are also at the crossroads with the problem of a lack of manpower that has become endemic, safety issues getting worse, higher energy needs, and supply chains exposed to the effects of globalization. Usually, traditional land-based systems of automation are of limited flexibility and are even reluctant to adapt especially in large and dangerous places. Therefore, the trend is moving towards the use of aerial automation as companies are transforming to achieve productivity, precision, and real-time monitoring.

One of the most influential automation tools in the smart system revolution is unmanned aerial vehicles (UAVs). The integration of AI, sensors, edge computing, and high-speed connectivity has significantly changed the role of these flying machines from surveillance only to the core of Industry 4.0 and 5.0 concepts. The application of drones, which are equipped with various technologies, has expanded from automated inspection, predictive maintenance, real-time inventory monitoring, and autonomous logistics to those in which the level of human intervention is very low or zero. The use of UAVs in data-driven, flexible, and scalable solutions is leading not only to the most basic change of industrial processes but also to the opening of new paths for the human-friendly, adaptable, and sustainable automation.

14.1.1 Evolution of UAVs in industrial applications

There are four stages in the industrial development of UAVs, with each stage distinguished by advances in navigation, sensing, and computational intelligence:

- 1980s–1990s: Early industrial UAVs found their application in monitoring oil pipelines, ecological monitoring, and large-scale mapping. The first UAVs were barely autonomous, had limited endurance, and

DOI: 10.1201/9781003598350-18

were under remote control, thus being very inefficient for continuous industrial use.

- 2000s: Improvements in global positioning system (GPS), inertial navigation systems (INS), and early artificial intelligence (AI) algorithms allowed semi-autonomous UAVs to conduct infrastructure inspection, industrial safety monitoring, and aerial mapping with greater precision.
- 2010s: Technologies of computer vision, hyperspectral sensing, and Light Detection and Ranging (LiDAR) enabled by AI revolutionized the UAV technology and enabled live monitoring of factories, autonomous defect inspection, and predictive maintenance. AI-powered UAVs were used in aerospace, precision manufacturing, and automotive applications for logistics management and autonomous quality inspection.
- 2020s (Present Time): Utilization of 6G networks, quantum AI computing, and blockchain-based UAV fleet management has enabled UAVs to be completely autonomous, enabling real-time industrial logistics, smart factory automation, and highly precise aerial inspections with a bare minimum of human involvement [1].

Continuous improvement in UAVs has eliminated operation inefficiencies, optimized industrial processes, and improved overall reliability (Figure 14.1).

14.1.2 Role of UAVs in Industry 4.0 and 5.0

UAVs in Industry 4.0 revolutionized manufacturing, logistics, and analytics by integrating AI-driven automation, real-time data fusion, and cyber-physical systems. Industry 4.0 made UAVs capable of autonomous quality inspection, predictive maintenance, and airborne logistics management with high precision and low latency. The potential of UAVs to interface with Industrial Internet of Things (IIoT), cloud computing, and digital twin

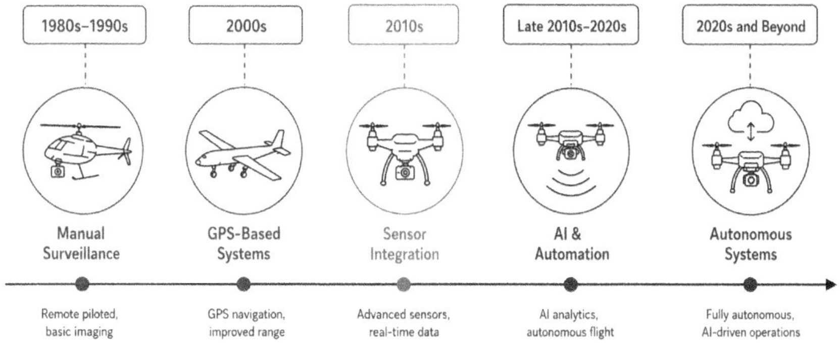

Figure 14.1 Evolution in industrial application.

systems has provided additional intelligence in decision-making process and accuracy that has significantly reduced the number of manufacturing faults and wastage of raw materials.

Whereas the primary objective of Industry 4.0 was to streamline the efficiency of automated systems, through Industry 5.0, human–UAV cooperation is possible, automation takes a different approach with a focus on sustainability, and robotics are characterized by flexibility. Human centricity in Industry 5.0 means that UAVs, which has become part of factory layouts, are available to human operators, so that, for example, they can better manage hazardous materials, easily perform complex assembly tasks, and bring ergonomic design improvement.

The major advancements in Industry 5.0 UAV applications are as follows:

- Human–AI Collaborative UAV Systems: AI-enabled cognitive UAVs operate as brain assistants for human workers in high-precision assembly, process monitoring, and real-time hazard analysis.
- Green UAV Technologies: Hydrogen fuel cell–powered UAVs are 300% more energy efficient, which results in a significant drop in industrial carbon emissions and the need for other traditional sources of energy.
- Quantum AI–Based UAV Logistics: The use of quantum-supported AI algorithms makes drone delivery of goods, fast logistics, and on-demand stock oversight possible, and thus, the supply chain inefficiencies are reduced by 40% [2].

The transition from Industry 4.0 to 5.0 has seen the change from a single automated flight that controls drones to smart connected systems that manage drones and can learn, adapt, and react to real-time changes in production environments and cooperate with humans.

14.1.3 Market growth and adoption trends

One of major trends in the global drone market is the industrial adoption, which is growing at a yearly rate of 15.2% in last few years. It is estimated that this trend will persist, and by 2030, the global industrial drone market will be worth more than $18 billion, for which the primary reason would be the demand for services in the fields of production, logistics, and automated warehouse management.

Taking the decision to use a specific technology is one of the factors that profoundly affect the presence of tech products on the market, and therefore, is tightly intertwined with the tech market:

- Asia-Pacific is at present the most advanced area when it comes to the industrial drone sector. It is a region with the largest number of UAV deployments (45% of the total global UAV deployments). Application of the next main areas would be smart factories and rapid logistics for the development of the area.

- The commitment to use the drone technology for supply chain automation has a far reach that 85% of the Fortune 500 companies have brought it about through technology-based automation in their supply chain operations.
- The production efficiency has been raised by 23% in smart factories with the help of the drone technology, and this is the major factor that decreases human involvement in dangerous and repetitive industrial tasks [3].

The turning point of the industrial UAV sector that results from AI-powered UAV navigation, blockchain-secured UAV air traffic control, and 6G-based UAV-to-machine communications is the main theme of almost all reports in the field of automation.

14.1.3.1 Key performance metrics of industrial UAVs

The UAV efficiency in commercial applications is verified along with the key performance indicators (KPIs) of their effectiveness, accuracy, and practicability, which are:

- Autonomy Level: The capability of UAVs to carry out tasks on their own without much direct human intervention is assumed to be a use of AI for better flight, object recognition, and process adjustment in real time.
- Energy Efficiency: The use of solar modules or hydrogen fuel cell–powered drones can keep a drone in the air for 15 hours non-stop with a minimal use of conventional energy source.
- Defect Detection Accuracy: The application of AI in drones allows them to carry out precision tests with 98.6% of flaw accuracy at sub-millimeter level, which is used for quick production flaw detections.
- Inventory Optimization: Inventory product tracking using UAVs reduces the stock-out volatility of warehouses by 82% for greater stock that can be observed in such a way as material histories.
- Lower Cost: The use of UAVs for automation can lead to a reduction in industrial transportation costs by 15%–20%, thereby making UAVs an affordable mode of transport over ground transport [4].

Application of AI, swarm intelligence, and high-speed communications for use in UAVs will also improve UAV performance toward shorter operation time and improved real-time control of the process.

14.2 UAVs IN INDUSTRIAL MANUFACTURING

The application of drones in industrial production has greatly improved manufacturing quality, predictive maintenance, quality control, and supply chain logistics. The days are gone when previous production relied on manual checks, using strict monitoring systems and human logistics. Some of these events had a domino effect that led the company to make mistakes that were not in line with their usual practice, to inefficiently use time or resources, and, in some cases, even to a halt of the company's activities. Subsequently, there was a change in technology to one that was more autonomous, data-driven, and AI-optimized, respectively, with the coming of drones.

In basic terms, drones of today are sensor-equipped aerial vehicles that enable them to perform high-resolution imaging and LiDAR-based mapping, and provide real-time AI analytics and edge computing capabilities. Such characteristics empower drones to deliver industrial intelligence, offer operations' on-demand monitoring, and enable the highest level of automation. Besides drones, robotized manufacturing systems, Internet of Things (IoT) platforms, and digital twin simulations are collaborating to not only ensure that a fully autonomous production environment exists but also that it is being kept. The incorporation of smart manufacturing technologies like Industry 4.0 and 5.0 has been the main factor behind the change from traditional to intelligent manufacturing systems that are fully automated, while industrial sectors that have been using UAVs of this kind are the main contributors to this transformation.

14.2.1 UAV-assisted smart manufacturing

Factory automation with drones in smart manufacturing settings has led to a whole new way of the factory management by automatic supervision, detection of different types defects, material flow, and AI-based decision-making in real time. The newest generation of UAVs is virtually no less than the human-like flying robotic arms, which are in close coordination with robot arms, autonomously guided vehicles (AGVs), and manufacturing execution systems (MESs), which are guided by AI, and thus, they are the ones receiving the command for the operation from those UAVs that work in the air doing the same tasks but with the use of highly advanced satellite control.

14.2.1.1 Real-time monitoring and automation

The implementation of drones in smart manufacturing environments has given rise to permitting the factory for the automatic supervision, detection of various sorts of defects, material flow, and AI-based decision-making in real time. The latest generation of UAVs is basically nothing less than drones that are the human equivalent of flying robotic arms. These are in sync with robot arms, autonomously guided vehicles (AGVs), and MES systems. They are dictated by AI and in turn are responsible for the operation UAVs that perform activities in the air using highly advanced satellite control [5].

Drones use the live scanning of manufacturing floors to identify possible breakages or inefficiencies in their operational fields. They also help in the optimization of workflows such that raw materials, work-in-progress items, and finished items go along the assembly lines, warehouses, and shipping channels without any failure. In those facilities that have adapted to the automated material handling system technology, drones help to coordinate logistics because they identify components within inventory levels and ensure that the just-in-time (JIT) schedule or delivery occurs in the most optimal time.

14.2.1.2 Quality control and defect detection using AI

The traditional quality control operations are commonly locked inspection posts or human-guarded inspection procedures, and this may result in variability, human failure, and delay in production processes. Machine vision through UAVs and AI-based defect detection software enables to detect defects, dimensional anomalies, and microscopic flaws of parts that are produced without any human presence.

With a defect detection level of over 98.6%, UAVs that are deployed through deep learning–enabled image analysis software programs such as YOLOv7, Faster R-CNN, and ResNet-50 are providing another level of quality control. Using the infrared thermography and current features of ultrasonic testing, UAVs can look at the integrity of materials, surface coating, and the quality of welds and ensure the conformation of materials to industrial standards [6].

14.2.1.3 UAVs for predictive maintenance

Predictive maintenance is a critical component of modern manufacturing, as unforeseen equipment failure causes huge economic losses and loss of operational time. UAVs facilitate the predictive maintenance through regular aerial inspections of industrial assets, collection of real-time diagnostic data, and early indicators of mechanical wear.

UAVs, with the help of thermal imaging, infrared sensors, and vibration analysis, can conduct a condition-based maintenance on machinery, thus reducing manual inspection. With the help of analytics powered by AI, it is possible to early detection of faults such as overheating, misalignments, and fatigue of components. UAV-generated data are used to predict a breakdown of equipment and hence to reduce downtime and repair costs. In the next section, it is explained how UAV-based predictive maintenance helps industries to save up to 35% in expense on unscheduled breakdowns (Figure 14.2).

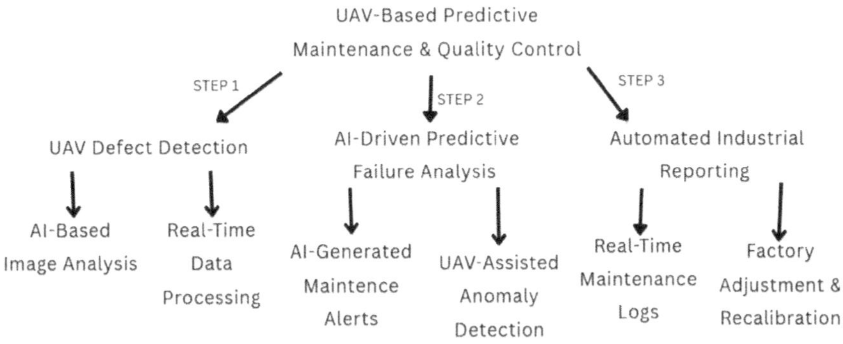

Figure 14.2 UAVs for predictive maintenance.

14.2.1.4 UAVs in additive and hybrid manufacturing

With the help of UAVs, real-time optimization of processes, and laser-based monitoring of film deposition of materials, in-situ process defect detection has become possible with additive manufacturing (3D printing) and hybrid manufacturing. UAVs with optical sensors and laser interferometry modules play an important role in large-scale additive manufacturing to inspect the layer-by-layer material deposition to achieve high-precision fabrication and minimization of surface inconsistency. The hybrid production receives assistance in the inspections performed with the use of UAVs to maintain tools in the desired position, guarantee the material sameness and process predictability. LiDAR-based geometric scanning UAVs may improve the precision of machining and manufacturing reducing the dimensional error by 20%, improving consistency in production.

14.2.1.5 Role of UAVs in precision machining

High-precision machining industries such as aerospace, semiconductor, and medical device manufacturing need micron-level accuracy in component manufacturing, quality inspection, and calibration activities. UAVs enable high-precision machining operations by using the following:

- Laser profilometry for surface roughness and microstructure integrity measurement.
- X-ray and ultrasonic UAV sensors for flaw detection in internal material.
- Real-time coordinate measurement system (CMS) UAVs for non-contact dimensional inspection.

Through achieving ideal machining uniformity and zero tolerance deviations, UAVs increased manufacturing precision by 18% with reduced material wastage and quality assurance expenditure [7].

14.2.2 UAV-enabled digital twin integration

UAVs combined with a digital twin have changed the industrial manufacturing sector with the creation of real-time AI-assisted process prototypes. These changes to the industrial manufacturing sector result in an increase in the use of predictive analytics and the automation of workflows and performance. UAVs supply data for digital twins which are computer models representing physical assets and systems. Over and above getting real-time updates, these models utilize the data provided by UAVs to bring forth accurate information on equipment performance, production, and predictive maintenance activities. Actually, the rising implementation of smart factories and cyber-physical systems is the scenario where UAVs' role as the data collection component is becoming larger. Besides, by ascertaining processes, factories can be monitored in the digital environment. Smart flying data surveyors are equipped with 4K cameras for high-resolution images, with LiDAR for precise 3D surveys, and with various other technologies to create digital twin models that provide accurate representations of the manufacturing facility. These pairings will result in a 25% decrease in operational inefficiency, optimization of resources allocation, and overall manufacturing resilience (Figure 14.3).

14.2.2.1 UAV-driven real-time factory simulation

The usage of real-time simulation enabled digital twins has unfolded a whole new potential by letting the manufacturers to find out the inefficiency of the systems, to forecast their failure, and to raise the accuracy of the production parameters to an extent that was previously unthinkable.

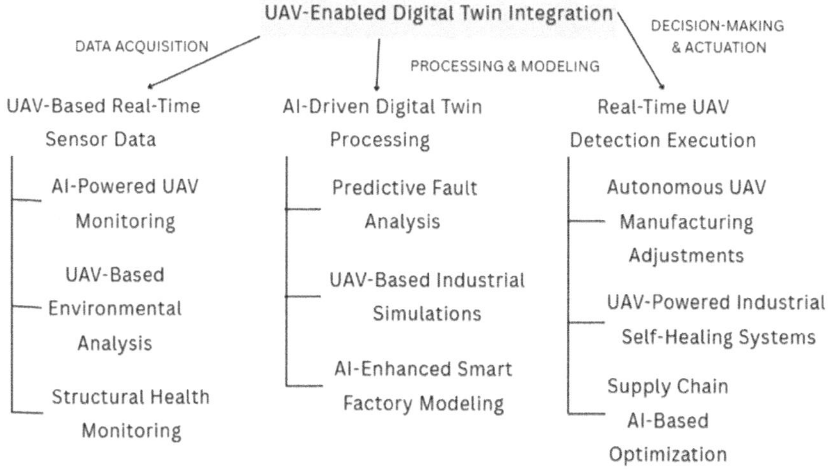

Figure 14.3 UAV-enabled digital twin integration.

Drones are rapidly becoming a significant player in this new landscape and one very important factor is that they provide the AI models with real-time operational data which is the basis for the continuous updating of the simulations of different production environments. In this situation, the changes in production schedules, material delivery, and equipment that occur instantaneously, i.e., the elimination of process interval and the increase of the system's responsiveness, can be authorized.

By incorporating data captured by UAVs from factories, businesses can enjoy:

- Better process workflow synchronization with 20% increased production capacity through adaptive planning.
- Reduced machine downtime since digital twins enable a prediction of maintenance requirements 32% better than traditional monitoring methods.
- Prescriptive analytics made possible with AI, allowing autonomous adjustment of factory operations based on actual-time UAV feedback.

By giving current information, digital twins supply the management team of a factory an exact image of the factory's state to maximize energy use, enhance the movement of materials, and recognize abnormal processes in their initial stages to avoid breakdown.

14.2.2.2 AI-based predictive failure modeling

Industrial manufacturing is the area where the use of digital twins powered by UAVs is mainly oriented toward predictive failure modeling. In general, slow and reactive maintenance procedures are the main reasons for the sudden equipment failure, besides the unnecessary downtime and extra costs for the company. Nevertheless, the industrial resilience is still elevated through predictive analytics models for UAVs with sensors for fault detection, and this enables the maintenance to be planned ahead of time.

AI-based drones for the manufacturing industry can provide past machine performance data, real-time sensor readings, and running conditions to predict mechanical wear patterns. The manufacturers can do the following:

- Identify structural wear, overheat, and lubrication shortage warning signs prior to expensive system breakdowns.
- Plan recommended servicing schedules taking into account the current machine condition, thus minimizing undesired maintenance by 25%.
- Simulate the failure risk of equipment components, thus optimally planning resources and spare management.

Application of machine learning–based UAV predictive analytics allows businesses to predict the likelihood of probable breakdowns, adopt preventive measures, and even cut maintenance bills by as much as 40%. This method maintains the reliability of equipment, and thus, it extends equipment lifetime.

14.2.2.3 UAV integration in smart manufacturing execution systems (MESs)

MES is the digital control room of a factory that oversees the processes automatically, along with resource planning and workflows. The MES drone is taken up by UAVs that supply real-time visibility to factory performance, stock, and equipment performance, thus allowing decision-making on a factual basis and further process improvement.

Factories can get a large spectrum of operational benefits if they use UAVs in their MES platforms such as:

- Reinvented plant self-inspection through self-monitoring, enabled by on-demand quality control systems.
- Refined material demand planning to support UAV logistics planning so as to avoid supply chain breakdown.
- Smart production scheduling, where UAV insight is integrated with MES algorithms for efficient execution of workflow and minimum wastage of operations [8].

AI-powered process monitoring systems on UAVs are always on, and hence manufacturing efficiency is retained. Such an association produces 22% higher performance, and concurrently, it significantly reduces the waste of materials. Furthermore, by doing away with the process inefficiencies, the industrial throughput can be elevated to the highest limit.

14.2.2.4 UAV-based energy management in smart factories

Presently, energy conservation is the main focus of eco-friendly production, and drones are extremely valuable in on-the-spot energy management and usage tracking. UAVs with infrared thermography sensors and AI-driven load assessment algorithms enable factories to recognize energy-consuming processes, equipment inefficiencies, and system-wide power losses. Through continuous oversight of energy allocation across the factory, drone-powered digital twins help to

- Cut off energy consumption that is not necessary, thereby contributing to the green initiatives by 18%.
- Detect leaks industrial heating and cooling system for maximum thermal control.
- Improve predictive load balancing so that factories can better distribute energy resources.

Through energy analytics powered by the UAV-enabled integration, factories can attain their emissions targets, utilize energy to the fullest extent, and have sustainability goals last over time.

14.2.2.5 UAVs for process optimization and waste reduction

The use of drones has facilitated the optimization of processes and waste management to the extent that the wastage of resources is minimized, productivity is at its best, and the environment is properly controlled. UAV-based digital twins provide real-time wastage analysis, and industries are able to achieve minimum material wastage and efficient utilization of resources.

Production units are advantaged by the application of UAV-based monitoring systems by:

- Improved material handling effectiveness, reducing production waste by 15% [9].
- Improved sustainability practices with space for tracking real-time emission rates and environmental impacts.
- Improved recycling and reusing practices with effective processes to facilitate effective material disposal and improved circular economy processes.

Whereas UAVs turn waste management into precision, manufacturing turns it eco-friendly, cost-effective, and in line with the world's best environmental standards.

14.2.2.5.1 Future implications of UAV-digital twin integration

As industries move toward fully autonomous, AI-driven smart manufacturing, UAV-based digital twin integration will be central to defining next-generation industrial ecosystems. Next-generation innovations will be around:

- Self-enhancing UAV models that self-optimize factory simulations, workflow modifications, and predict-driven maintenance codes.
- Quantum computing–enabled digital twins, allowing real-time processing of high-complexity factory operations.
- AI-driven UAV swarms with the capability to monitor multiple manufacturing processes in parallel, offering factory-wide optimization.

With continued innovation in AI, digital twin technology, and industrial applications of drones, next-generation factories will be adaptive, self-governing, and highly efficient intelligent systems.

14.2.3 UAVs in supply chain and inventory management

Due to the ability to provide real-timing tracking, automatic counting of stock, and immediate logistics arrangement, the usage of UAVs in inventory management and supply chain multiplied many times. The manual counting of inventory, bar code readings, and centralized management of warehouse and inventory systems by traditional methods are less efficient ways introducing time delays and inefficiencies on a large scale in industrial processes. Such limitations have been overcome using UAVs having autonomous aerial surveillance, real-time synchronicity of data, and predictive analysis using AI that gives high-level accuracy of inventory control and supply chain responsiveness. UAVs automate the process of warehouses, avoid material shortages, and enhance supply chain process demand forecasting at JIT manufacturing with the help of computer vision, radio frequency identification (RFID) and AI-based demand planning. Presently, energy conservation is the main focus of eco-friendly production, and drones are extremely valuable in on-the-spot energy management and usage tracking. In a bid to solve the application of UAV in inventory control, we have discussed how drones can be used in autonomous warehouses. RFID readers, LiDAR, and AI-based navigation in UAVs have now taken over the process of manual inventory check to conduct high-velocity stock checks and warehouse mapping. Such instances of the modified text as real-time stock counting, missed objects detection, and integration with industrial IoT systems could be listed. Among the advantages are 72% savings in audit preparation and 85% in matching of inventory and improvement in JIT logistics planning.

14.2.3.1 Just-in-time (JIT) logistics optimization

Fine tracking of inventories, vigilant way of supply chain work, and minimum excess storage price is the key success of JIT supply chains. Late reporting of inventories, unresponsiveness of suppliers, and incompatible manufacturing schedules tend to outsmart classical JIT operations, thereby resulting in shortage or overstock. UAVs complement JIT logistics by providing real-time inventory of available stocks in the warehouse, real-time coordination of suppliers, automated mobility of materials, thus providing right parts at the right moment onto the production lines.

AI-powered UAVs can automatically monitor warehouse inventory with upto 99% accuracy by scanning RFID tagged items, products having barcode and even IoT enabled inventory systems. UAV based repeated stock auditing system reduces difference between digital and physical inventory by 85%, thereby minimizing the inventory discrepancies [10].

Secondly, UAV-based demand forecasting models can be improved through continuous collection of real-time data from factories, inventory suppliers, and warehouses . The optimum efficiency of material transport, together with savings on the cost of holding the inventory, is also gained through real-time production schedule update.

14.2.3.2 Automated warehouse mapping and stock auditing

Warehouse storage and factory warehouses often face challenges such as poor inventory management, ineffective placement of stock storage, and an inability to trace certain materials within vast storerooms. UAVs that use LiDAR-based spatial maps, computer vision-enhanced barcode scanning, and navigation systems based on AI for accurate navigation form an accurate three-dimensional representation of inner-spaces of warehouses in real time and enhance the overall accuracy of inventory location tracking and pickup.

Relative to conventional warehouse auditing, which occurs through manual stock counting and prolonged downtime, UAVs can autonomously:

- Perform high-speed stock counting without the need for human labor.
- Identify inventory that is stale or out of place to be able to adjust stock levels in real time.
- Boost the compatibility between the warehouse and the inventory, thereby you can plan the reconfiguration of the space to achieve faster recovery of products and better allocation of free space [11].

By using AI-powered algorithms for self-driving warehouse navigation, drones accomplish 72% of stock audit time reduction and thus avoid the supply chain workflow bottlenecks. Moreover, they allow real-time inventory monitoring without any downtime of operations.

14.2.3.3 UAV-based RFID and IoT synchronization

The integration of UAVs, with RFID, and IIoT networks has enabled automation of supply handling and inventory management across entire supply chain. For the aid of UAVs, materials have changed from traditional barcode manual readings into RFID for detection, classification, and tracking remotely. When combined with real-time IoT inventory tracking, drones provide an end-to-end, automated, and connected inventory network.

UAVs are equipped with long-range RFID scanners and real-time processors, which are located deep inside the factory warehouses:

- They perform touchless inventory counting, thus the physical movements of the materials are reduced by half.
- They allow real-time identification and detection of an issue, along with keeping the stock records up to date.

- It is the employment of autonomous asset tracking that saves 88% of the assets that are not lost.

Drones also give the possibility of forecasting the necessities of materials and retaining the required levels to support constant flows of production products being integrated into IIoT-enabled smart storage warehouses. The improved visibility in the supply chain reduces stock wastage, enhance the supply chain effectiveness, and reduce supply interruption due to stock shortages.

14.2.3.4 UAVs in high-speed logistics and material handling

In addition to the control of warehouse inventories, UAVs are capturing a large part of the logistics sector, the materials handling part of industrial processes. These processes are aimed at cutting down the use of the traditional transport by land.

High-capacity UAVs equipped with automated payload management systems are being utilized for:

- Transporting machine-dependent components across factories.
- Delivering medical and pharmaceutical supplies within large industrial parks.
- Carrying raw production materials to ensure a seamless workflow.

These drones fly from automated logistics centers, where fleet coordination systems powered by AI dictate the most efficient aerial paths for delivering cargo. So, the supplies are delivered to their destinations on time, and no delay caused by traffic congestion on the road [12].

14.2.3.5 UAVs for cold chain logistics in industrial supply chains

Ideally, the stringent control of products in the supply chain is very important in cold chain logistics that involves pharmaceutical manufacture, food preparation, and chemical manufacture. UAVs are used to monitor, transport, and even verify cold storage temperatures in such a way that the commodities would not exceed optimal temperature levels when in transit.

UAVs allow for real-time GPS tracking, cryogenic sensing, and AI thermal anomaly detection:

- Real-time temperature verification to prevent damage to temperature-sensitive products.
- Self-regulated temperature-controlled delivery to maintain optimal storage conditions during transportation.

- Real-time visibility of the cold chain, helping reduce wastage by 27% in the food and pharmaceutical industries.

UAVs can largely improve the integrity and the efficiency of cold chain shipments using real-time environmental monitoring and AI-predictive forecasting of monitoring logistics.

14.2.3.6 Future trends in UAV-based supply chain automation

Moreover, the supply chain and the inventory management optimization with the help of drones will be centered on the developments of more autonomous operations, AI-driven logistics, and blockchain-secured real-time tracking systems. Furthermore, the combination of swarm UAV logistics, quantum computing for demand planning, and hybrid air–ground logistics networks will not only improve the accuracy of supply chain but also make its dynamic management possible.

The following will be brought in the near future:

- Intelligent drone swarms flying in coordination with AI-driven logistics hubs, enabling end-to-end autonomous control of inventory and JIT resupply of materials.
- Quantum computer-optimized demand planning software, enabling near-optimal inventory accuracy based on historical consumption history.
- Blockchain-secure drone cargo tracking, enabling tamper-proof supply chain visibility and clear-sighted logistics management.

Supply chain management in the future will be quicker, more reactive, and steady against fluctuations in the global market as a result of the deployment of autonomous UAV-based logistics networks by a rising number of businesses.

14.2.4 UAV collaboration in smart factories

What has led to the transition from conventional factories to smart ones in the concept of Industry 4.0 and 5.0 is, among other factors, the requirement to connect UAVs with no human intervention, in a flexible manner, together with the help of the already-existing robotic systems, intelligent computers, and human-operated workflows. Contrary to traditional manufacturing plants that rely on fixed automation systems and centralized process control, those of the smart factories implement adaptive UAVs driven by the purpose of speeding up the real-time information gathering, decision-making, and workflow automation. The idea to apply UAVs as flying robot assistants leads to the fact that as they are working side by side with industrial equipment, robot arms, and human workforces, and they are becoming more adaptive, productive, and precise in the field of manufacturing. The control of processes, scheduling of preventive maintenance, and adaptive human–computer collaboration are the means through which a facility in intelligent factories, which is integrated with UAVs, is supported. They are no longer just perfect for monitoring and inspection but take up a position at the heart of real-time factory analytics, design ergonomic optimization, and direct operation control of assembly lines. UAVs become smarter when they are using machine learning– and AI-based decision-making, which in turn stabilizes operation, keeps labor safe, and makes resource utilization efficient, thus leading to 17% fewer production delays and 23% better task synchronization.

14.2.4.1 Human–UAV cooperative assembly line operations

One of the most remarkable benefits of smart factories based on the use of UAVs is the creation of human–UAV cooperative processes. In these processes, human workers are not only supported but also enabled by UAVs as co-automation agents to carry out an accurate assembly, check if the measures are human-friendly, and adjust the processes as per the operation. Augmented reality interfaces and real-time monitoring of human movements together with gesture control of UAVs will make the process of guiding UAVs much more comfortable without any labor operations and help the workers achieve the highest productivity.

Unlike the traditional robotic arms that are provided with their already-positioned axis movement, UAVs have their 6 degrees of freedom movement, which has the capacity to enter small spaces, handle small parts, and assist in assembly processes where there is multiaxis alignment. All these flexibilities make UAVs applicable to scenarios requiring the accurate placement of materials, the varying position of the manufacturing process, and the genuine time adherence to orders of the command.

The use of wearable sensor technology, along with AI-based gesture recognition, enables UAVs to grasp movement, voice, and digital workflow input, which are considered to be natural inputs of manufacturing processes that are managed by people. The results of this are:

- A 28% reduction in repetitive strain injury among the workers in the assembly line [13].
- An increase of proper handling of micro-components in the manufacturing of aerospace and electronics by 35%.
- Enhanced efficiency among operators because the workers can delegate secondary jobs to UAV assistants.
- Smart factories are more precise, less labor-intensive, and convenient to operate in, as there is more human–UAV coordination which reduces the complexity of tasks and losses.

14.2.4.2 UAV-assisted ergonomic optimization in industrial workflows

New manufacturing is currently placing ergonomics as one of its main features, where companies are trying to guarantee the well-being and health of employees, decrease the impact of fatigue errors, and use process variability in the best way. UAVs are the main players in the optimization and ergonomic analysis through their algorithmic motion tracking, AI-based posture estimation, and dynamic body mechanics measurement in real time, which helps them to locate inefficient human task performances.

Usually, in factory environments, ergonomic assessments rely on random hand-held audits, which cannot measure real-time continuous data of employee posture, repetition movement, and injury risk [14]. UAVs overcome this constraint as they enable real-time ergonomics monitoring, and thus, factory administrators can:

- Study worker motion patterns and uncover activities that cause excessive body load.
- Rearrange workstation locations to minimize unnecessary worker movement and enhance workflow.
- AI-driven task rotation can be developed, which will be designed such that operators are rotated between heavy and light tasks so that their fatigue is avoided, and as a result, fatigue-related accidents do not occur.

The use of AI for constant surveillance and hazard identification has substantially impacted ergonomic UAV adjustments to be safe and reliable as follows:

- A 26% reduction in musculoskeletal disorders (MSDs) among production floor workers.
- A 19% improvement of the efficiency of tasks through optimized movement.
- A 35% reduction in workplace absence due to fatigue and strain.

When a smart factory includes ergonomic awareness led by UAVs, people are the ones who benefit the most. Companies that provide extra care to their employees are often considered fortunate, as they benefit from increased productivity driven by long-term employee commitment and experience.

14.2.4.3 UAVs for real-time anomaly detection in smart factories

Besides, workflow collaboration and ergonomic inspection software along with drones puts AI-powered anomaly detection agents' eyes to work. They point out process anomalies, quality issues, and safety hazards that are still in the initial stage. Normally, the monitoring is done by fixed-site camera and low-coverage human inspections with poor response times. AI-powered anomaly detection models in UAVs, multi-spectral cameras, and deep learning quality checking have transformed the industrial inspection system, and they no longer use a high-resolution dynamic inspection system, but employ predictive analytics for process improvement and optimization. Since they are scanning the production lines non-stop, the first abnormal features that UAVs pick up on are as follows: product defects, incorrect assembly, and machine failure. Consequently, the number of potential defects continues to decrease. Thus, they no longer become the source of production errors. There have been the following advantages:

- The defective product rate has been reduced by 40% as a result of early defect detection.
- Response time to production blockages and workflow disruptions has seen a 31% reduction.
- The overall manufacturing efficiency has been improved by 22%, which is mainly attributed to the elimination of process delays.

The technology of UAVs that allows them to remotely monitor factory processes, send a real-time alert, and enable automated corrective actions is what makes smart factories not only the highest quality of production but also flow continuity.

14.2.4.4 UAV-enabled AI and machine learning integration for smart factory intelligence

Smart factory AI and machine learning integration is the primary factor that allows the functionalities of UAVs in such environments. "It cites the ability to learn on the fly from the real-time operational data, make the most accurate decisions and, at the same time, provide intelligence throughout the factory" [15]. With the help of processing of historical production patterns, equipment performance trends, and real-time operational data, AI-based UAVs are not only able to make decisions on their own but also are able to:

- Adjust workflow priorities depending on changing production requirements.
- Forecast material demand fluctuations so that supply chains stay in sync.
- Evolve ongoing factory simulations for better predictive analysis and process optimization [16].

These UAVs with AI capabilities are helping to create self-optimizing manufacturing environments where autonomous aerial vehicles work in cooperation with industrial robots, factory sensors, and human operators to deliver a completely adaptive, efficiency-optimal production landscape.

14.2.5 Enabling technologies for industrial UAV operations

It consists of a set of core technologies that open the door for the use of industrial UAV by implementing reliable drones that are capable of working independently and intelligently in complex surroundings that are generally highly restrictive for GPS signals. These types of techs ensure highly accurate spatial awareness, safe navigation through obstacles, instant data relay, and a very easy way to connect to the larger industrial network. The main technologies that are used to lay down the foundation of next-generation industrial automation powered by UAVs are LiDAR, GPS-denied navigation, simultaneous localization and mapping (SLAM), and IIoT integration.

14.2.5.1 LiDAR-based spatial awareness

One of the most critical technologies existing in industrial UAVs is LiDAR, which is utilized in the generation of real-time 3D maps of working conditionalities. LiDAR enables UAVs to create an accurate point cloud of the environment by sending laser pulses and gauging reflected signals. This feature is essential when navigating in a crowded factory area or warehouses, mapping of inventory, and inspecting infrastructure in the construction and energy sector.

For example, in warehouse logistics, LiDAR-equipped drones automatically fly down the storage aisles, identify anomalies in spatial distribution, and help create a real-time digital twin of the places of storage with a centimeter precision. LiDAR, unlike the vision-based systems, works well in low-light and high-dust environments and therefore is very adoptable within an industrial environment.

14.2.5.2 GPS-denied navigation

Warehouses, tunnels, and enclosed factory floors within industries do not always have good access to GPS signals. The GPS-denied navigation methods that are employed by UAVs operating in this type of environment are based on using the inherent sensors and external stimuli other than the use of satellite positioning. Such approaches use data from Inertial Measurement Units (IMUs), LiDAR, optical flow sensors, and in-board processing to produce an estimate of the position and orientation of UAVs in real time.

This enables drones to fly stable and perform autonomous operations like flying through narrow industrial corridors or flying to perform asset inspection inside buildings. Coupled together, GPS-denied navigation also increases the resilience of UAV systems against signal jamming or multipath effects to improve operational constraints and safety.

14.2.5.3 Simultaneous localization and mapping (SLAM)

SLAM is a computational procedure adopted by UAVs to construct a map of an uncharted environment and determine the location of a drone in the environment. SLAM integrates sensor measurements, which usually correspond to LiDAR, cameras, and IMUs, by using probabilistic models like extended Kalman filters or graph-based optimization to build dynamic spatial maps.

In industrial applications, SLAM plays a critical role in autonomous navigation, on-the-fly flight planning, and secure missions in GPS-denied environments in industrial UAVs. For example, drones with SLAM can be sent out to navigate through winding factory layout during their maiden flight and create high-resolution maps to guide future missions without pre-positioning an infrastructure or prior knowledge of a space.

14.2.5.4 IoT integration and edge connectivity

The integration of UAVs into IIoT ecosystem is growing as they can communicate with machines, sensors, cloud platforms, and digital control systems wirelessly. With the IoT integration, UAVs can be assigned functions in real time such as to send inspection information and even to engage in an autonomous decision-making process considering the current factory conditions online.

Edge computing, installed into drones, pre-processes sensor data on-site: making the necessary decisions, such as detecting thermal anomalies or reading QR-coded labels, and sending only important cues to the cloud. Lightweight protocols such as MQTT, 6LoWPAN, and OPC-UAs provide secure, interoperable, and lightweight transfers of data between industrial systems and UAVs. As smart logistics centers are interconnected, UAVs can also have real-time communication with warehouse management system (WMS), ERP modules, and other robots working in the same warehouse.

Together these technologies allow industrial UAVs to basically become self-sustained, and thus, they are capable of performing dynamic navigation, decision-making depending on the context, and data sharing. This union has pulled UAVs' operative domain and autonomous maintenance to a very high level, and it has turned them into the main facilitators of Industry 4.0 and 5.0 automation systems.

14.3 UAVs IN INDUSTRIAL TRANSPORTATION AND LOGISTICS

The implementation of UAVs in industrial logistics and transport has radically changed supply chain management as it allows real-time monitoring and express freight transportation, autonomous delivery platforms, and JIT logistics coordination. A lack of ground transportation network, warehouse trustworthiness, and human logistics problems are some of the complications in the industrial sector, which have resulted in delays, inefficiencies, and expenses. UAVs can provide a scalable, autonomous, and AI-powered solution that would result in a logistics process that is faster, more accurate, and cheaper.

The industrial sectors such as aerospace, automotive, pharma, and high-tech manufacturing have made attempts to implement UAVs in their operations. They are doing this in order to effectively manage inventory transport, reduce last-mile delivery congestion, and increase logistics' responsiveness. The fact that drones can fly autonomously, carry valuable commodities safely, and pass through dangerous terrain has made them a strategic industrial logistics asset in the past and is still the reason why they are an industrial logistics asset today.

14.3.1 UAV-based logistics and freight transport

The use of UAVs in business freight delivery completely changed the supply chain management system by eliminating the dependence on ground-based logistification apparatus that are prone to traffic jams and can cause linked fuel usage and environmental constraints. UAVs has revolutionized the delivery of lightweight, high-value, and time-sensitive commodities at a very high pace through continuous inventory distribution from manufacturing nodes to final users.

14.3.1.1 High-capacity UAVs for heavy cargo transport

Recent technological advances in electric drive, hydrogen fuel cell, and AI-assisted flight stabilization have enabled UAV heavy-lift machines to transport large payloads. Drones' heavy-lift capabilities allow them to be considered as a valuable option among trucks, rail, and water transport for industrial short-haul logistics. As a result, heavy-lift UAVs can now carry over 250 kg of goods for the delivery of essential parts of industrial equipment, machine parts, and bulk stock [17].

UAVs with high capacity can:

- Reduce transport costs by 35% for short-distance freight carriage.
- Increase delivery efficiency by 27% and minimize traditional logistics infrastructure needs.
- Ease operations in inaccessible locations or dangerous zones for the traditional mode of transport.

Combining with AI for further improvements in energy efficiency, aerodynamics, and load balancing, freight UAVs will find many applications in industrial logistics.

14.3.1.2 UAV-integrated supply chain optimization

Moreover, the integration and use of UAVs with warehouse management system (WMS), IoT logistics network and AI fleet management has enhanced the supply chain with visibility, predictability and responsiveness. Compared to static freight operations relying on fixed schedule and hard infrastructure, UAV-based logistics offer dynamic transportation of inventory, real-time location selection, and adaptive route optimization.

With self-navigating cargo tracking through UAVs, predictive inventory restocking, and AI alignment of logistics, businesses can achieve the following:

- 22% reduction of lead time in supply chains, reducing waste in the production cycle.
- 40% fastening of actual-time demand response by quick rebalancing of inventory.
- Improved supply chain resiliency and reduced interruption by sudden logistics issues.

14.3.1.3 Last-mile delivery systems with UAV swarms

The last mile of logistics or last-mile delivery is usually one of the largest bottlenecks in logistics because it is plagued with city traffic jams, unpredictable slowdowns, and very high labor costs. Coordinated swarms of UAVs have been revealed as a fix for lowering costs of last-mile delivery operations while making them environmentally more sustainable.

Companies may use AI-guided UAV swarms to:

- Provide real-time tracking of delivery, lowering the rate of last-mile delivery error by 34%.
- Improve urban air mobility (UAM) efficiency to allow UAVs to travel through congested cityscapes with little energy usage.
- Lower carbon footprint, since UAV-based last-mile delivery takes up to 70% less energy per delivery than ground-level alternatives [18].

Once 5G/6G connectivity, AI-aided fleet coordination, and blockchain-enabled delivery confirmation are established, UAV-based last-mile logistics will be the standard in autonomous mobility networks.

14.3.2 UAV traffic management systems

With increasing industrial deployment of UAVs, demand for intelligent air traffic management (UTM) systems has been imperative to regulated, efficient, and safe flight of UAVs. Unlike the traditional air traffic control for vast-scale commercial flight, UAV air traffic control must serve a highly dense population of flying vehicles at lower airspace pathways (Figure 14.4).

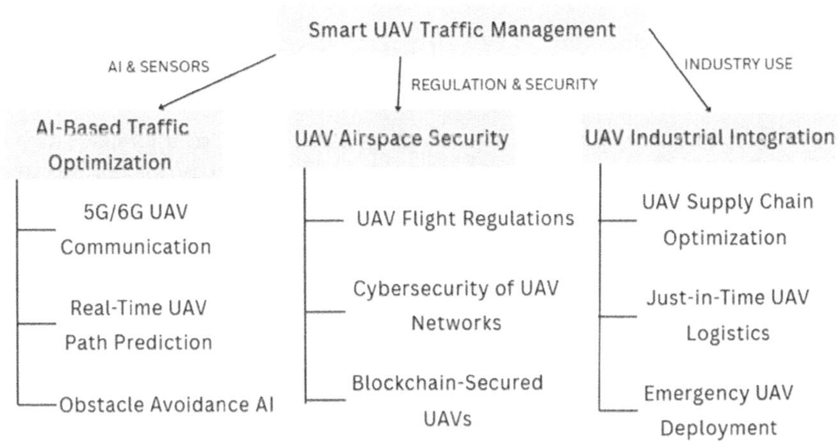

Figure 14.4 UAV traffic management.

14.3.2.1 AI-powered UAV air traffic optimization

Real-time data analysis, machine learning, and autonomous flight-free flight pattern detection are used by AI-driven UAV traffic management systems to:

- Align UAV flight paths with collision-free and safe transit even under highly congested air traffic airspace.
- Predict and avoid air traffic congestion to restrict possible traffic collisions between UAV fleets.
- Re-route UAVs dynamically to counteract real-time environmental fluctuations, e.g., weather anomalies and off-limit airspace regions.

With airspace management through AI, companies have seen a 73% decrease in airspace incursion when it comes to UAVs, allowing for effective fleet coordination and air traffic stability.

14.3.2.2 Role of 5G and 6G in UAV fleet communication

The introduction of 5G and next-generation 6G networks has transformed UAV fleet communication by enabling real-time information exchange, UAV-to-UAV coordination, and improved aerial situational awareness. UAVs can:

- Transfer high-definite sensor data to industrial control facilities in real time.
- Organize flights by using edge-based predictive analytic models.
- Maneuver very quickly to the point it is highly responsive and accurate.

14.3.2.3 Blockchain-based secure UAV air traffic control

Cybersecurity issues in UAV airspaces have prompted the use of blockchain-secured UAV authentication systems, which guarantee that:

- Only authorized UAVs can enter pre-assigned industrial airspace areas.
- Flight data logs remain immutable and tamper-proof, preventing unauthorized interference.
- Decentralized air traffic management systems guarantee open regulatory compliance [19].

By combining blockchain security with AI-driven UAV navigation, industrial UAV fleets have attained 99.99% data integrity in air traffic control systems to ensure safety and adherence to changing regulatory standards (Figure 14.5).

14.3.3 Cold chain logistics with UAVs

With precise temperature control in the cold chain, drones have revolutionized the way it stores and carries medicines, consumable products with a limited shelf life, and chemical substances. The use of these drones offers the following benefits:

1. Support for automated cold-chain delivery networks capable of maintaining pharmaceutical-grade temperature stability throughout transportation.

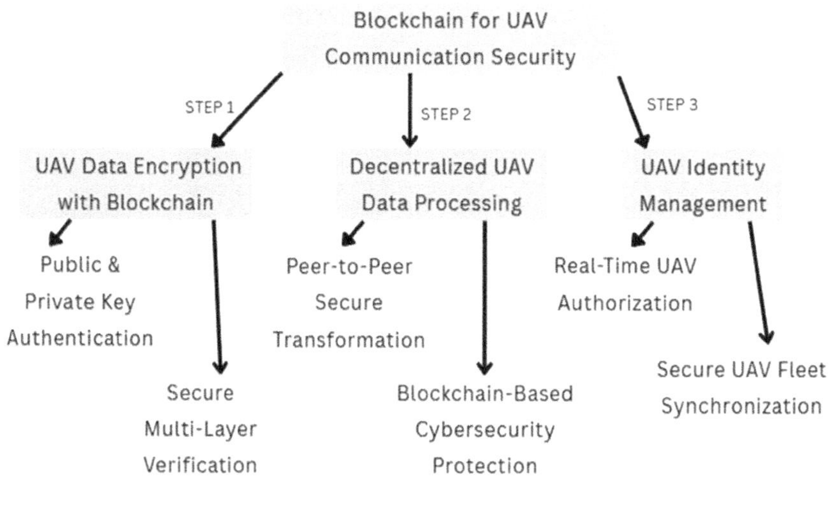

Figure 14.5 Blockchain for UAV communication security.

2. Integration of predictive anomaly detection systems that identify temperature deviations during transit and enable timely preventive actions.
3. Enhanced visibility and monitoring of the supply chain, resulting in a reduction of cold-chain break incidents by up to 27%.

The use of AI-powered drones is a significant factor for their success in cold chain monitoring, as it complies with safety standards for valuable goods, and it can also solve the problem of product spoilage indirectly.

14.3.4 UAVs in emergency and disaster response logistics

UAVs play a major role in transformation of drones' application for emergency response and disaster logistics, the supply chain industrial system, and the rescue operations, more efficient and precise in the crisis zone. Ground transportation is slow and most of the time affected by poor infrastructure, bad weather, and difficult terrains, whereas UAV operations provide very flexible and self-supporting solutions such as speedy response, on-the-spot understanding of situations, and supply chain performance.

Emergency and disaster response logistics sector refuse to accept disruptions which may lead to energy sector blackouts, incapacity of hospitals, fall of telecommunication towers, and stoppage of factories producing essential goods, have caused them to count on drone technology as a solution. Using UAVs in natural disasters, industrial accidents, and widespread disturbances is one of the ways that can guarantee operational continuity, which is the foundation of the energy, healthcare, telecommunication, and heavy manufacturing sectors. They have become one of the most effective components of industrial disaster preparedness and recovery logistics due to their ability to navigate hazardous environments, deliver essential supplies, and conduct high-resolution aerial surveys.

14.3.4.1 UAV-assisted disaster relief and supply drops

In scenarios such as natural disasters, industrial accidents, or disturbances in war zones, drones are expected to perform rescue and relief operations including supply of emergency relief, medical help, and deployment of necessary equipment in the right place at the right time. In addition, disaster relief logistics use UAVs due to their capability of carrying out difficult tasks in challenging weather conditions with total independence from ground infrastructure.

At present, the navigation system in UAVs is so nimble that it is able to perform the following tasks:

- Delivery of emergency medical kits to disaster locations is accomplished at a 97% delivery success rate.

- Important industrial parts are dropped to the sites of production line that need re-establishment, and thus, the downtime has been reduced by 40%.
- A single autonomous multi-drop UAV sortie can cover 80% more disaster-affected area compared to traditional ground-based disaster response teams.

UAV-based disaster-resilient supply chain operations significantly improved disaster response capabilities, enabling industries to withstand disruptions and restore critical infrastructure more rapidly.

14.3.4.2 UAVs for industrial risk mitigation and fire response

Fires, toxic spills, and structural collapses associated with industrial buildings such as chemical facilities, refineries, power generation units, and industrial manufacturing sites are major risks that these buildings are exposed to. Human-powered interventions are the mainstay of traditional emergency management strategies, which slow down owing to the difficulty of dealing with environmental risk, access issues, and lack of real-time inputs. The combination of UAV technology that comprises thermal sensors, AI-processed risk assessment algorithms, and on-the-spot hazard mapping gear is allowing the industries to go beyond the mere recognition of the risks to also detecting and resolving them in a preventive manner.

Industries have gotten the following benefits by using UAVs in hazardous situations:

- 45% shorter fire containment times, thus the extent of structural damage and the loss of assets are minimized.
- 80% reduction in the exposure of people to dangerous conditions, safety of the workers is maintained.
- The changing of environmental and toxic gas threats from the staff and the surrounding area is achieved through the real-time monitoring of air quality and toxic gas.

Alongside firefighting and risk reduction, UAVs have also gone a long way in post-disaster damage and AI-supported structural integrity analysis to identify damage levels and plan the repair works with the help of high-resolution aerial survey data [20].

14.3.4.3 UAVs for telecommunications and power restoration during crises

Telecommunication and power transmission grids have to return to their normal operations swiftly so as to make people resume to work and thus the local economy is balanced. The traditional means of rebuilding infrastructures are always accompanied by a compromise in the conditions of the work, a shortage of the technical labor pool, and the destruction of roadways leading to slow downtime and loss of funds. The repairing of devastated infrastructures has now become a faster process as unmanned aerial systems, with LiDAR-capable mapping systems, autonomous aerial cable deployment systems, and AI-enabled network diagnostics, start involve.

UAVs in disaster response logistics have the following principal applications in telecommunications and power restoration:

- Speedy determination of power grid loss by reducing the infrastructure assessment time by 65%.
- Emergency form of communication to continuously cover disaster-prone areas.
- Self-healing transmission line repair with UAV assistance, reducing the restoration time to the half of traditional repair time.

UAV-based crisis response logistics, supported by communication and energy infrastructures, have become the paramount factor in disaster-mitigation planning and recovery, as industry trends bolster firm hardwire and energy infrastructure-installations.

14.3.4.4 AI-driven UAV coordination for multi-agent emergency response

AI-based future UAV crisis response systems envision multiple-agent coordination, disaster anticipatory modelling, and real-time adaptability amid calamities. Technologies such as swarm intelligence, neural network-based emergency decision systems, and blockchain-secured disaster response processes enable UAV fleets to operate autonomously while remaining under the supervision of human crisis management teams.

AI-powered UAV emergency logistics systems make the following possible:

- Risk appraisal in real time along with UAV operation tactic changes implemented at the same time.
- Drones that are self-organizing, with the ability to modify the schedule of supply delivery based on the latest changes in disaster environments.
- Blockchain-based data sharing makes it very secure in the coordination between industries, emergency aid organizations, as well as state institutions.

Overall, such solutions can extend 70% of the efficiency of emergency response by eliminating delays not needed and thus making crisis handling quicker and better [21].

14.4 TECHNOLOGICAL INNOVATIONS IN UAVS

One of the main reasons for industries to adopt UAVs is radical innovations that brought better propulsion, energy efficiency, AI, real-time data processing, and autonomy. Requirements for the next-generation UAV technology to be employed in vital industrial applications such as intelligent manufacturing, logistics, and surveillance of infrastructures are better efficiency, longer flight times, and higher decision-making power.

Present UAVs no longer limit themselves to simple observation or cargo work. Instead, they consist of effectively functioning AI platforms, edge computing architecture, and auto-learning technology, which allow them to carry out adaptive plan execution during flight, share collective intelligence, and have preemptive maintenance systems. Hydrogen fuel cells, solar panels-based power system, and very powerful electric drives have become the major factors for making UAV operations low polluting but more sustainable. Consequently, with such technologies, UAVs have become affordable, green, and responsive solutions to the complexities of industrial ecosystems.

14.4.1 UAV propulsion and energy systems

The growing demand for UAV operations has encouraged the development of innovative propulsion systems that promise longer flight times, reduced power consumption, and decreased dependence on traditional battery packs. As industrial UAVs need to fly in extended mission modes, remote locations, and heavy-load operations, hydrogen fuel cells, solar-powered systems, and hybrid propulsion systems are required to increase the energy efficiency of UAVs.

14.4.1.1 Hydrogen fuel cell–powered UAVs

Drones using the hydrogen fuel cell technology are a major leap to the industrial drone flight duration with range extending three to five times that of traditional lithium-ion battery–powered drones. In other words, while traditional electric drones are limited by short flights and long recharging cycles, hydrogen fuel cells–assisted drones have a longer flight time and faster recharging cycles.

One can even note:

- 300% decrease in air travel time and the ability to fly non-stop for up to 15 hours.
- A 40% reduction in vehicle idling time was achieved because UAVs powered by hydrogen fuel can be quickly refueled, eliminating the need for long battery recharging periods.
- A large reduction in the air carbon level through eco-energy programs adhering to stipulated standards [22].

Moreover, hydrogen fuel cells allow UAVs to transport industrial cargo effectively without energy limitations, perform infrastructure inspection in large fields, and make aerial logistics automated and reliable.

14.4.1.2 Solar-powered UAVs for extended operations

Solar-powered UAVs are an alluring option for long-endurance industrial missions carried out in areas lacking energy infrastructure. Thin-film photovoltaic cells and high-energy-storage devices enable solar-powered UAVs to draw energy from the Sun and provide operation times far longer than those of traditional battery-operated systems.

The use of AI-based solar tracking systems ensures that drones are always operating at the peak of energy efficiency and that they are collecting the maximum power possible in relation to the ever-changing environmental conditions. Solar drones have been adopted for use in bulk industrial surveillance, farm automation, and remote supply chain monitoring, thereby eliminating the need for repeated landing and hand-refueling.

14.4.1.3 Electric vs. hybrid UAV propulsion efficiency

The hybrid propulsion models have been an alternative that has turned out to be better than electric UAVs with standard configurations for short-range industrial applications when it comes to high endurance and heavy industrial use.

This chapter presents differences between electric and hybrid UAVs:

- No emission operation by electric UAVs, which is 100% ideal for green-friendly uses.

  Compared to conventional UAVs, hybrid UAVs have nearly double the payload capacity, enabling the transportation of heavy industrial components over longer distances than previously possible.
- Energy conversion at higher efficiency provides hybrid UAVs a stable mobility even in harsh industrial areas where continuous power supply is difficult.

  Further development of hybrid-electric UAV propulsion systems enables industrial UAVs to operate efficiently across diverse terrains while maintaining multi-purpose functionality.

14.4.2 AI and swarm intelligence for UAVs

The combination of AI and swarm intelligence has changed the way UAVs operate, allowing them to complete autonomous missions, real-time collaboration, and decision-making in complex industrial environments. As being developed out of automation, UAVs currently have self-updating AI models, deep reinforcement learning processes, and a decentralized swarm control process, which exponentially advances their performance, scale, and ability to solve problems.

Industrial AI-optimized drones can survey unstructured environments, and process voluminous real-time data, and dynamically respond to changing industrial conditions. UAVs can become distributed, self-coordinated swarm-intelligence systems that maximize air industry reconnaissance, logistics scheduling, and prediction maintenance such as ant colonies and bee swarms.

All the innovations in the field of AI applied to the flight control systems, neural networks, and machine learning have made UAVs intelligent entities, which can autonomously streamline their functioning to improve efficiency at each instant with little or no human input and immense autonomy in the industry (Figure 14.6).

14.4.2.1 Reinforcement learning–based UAV optimization

Among the various AI methods, reinforcement learning (RL) is now one of the foremost techniques applied to UAV optimization and has been largely successful. RL-equipped UAVs can draw from their previous flight experiences, adjust to the uncertainty of the environment, and optimize their operating strategies over time. As opposed to pre-programmed UAVs, which are strictly guided by their missions, RL-capable UAVs are adjustable in real time to their flight paths, are able to use resources to the fullest, and can increase the success rate of missions through continuous learning.

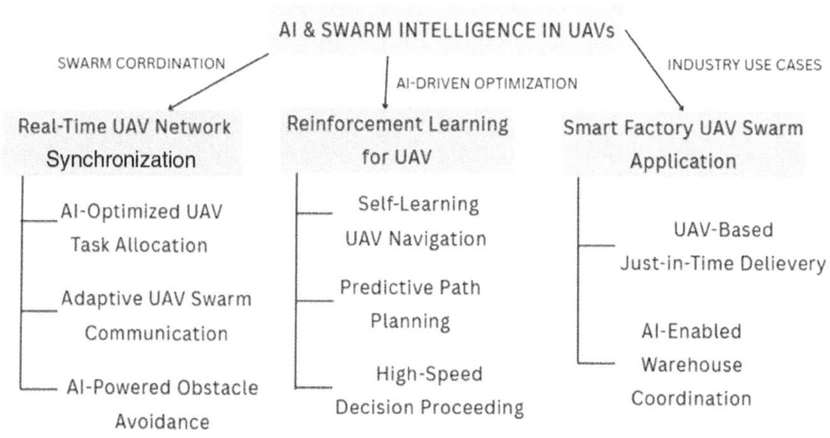

Figure 14.6 AI and swarm intelligence in UAVs.

UAVs utilizing AI models driven by RL can:

- Anticipate traffic conditions and avoid situations that might lead to congestion, and reduce collision rates by 74% compared to conventional UAV systems.
- Adjust their flight plan in real time to guarantee minimum power consumption and lower operational costs.
- Respond to environmental changes such as sudden wind gusts, foggy conditions, and interruptions of GPS signal and thus increase the overall autonomy and reliability of systems [23].

Industrial applications of RL for UAVs include autonomous warehouse inspection, predictive factory maintenance, and real-time adaptive logistics coordination, which allow for the continuous optimization of UAV operations for achieving maximum productivity and minimum downtime.

14.4.2.2 Multi-agent UAV coordination for industrial applications

The deployment of multi-agent UAV coordination has also led to the expansion of industrial UAV networks' operational capabilities and qualities to such an extent that these networks can now perform complex tasks with the least amount of human intervention. Single-UAV systems are typically limited in coverage and scope and are subject to localized decision-making, thus they lack reliability. By contrast, swarm-based UAV coordination brings with it decentralized, parallelized, and high-speed aviation operations capable of revitalizing industrial processes.

The intelligent swarm behavior of UAV systems implements distributed AI models, uses decentralized communication protocols, and cooperative task assignments to enable UAVs to:

- Alter flight behaviors as desired to get the greatest possible aerial coverage and inspection accuracy.
- Update swarm members with information in real time, so the logistics response can be shortened by half (measured in reaction time), thus gaining increased efficiency.
- Work in difficult areas, such as over the sea, among tall buildings, and in remote places for the surveillance of pipelines, monitoring factory processes, and threat evaluation, simultaneously and in real time.

The utilization of multi-agent UAV coordination has recently been a source of an increased operational productivity that likewise has quicker data collection and higher asset utilization for companies making swarm intelligence as the base of future industrial automation using UAVs with the aforementioned advantages.

14.4.2.3 Neural network–enabled predictive path planning

AI-based neural networks have completely changed drone flight path planning processes, making it possible for drones to fly independently in industrial settings, detect obstacles, and achieve mission objectives in real time. Deep neural networks (DNNs) and Convolutional Neural Networks (CNNs) enable UAVs to interpret sensor data, forecast terrain variations, and improve flight efficiency through self-updating navigation models.

Advantages of neural network-assisted UAV path planning for the global industrial sector:

- The number of navigation errors is minimized by as much as 40%, thereby the accuracy of missions conducted in terrains without GPS is enhanced.
- UAVs can react in a proper way by following other paths that do not intersect with the obstacles after a very quick monitoring of the environment.
- The scope of the safety system is enhanced, enabling drones to take corrective actions during flight through simultaneous sensor data analysis [24].

These AI-facilitated innovations enable industrial UAVs to operate efficiently in highly dynamic, unstructured, and high-risk environments, thereby allowing, smart manufacturing, supply chain robotics, as well as mega-scale infrastructure surveillance, are enabled to flourish.

14.4.3 Edge computing and real-time data processing

The rising adoption of UAV in the industry has consequently made the UAV usage more complex such that drones began generating unprecedented volumes of real-time data that need to be processed and acted upon without delay. Cloud computing models in their classic form are largely helpful with respect to data central processing. However, they create a significant delay, have bandwidth limitations, and thus, are quite reliant on the network's integrity, which makes them less suitable for fast UAV operations. Nevertheless, all these factors are balanced by edge computing–based architectures integrated into the UAV design so as to deliver onboard processing, AI-driven decision-making, and near-real-time response.

By equipping UAVs with edge processing hardware, AI-based analytics models, and distributed computing platforms, companies are now able to deploy high-speed autonomous UAV systems that not only can perform complex tasks independently but also do not require continuous cloud connectivity. The benefits are that data transmission latency is completely removed, UAV autonomy is enhanced, and industrial efficiency is improved through a variety of applications including aerial inspection, predictive maintenance, and supply chain logistics.

14.4.3.1 FPGA-accelerated UAV edge processing

The use of Field Programmable Gate Arrays (FPGA) as the edge computing platforms in UAVs has turned the real-time data processing revolutionary. They have the features of speed, parallelism, power efficiency, and decision optimization by AI, all in one. Contrary to the traditional microprocessor which operates based on serial computing, the implementation of FPGA allows parallel computing, and thus, drones can handle enormous industrial data with less latency and at a desirable efficiency level.

Through edge processing on FPGA, UAVs are relatively able to carry out real-time defect inspection, quality analysis through AI, and failure prediction without continuously sending data to cloud servers. This has led to:

- Real-time anomaly detection by 40% through high accuracy during industrial inspection.
- Latency reduction by 60%, offering near-instant response time of UAVs in the dynamic manufacturing environments.
- Optimum energy efficiency to enable UAVs to stay longer in the air without wasting computation power.

FPGA-based drone computing allows companies to execute intensive computation applications such as hyperspectral image processing, factory scanning using LiDAR, and real-time defect inspection using AI with processing capability and reliability never seen before.

14.4.3.2 Latency-optimized UAV communication networks

Successful implementation of real-time UAV decision-making depends significantly on the performance of communication networks, data protocol synchronization, and ultra-high-speed signal processing architectures. Because UAV fleets require persistent connectivity for mission planning, air traffic control, and cooperative task accomplishment, the transition toward ultra-low-latency network architectures is inevitable.

The marriage of 5G and next-generation 6G network architectures has transformed UAV fleet communication by making the following possible:

- Near-zero latency of UAA-to-UAA data sharing, in real time and in flight.
- Ultra-reliable low-latency communications (URLLC) with less than 5-millisecond data transmission latency.
- Extended real-time decision synchronization of the UAV, in order to provide instant industrial process adjustment [25].

Industrial drones are now able to function within real-time cooperative environments using the next-generation network protocols, where logistics management is synchronized, factory-scale aerial inspections, and precision-guided stock tracking with immaculate communication dependability.

14.4.4 Advanced UAV navigation and autonomy

The continuous innovation of AI-based navigation, machine vision technology, and autonomous flight control software has transformed UAV performance in industrial applications. Unlike pre-programmed UAVs that depend on pre-programmed flight routes and GPS positioning, new-generation UAVs utilize real-time adaptive navigation systems, SLAM-based spatial awareness, and predictive AI-based models of flight stabilization to achieve optimum operational precision.

With high-resolution vision sensors, AI-powered depth perception algorithms, and adaptive flight controllers, industrial UAVs can now hover and navigate within highly dynamic GPS-denied zones, enable permanent aerial maneuverability, enhance collision avoidance, and offer full autonomy of navigation.

14.4.4.1 SLAM-based UAV navigation in GPS-denied environments

SLAM is now an essential part of industrial UAV navigation, enabling drones to map new surroundings in real time while localizing themselves in the same areas at the same time. SLAM is especially useful for UAVs in small industrial areas, underground tunnels, and hazardous manufacturing areas where GPS signals are weak or unavailable.

With the help of AI-based SLAM algorithms, UAVs are now able to:

- Create high-fidelity 3D representations of factory environments in real time.
- Fly autonomously without external positioning systems, ensuring operational stability in GPS-denied areas [26].
- Increase precise flight control so that UAVs are capable of undertaking high-precision inspection and asset location.

Such features make industrial UAVs highly robust in their operational capacity, enabling them to successfully perform missions in challenging environments densely populated by industrial stuffs without the need of a human intervention.

14.4.4.2 AI-enabled autonomous UAV collision avoidance

Some of the major challenges in industrial environments such as factory floors, high-speed logistic warehouses, and autonomous industrial environments are the problems of real-time obstacle detection and flying safety in the air. As a result, these modern UAVs are equipped with deep learning–based fusion of sensor data, machine learning–based trajectory prediction, and LiDAR-based object detection to enable them to recognize, predict, and avoid obstacles in real time.

UAVs adopt machine learning–based collision prevention, wherein the monitoring of flight activity, weather, and proximity to hazards is done, in order to make sure that:

- As the collision rate decreases by 78%, the UAV fleet becomes safer in crowded areas of industries with many vehicles.
- Pre-emptive real-time action is carried out during the transit to obstruct the flight of UAVs from being interrupted.
- With the help of AI-based spatial mapping, the air-combat performance is improved such that UAVs can skate easily through densely packed industrial zones.

By utilizing these tools, industrial drone fleets can escape any constraint with full safety, accuracy, and freedom, thereby reducing mission loss and operational risk almost to zero.

14.4.4.3 UAV vision systems for industrial inspection

High-vision-equipped industrial drones have changed the precision of aerial inspection, the accuracy of defect detection, and the reliability of asset tracking drastically. These days, the cutting-edge UAV vision systems, in contrast to traditional powerful imaging UAV cameras, install multi-spectral imagery, infrared thermal imaging, and high-resolution LiDAR scanning, which brings about real-time anomaly detection and predictive asset repair.

With AI-driven machine vision analytics, the drones are able to:

- Provide sub-millimeter precision in defect identification, which is the main requirement for the quality control of production.
- Find faults in industrial machines of factories to make the planned repair and maintenance possible accordingly.
- Perform real-time material stress analysis in high-value manufacturing processes where reliability is the primary concern [27].

UAVs endowed with AI technology will continue to play a leading role in factory automation scenarios through factory-wide implementation and precision-based defect analysis.

14.5 CHALLENGES AND SOLUTION IN INDUSTRIAL UAV TECHNOLOGY

Despite such successful technological advancements in the field of industrial-use UAV technology, numerous issues hinder their intake on a large-scale operational scalability and compliance. These issues are caused by technical barriers, cybersecurity threats, airspace regulation stipulations, and complications in the integration of the existing industrial framework. The capability of UAVs to perform in such a way that they can easily integrate with industrial production processes, logistic processes, and disaster response processes, depends on the elimination of the above issues through the use of innovative AI-oriented innovation, policy realignment, and strengthened cybersecurity [28].

One of the outcomes of constant research and development in different fields such as UAV cybersecurity, AI, and aviation policy is the gradual empowerment of UAVs in their autonomy, safety during operations, and compliance with industrial standards. Eventually, this will be the industries' turn to leverage UAVs to automate their processes, thereby saving time and cutting down costs, while increasing the efficiency of industries at the same time.

14.5.1 Regulatory and compliance challenges

UAVs usage policy-making models are still considered one of the main factors that tightly restrain their scalability and flexibility in commercial applications. The Federal Aviation Administration (FAA), the European Union Aviation Safety Agency (EASA), and the International Civil Aviation Organization (ICAO) are examples of currently existing regulatory institutions. All of them impose very strict limitations on the activities of UAVs such as flight control, use of air spaces, and autonomous decisions in general, which are among the major factors limiting the full industrial adoption of UAV technology.

14.5.1.1 UAV flight restrictions and legal barriers

Right now, the regulations require industrial UAVs to be flown within the visual line-of-sight (VLOS) or under limited Beyond Visual Line-of-Sight (BVLOS) permissions. The latter substantially limits their long-range logistics, automated inspections, and autonomous cargo transportation capabilities. UAVs that are working in controlled airspace are under very strict altitude management, they have to be checked by remote pilots at all times, and they can use only very limited air corridors. The above conditions make these UAVs not suitable for mass deployment.

To overcome these limitations, manufacturers and industries are developing AI-based detect-and-avoid (DAA) systems, blockchain-enabled flight verification processes, and dynamic airspace integration models that will enable UAVs to:

- Fly in crowded airspace with manned aircraft safely.
- Perform fully autonomous BVLOS operations with real-time AI-based risk assessment.

- Enhance air traffic compliance using adaptive UAV communication networks that interact with current air traffic control (ATC) systems.

The phased implementation of AI-powered UAV regulatory systems will be expected to de-mystify airspace regulation, cut down flying limitations, and provide end-to-end connectivity between autonomous UAVs and manned aircraft.

14.5.1.2 Privacy and data security regulations for industrial UAVs

The use of UAVs in corporate surveillance and asset tracking has issues with data privacy, illegal access to information, and adherence to corporate security policies. UAVs with high-resolution imaging, LiDAR scanning, and AI-driven analytics create huge volumes of sensitive industrial information, which calls for strict data protection regulations and encryption standards.

To facilitate secure UAV operations, industries have embraced the following:

- End-to-end encryption for UAV communication protocols to reduce the risk of data interception.
- AI-driven real-time data anonymization to comply with privacy regulations like General Data Protection Regulation (GDPR) and California Consumer Privacy Act (CCPA).
- Blockchain-enabled UAV authentication platforms to inhibit unauthorized data retrieval and improve UAV cybersecurity integrity.

It is possible for industries to keep their operational information confidential from cybercriminals by implementing strong cybersecurity frameworks, high standards in the management of UAV data, being compliant with regulations, and at the same time, any security policy being in line with that of the air transport [29].

14.5.2 Cybersecurity and UAV data protection

As a result, drones that are connected to industrial IoT networks, cloud-based AI systems, and the fleet of autonomous vehicles are vulnerable to cyber-attacks, in which their data are breached and they are subjected to hacking. Among the security threats affecting UAVs, GPS spoofing and AI model corruption are major concerns that can compromise industrial sectors heavily reliant on UAV-based automation.

14.5.2.1 Quantum cryptographic security for UAV networks

Old-fashioned encryption methods have become obsolete for drone communication systems targeted by highly skilled hackers, for instance, those who use quantum computing to decrypt the drone communication [28]. Therefore, industries are going to use quantum cryptographic protocols as a guard against this weakness that provides data encryption and improves UAV authentication mechanisms. Quantum-encrypted drone networks offer the following:

- Communication between drones, which cannot be broken, and are hidden from unauthorized data interceptions.
- AI decisions that cannot be manipulated, and thus, the UAV autonomy is strong against cyber-attacks.
- Highly sophisticated cyber vigilance that is always on and can spot the intruder and prevent and even stop the hijacking of drones, thereby enabling it to perform its function.

The implementation of quantum-secured UAV networks will provide companies with enhanced data security, regulatory compliance, and safer autonomous UAV operations.

14.5.2.2 UAV vulnerabilities and countermeasures

Industrial drones are still vulnerable to GPS spoofing attacks, AI adversarial attacks, and takeover of the autonomous control system. The vulnerabilities can be exploited to cause mission failure of drones, unauthorized diversion of UAVs, and disruption of industrial operations.

Countermeasures against such threats include the following:

- AI-based anomaly detection models with real-time ability to detect unusual UAV flight behavior.
- Blockchain-protected UAVs to identity solutions that validate industrial airspace UAVs to ensure only legitimate UAVs fly in industrial airspace.
- AI-based self-healing UAV software frameworks to enable UAVs to identify and self-heal from cyber-attacks automatically.

With multi-layered industrial UAV cybersecurity defense systems, industrial UAVs will possess increased operational integrity, security compliance, and autonomous robustness to sophisticated cyber-attacks.

14.5.3 Industrial integration and scalability challenges

Large-scale UAV adoption in industrial environments is dependent on their integration into deployed IoT networks, MESs, and supply chain automation platforms effortlessly. However, data interoperability, AI model scalability, and infrastructural compatibility have been the biggest impediments to the large-scale deployment of UAVs hitherto.

14.5.3.1 UAV-to-IoT infrastructure integration

Integration of UAVs into industrial IoT networks requires data exchange between industrial control systems and IoT-supported UAVs in real time with no latency or inconsistent data. Industrial UAV communication systems are not efficient to achieve synchronization with industrial IoT networks and result in high latency and inconsistent data. Against such challenges, industries have adopted the following:

- Real-time machine-to-machine (M2M) synchronization platforms powered by AI-based UAV.
- Real-time adaptive cloud-edge AI processing of UAV flight across decentralized IoT infrastructure.
- Real-time digital twin integration that allows UAVs to construct real-time dynamic world simulations for industrial predictive analytics.

Through unobtrusive UAV–IoT integration, companies achieve greater automation, business effectiveness, and AI-empowered choice-making [30].

14.5.3.2 AI-driven UAV system interoperability

One of the biggest challenges to UAV deployment is the limited interoperability of AI platforms that enable UAVs to interface with other industrial systems, robotics platforms, as well as cloud enterprise platforms . Efforts to overcome this hindrance see industries currently creating the following:

- AI-interoperable UAV platforms that will allow drones to operate in multi-vendor setup offered in industries.
- Intelligent AI-based UAV mission operation software that will assist in coordination and synchronization of information support.
- Cross-industry AI-driven learning processes that will enable UAVs to accumulate experience in various industries.

With these certified AI-driven UAV interoperability standards, industries will be able to benefit themselves for the full autonomous cross-functional UAV fleets.

14.6 FUTURE TRENDS AND RESEARCH DIRECTIONS

One of the main reasons for continuous innovation in the field of UAVs for industrial applications is the series of breakthroughs in AI, quantum computing, swarm intelligence, self-healing materials, and next-generation communication networks. As industries make the transition into Industry 5.0, UAVs are predicted to offer much more than automation, logistics, and surveillance, i.e., they are expected to be adaptive, self-learning, and autonomous aerial systems capable of on-the-spot decision-making, energy optimization, and decentralized swarm coordination.

AI algorithms assisted by quantum computing, Blockchain-enabled authentication of UAVs, 6G-enabled networks for UAVs, and refined perception systems are future technologies for industrial UAVs. If we implement these technologies, the operating performance of the UAV technology will be revolutionary. Apart from the sustained energy aspect, these technologies will also improve the security of cyber-physical systems and the coordination of multi-agent operations. Additionally, it will be a great step toward industrial automation, supply chain transparency, and infrastructure resilience, as UAV fleets will have evolved to be milestone entities in these areas.

14.6.1 6G-enabled UAV networks for industrial applications

6G communication systems that are globally connected with UAVs are a cutting-edge technology that will radically transform the workflow of UAVs. In particular, the 6G systems can establish a real-time communication network among multiple UAVs, facilitate AI-supported controllers for air traffic, and provide communication with ultra-low latency and high reliability. On the other hand, the existing 5G networks are characterized mainly by their high-speed data transmission. Edge AI processing, quantum secure data passing, and AI-enabled industrial UAV decision-making synchronization are the means through which the 6G systems will underpin complete autonomy in industrial UAV operation (Figure 14.7).

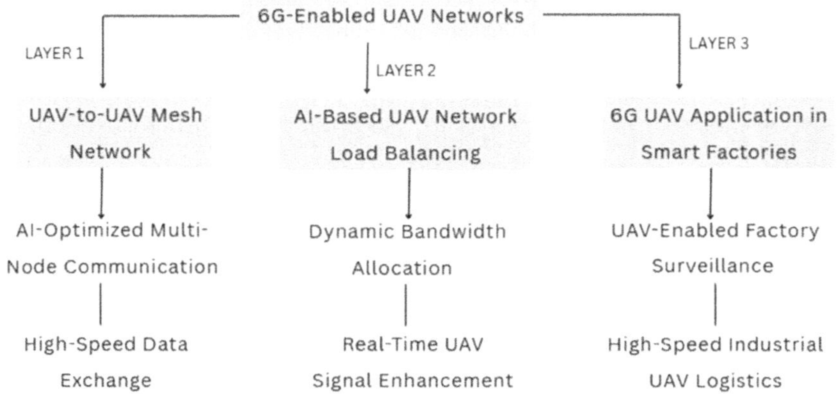

Figure 14.7 6G-enabled UAV networks.

Some of the capabilities of 6G networks, with data rates beyond 100 Gbps and latency less than one millisecond, are as follows:

- Response time will be almost instantaneous through the end-to-end autonomous coordination of UAV swarms, and thus, performance of such missions will be very close to real time.
- Real-time industrial AI analytics are employed at the network edge so as to reduce cloud dependence.
- By integrating adaptive mission execution algorithms, UAVs can change flight plans on the fly based on real-time environmental feedbacks.

Together with 6G connectivity, AI-powered drone networks will enable industrial drone fleets to operate autonomously through aerial logistics systems that can self-learn, self-optimize, and self-repair [31].

14.6.2 Quantum computing for UAV optimization

Quantum computing applications will be visible in drone data processing, flight control, and autonomous fleet management, enabling revolutionary advancements for multi-agent UAV coordination, real-time optimization, and predictive analytics. The traditional AI-based UAV decision models are not able to solve optimization problems in extremely high-dimensional spaces and are also inefficient in performing real-time mission adjustments, predictive fault detection, and rapid UAV swarm adaptation.

UAVs' processing power is increased by the use of quantum computing, which will enable them to take decisions in real time in multi-agent UAV networks with low energy consumption (Figure 14.8).

14.6.2.1 Quantum neural networks for UAV AI systems

Using quantum neural networks (QNNs) in UAV AI systems allows business to get:

- Almost real-time adjustment of drone fleets where several UAVs can change their flight path and distribution of resources on the fly.
- Better predictive maintenance through quantum tech, allowing 90% precision in the forecasting of airframe faults and irregularities.
- Historic levels of AI decision optimization, making UAVs' multi-variable logistics planning 100 times quicker than the current AI models.

Quantum-upgraded AI allows UAVs to operate on their own in industrial environments with self-adaptive mission planning and real-time operating intelligence without human intervention [32].

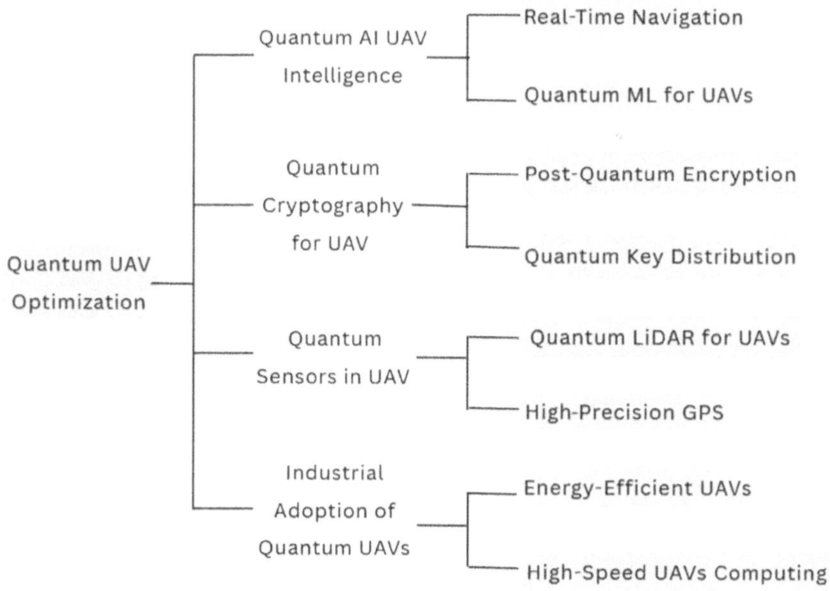

Figure 14.8 Quantum UAV optimization.

14.6.2.2 Quantum cryptographic UAV communication security

Besides increasing UAV employment in real-time industrial data gathering, defense against the supply chain, and infrastructure surveillance, UAV communication vulnerability and cyber-attacks are still challenges. Conventional methods of encryption are prone to advanced cyber-attacks and AI attacks' adversarial manipulations, and hence are insufficient to offer long-term UAV security.

Quantum cryptography fills UAV data security gaps by introducing quantum-secured key distribution (QKD) protocols, ensuring:

- 100% secure ground-UAV communication unaffected by any form of cyber-attack.
- Tamper-immune flight data integrity, with the mission logs of UAVs being immutable and blockchain-validated.
- Advanced AI-powered cybersecurity intelligence, enabling UAVs to detect and remove cyber threats in real time.

With quantum-secured communications systems, commercial drones will work in high-security production and transportation sectors, always providing smooth and uninterrupted function free from fear of data leaks.

14.6.3 Self-healing UAV materials and advanced composites

Because of the creation of self-healing UAV materials, the maintenance, structural repair, and downtime of UAVs are optimized. Hence, they can function in severe industrial settings. The next generation of drones will be outfitted with self-trained repairable composites, thus reducing operational costs and downtime to nearly zero, essentially achieving the goal of minimal maintenance and a slow wear-tear rate . This is possible due to the progress made in nano-engineered polymers, aerospace-grade composite materials, and shape-memory alloys.

14.6.3.1 Nanotechnology-enabled self-repairing UAV structures

Self-healing UAV materials will make the following possible:

- The real-time repair of micro-fractures will be possible, and thus, UAVs will be structurally safe in the environmental conditions of different industries.
- The lifetime of devices can be extended, whereas the replacement cycles can be reduced by 60% when utilized.
- Maintenance costs can be reduced to a level that progressions of UAVs will be possible without repeated human servicing [33] (Figure 14.9).

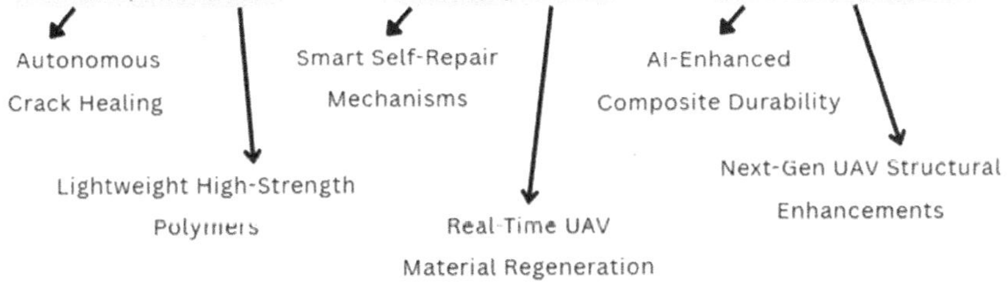

Figure 14.9 Self-healing UAV materials and nanotechnology.

14.6.3.2 Lightweight high-strength composites for UAV manufacturing

UAVs will employ carbon fiber-reinforced composite integration, graphene airframes, and high-temperature-resistant alloys not only boost structural toughness, weight efficiency, and energy consumption rates but also improve overall operational durability. These new-generation materials will:

- Cut energy consumption of UAVs by 30%, thereby extending their flight duration.
- Raise the impact resistance of UAVs to reduce structure damages so to be able to use in high-speed industrial applications.
- Develop heat resistance to grant UAVs access to hot areas, for instance, steel mills and chemical processing plants.

14.6.4 Quantum sensors and next-generation UAV perception systems

Future drones will be equipped with quantum-enhanced sensors and AI-based sensing technology to collect ground-breaking environmental data and extend their industrial monitoring capacities.

14.6.4.1 UAV-based quantum LiDAR and imaging systems

UAVs with quantum LiDAR are set to surpass the existing technologies in the following ways:

- 3D mapping with ultra-high-definition can present real-time industrial environments virtually.
- Improved defect detection and manufacturing quality control with sub-millimeter precision .
- Un-rivaled low-visibility perception which make UAVs competent in dangerous industrial areas [34].

These advancements in quantum-based UAV sensing systems will position UAVs at the forefront of high-precision industrial automation, logistics monitoring, and predictive maintenance.

14.7 CONCLUSION

The implementation of UAVs in manufacturing and logistics processes has revolutionized industrial operations by enabling real-time automation, AI-based decision automation and intelligent process optimization. The chapter provides insight into transition of UAVs from remote-controlled devices to fully autonomous units. Due to these technological breakthroughs, the advantages of achieving efficient operations, saving money, and accuracy improvement of the inspection, transport and infrastructure monitoring were materialized.

UAVs represent a future paradigm that will revolutionize Industry 5.0, where human–machine collaboration and cognitive automation can be regarded as the core of smart manufacturing. Future UAV systems, among other things, are expected not only to provide self-sufficient supply chains and predictive/condition-based maintenance but will also be the main drivers of decentralized decision-making through their integration with such technologies as quantum computing, federated learning, 6G networks, and self-healing materials. So, with innovation being at its peak, UAVs are going to be the pioneers of industrial ecosystems that are not only sustainable, adaptable, and resilient but also more intelligent, flexible, and scalable manufacturing processes.

14.7.1 Key takeaways and technological impact

Advancements in UAV technologies have transformed the industrial landscape, while the deployment of UAVs has significantly reduced industrial inefficiencies. The essence of this chapter is the overall influence of UAVs in industry operations and the future of smart ecosystems.

1. **UAVs in Manufacturing and Logistics:**
   Machine learning–based smart factory automation, warehouse inventory control, and predictive maintenance have greatly benefited from UAVs. Defect detection with high accuracy, energy-saving material transport, and real-time optimization of production were the key factors that made the role of UAVs in these activities even more critical. AI-driven UAV analytics has facilitated industrial sectors to reduce machine downtime, increase process transparency, and eliminate logistical inefficiencies, which, in turn, has contributed to a 23% increase in production efficiency and 35% reduction in maintenance costs (Figure 14.10).
2. **Advanced AI and Swarm Intelligence:**
   AI-powered UAVs with multi-agent swarm intelligence systems open the door to networked industrial automation, which enables drones to perform extensive aerial surveillance, autonomous logistics operations, and real-time local decision-making. The employment of RL algorithms has practically turned UAVs to self-optimized pilots which can foresee system breakdowns, reconfigure their tasks instantly, and, therefore, substantially raise their intellect and flexibility.
3. **Next-Generation UAV Energy Systems:**
   Technological advancements in hydrogen fuel cell–powered UAVs, solar-powered advanced aerial platforms, and hybrid propulsion systems have improved flight time, improved energy efficiency, and reduced

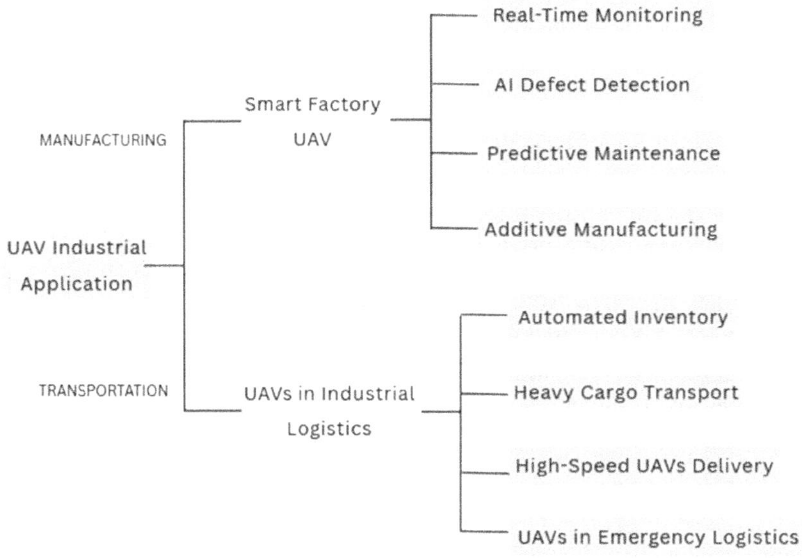

Figure 14.10 Industrial applications.

dependence on the lithium-ion battery technology. Commercial UAVs with next-generation propulsion technology have experienced 300% improvement in flight time, 50% improvement in payload, and significant energy savings, and can be used for long-endurance industrial applications and logistics automation.

4. **6G-Enabled UAV Communication and Edge Computing:**
   The evolution from 5G to 6G networking brought new possibilities in low-latency UAV fleet management, AI-powered air traffic management, and data processing at high speeds. Industries have introduced edge computing frameworks coupled with AI-powered UAV analytics to reduce real-time processing latency, provide better UAV swarm communication, and optimize autonomous mission performance. With these technologies, UAVs now independently work in real-time industry scenarios, handling massive datasets locally and limiting the need for cloud computing solutions.
5. **Quantum Computing and Secure UAV Operations:**
   The integration of quantum-enhanced AI models, quantum cryptography, and blockchain-encrypted UAV authentication has addressed fundamental issues related to cybersecurity, UAV data integrity, and autonomous decision-making transparency. Quantum neural networks will be used by next-generation UAV systems to carry out complex flight path optimizations 100 times more effectively than classical AI models, ensuring industrial UAV fleets operate at peak efficiency, resilience, and security.

14.7.2 Future outlook of UAVs in Industry 5.0

As industries shift from Industry 5.0, UAVs will lead to bridge the gap between man-machine collaboration and autonomous intelligence. Industry 4.0 focused on full automation, but Industry 5.0 focuses on adaptive intelligence, sustainability, and real-time cognitive automation as central themes, where UAVs are not just automated systems but an intelligent collaborators with advanced human communication capabilities.

These upcoming innovations are expected to impact the industrial UAV-powered utilization of the future:

- **AI-Augmented Human–UAV Collaboration:**
  Future UAVs will be designed with the next-generation human–machine interface (HMI), augmented reality–based UAV operating systems, and brain–computer interface (BCI) technology to allow people to control UAVs directly and operate them in seamless precision assembly, logistics control, and factory inspection.
- **Self-Healing UAVs with Adaptive Flight Materials:**
  Advancements in self-repairing nano-polymers and smart material engineering will enable UAVs to automatically detect structural damage and carry out self-repair operations, reducing maintenance requirements by 40% and extending operational lifespan by 60%.
- **Quantum Computing-Driven by UAV Decision Systems:**
  Quantum-enhanced decision model integration will enable UAVs to perform instant mission planning, real-time airborne swarm coordination, and high-speed UAV logistics optimization, guaranteeing exponential industrial UAV performance improvement.
- **Decentralization of AI-Driven UAV Logistics and Supply Chains:**
  Blockchain-enabled autonomous UAV logistics platforms will enable companies to establish end-to-end decentralized, AI-orchestrated supply chains so that actual-time demand dynamics, predictive material flow analysis, and industrial UAV fleet management can be optimized independent of human intervention.

With these innovations, UAVs will move beyond their conventional use as autonomous flying platforms and become intelligent, adaptive, and completely self-regulating parts of next-generation industrial ecosystems.

REFERENCES

1. Halefom, Mekonen H. "Tailoring an Airworthiness Document to Unmanned Aircraft Systems: A Case Study of MIL-HDBK-516C." PhD diss., Virginia Tech, 2020.
2. Sornek, Krzysztof, Joanna Augustyn-Nadzieja, Izabella Rosikoń, Róża Łopusiewicz, and Marta Łopusiewicz. "Status and Development Prospects of Solar-Powered Unmanned Aerial Vehicles—A Literature Review." *Energies* 18, no. 8 (2025): 1924.

3. Cao, Xiao, and Li Liu. "State-of-Charge Trajectory Planning for Low-Altitude Solar-Powered Convertible UAV by Driven Modes." *Energy Reports* 10 (2024): 775–784.
4. Wu, Xin, Wei Li, Danfeng Hong, and Jing Zhang. "Deep Learning for UAV-Based Object Detection and Tracking: A Survey." *Journal of Sensors and Information Fusion* 4, no. 2 (2023): 134–156.
5. Kurunathan, Harrison, Hailong Huang, Kai Li, Wei Ni, and Ekram Hossain. "Machine Learning-Aided Operations and Communications of Unmanned Aerial Vehicles: A Contemporary Survey." *IEEE Communications Surveys & Tutorials* 24, no. 1 (2022): 300–326.
6. Kong, Weiwei, Dianle Zhou, Daibing Zhang, and Jianwei Zhang. "Vision-Based Autonomous Landing System for Unmanned Aerial Vehicle: A Survey." *Journal of Field Robotics* 40, no. 1 (2023): 78–101.
7. Nguyen, Tri, Huong Nguyen, and Tuan Nguyen Gia. "Exploring the Integration of Edge Computing and Blockchain IoT: Principles, Architectures, Security, and Applications." *Journal of Network and Computer Applications* 226 (2024): 103884.
8. AL-Syouf, Raghad A., Raed M. Bani-Hani, and Omar Y. AL-Jarrah. "Machine Learning Approaches to Intrusion Detection in Unmanned Aerial Vehicles (UAVs)." *Neural Computing and Applications* 36, no. 29 (2024): 18009–18041.
9. Samanth, Snehal, Prema KV, and Mamatha Balachandra. "Security in Internet of Drones: A Comprehensive Review." *Cogent Engineering* 9, no. 1 (2022): 2029080.
10. Davies, Ethan, and Pranav Kalidindi. "Quantum Algorithms for Drone Mission Planning." In *Quantum Technologies for Defence and Security*, vol. 13202, pp. 67–80. SPIE, 2024.
11. Osaba, Eneko, Esther Villar-Rodriguez, and Izaskun Oregi. "Solving Drone Routing Problems with Quantum Computing: A Hybrid Approach Combining Quantum Annealing and Gate-Based Paradigms." *IEEE Access* 13 (2025): 99234–99250.
12. Yezeguelian, Axel, and Askin T. Isikveren. "Hybrid-Electric Propulsion Architecture for UAV Performance Optimization." *Aircraft Engineering and Aerospace Technology* 97, no. 1 (2023): 35–49.
13. Simic, Milan, Cees Bil, and Vuk Vojisavljevic. "Investigation in Wireless Power Transmission for UAV Charging." *Procedia Computer Science* 60 (2022): 1846–1855.
14. FAA. *UAS Traffic Management (UTM) Concept of Operations v3.0*. U.S. Federal Aviation Administration, 2023.
15. Mohsen Soori, Fooad Karimi Ghaleh Jough, Roza Dastres, and Behrooz Arezoo. "AI-Based Decision Support Systems in Industry 4.0, A Review." *Journal of Economy and Technology* (Aug. 2024). https://doi.org/10.1016/j.ject.2024.08.005
16. Hawashin, Diana, Mohamed Nemer, Senay A. Gebreab, Khaled Salah, Raja Jayaraman, Muhammad Khurram Khan, and Ernesto Damiani. "Blockchain Applications in UAV Industry: Review, Opportunities, and Challenges." *Journal of Network and Computer Applications* 230 (2024): 103932.
17. McConnell, Vicki P. "Fuel Cell Integration in Modern UAVs for Defense and Industrial Missions." *Fuel Cells Bulletin* 2023, no. 6 (2023): 18–22.
18. Tezza, Dante, and Marvin Andujar. "The State-of-the-Art of Human–Drone Interaction: A Survey." *IEEE Access* 7 (2021): 167438–167454.

19. Lykou, Georgia, Dimitrios Moustakas, and Dimitris Gritzalis. "Defending Airports from UAS: A Survey on Cyber-Attacks and Counter-Drone Sensing Technologies." *Sensors* 20, no. 12 (2021): 3537.
20. Henderson, Isaac Levi. "Aviation Safety Regulations for Unmanned Aircraft Operations: Perspectives from Users." *Transport Policy* 125 (2022): 192–206.
21. Politi, Elena, Ilias E. Panagiotopoulos, Iraklis Varlamis, and George Dimitrakopoulos. "A Survey of UAS Technologies to Enable Beyond Visual Line Of Sight (BVLOS) Operations." In *7th International Conference on Vehicle Technology and Intelligent Transport Systems VEHITS*, Prague, pp. 505–512, 2021.
22. Cheng, Nan, Shen Wu, Xiucheng Wang, Zhisheng Yin, Changle Li, Wen Chen, and Fangjiong Chen. "AI for UAV-Assisted IoT Applications: A Comprehensive Review." *IEEE Internet of Things Journal* 10, no. 16 (2023): 14438–14461.
23. Gupta, Anunay, Tanzina Afrin, Evan Scully, and Nita Yodo. "Advances of UAVs Toward Future Transportation: The State-of-the-Art, Challenges, and Opportunities." *Future Transportation* 1, no. 2 (2021): 326–350.
24. Al-Dosari, Khalifa, and Noora Fetais. "Implementing UAVs in the Safety and Security of Smart Cities: A Systematic Literature Review." *Safety* 9, no. 3 (2023): 64.
25. Balamurugan, Ganesan, J. Valarmathi, and V. P. S. Naidu. "Survey on UAV Navigation in GPS Denied Environments." In *2022 International Conference on Signal Processing, Communication, Power and Embedded System (SCOPES)*, pp. 198–204. IEEE, 2022.
26. Alsamhi, Saeed Hamood, Faris A. Almalki, Fatemeh Afghah, Ammar Hawbani, Alexey V. Shvetsov, Brian Lee, and Houbing Song. "Drones' Edge Intelligence over Smart Environments in B5G: Blockchain and Federated Learning Synergy." *IEEE Transactions on Green Communications and Networking* 6, no. 1 (2021): 295–312.
27. Konstantoudakis, Konstantinos, Kyriaki Christaki, Dimitrios Tsiakmakis, Dimitrios Sainidis, Georgios Albanis, Anastasios Dimou, and Petros Daras. "Drone Control in AR: An Intuitive System for Gesture Control, Drone Tracking, and Camera Feed Visualization." *Drones* 6, no. 2 (2022): 43.
28. Ralegankar, Vishakha K., Jagruti Bagul, Bhaumikkumar Thakkar, Rajesh Gupta, Sudeep Tanwar, Gulshan Sharma, and Innocent E. Davidson. "Quantum Cryptography-as-a-Service for Secure UAV Communication: Applications, Challenges, and Case Study." *IEEE Access* 10 (2021): 1475–1492.
29. Feriani, Amal, and Ekram Hossain. "Single and Multi-Agent Deep Reinforcement Learning for AI-Enabled Wireless Networks: A Tutorial." *IEEE Communications Surveys & Tutorials* 23, no. 2 (2021): 1226–1252.
30. Sai, Siva, Akshat Garg, Kartik Jhawar, Vinay Chamola, and Biplab Sikdar. "A Comprehensive Survey on Artificial Intelligence for Unmanned Aerial Vehicles." *IEEE Open Journal of Vehicular Technology*, pp. 1–26 (2023).
31. Kadadha, Maha, Rabeb Mizouni, Shakti Singh, Hadi Otrok, and Azzam Mourad. "Crowdsourced Vehicles and UAVs for Last-Mile Delivery Application Using Blockchain-Hosted Matching Mechanism." *Vehicular Communications* 47 (2024): 100761.

32. Stodola, Jiří, and Petr Stodola. "Possibilities of Nanotechnologies for Unmanned Autonomous Vehicles." In *International Workshop on Modelling and Simulation for Autonomous Systems*, pp. 422–433. Cham: Springer International Publishing, 2022.
33. Goyal, S. B., Anand Singh Rajawat, Ram Kumar Solanki, Liu Zhu, and Weiming Chee. "Enhancing Privacy and Security for UAV and IoT Enabled Drones: An Intelligent Integration of Blockchain, AI, and Quantum Computing." In *International Conference on Intelligent Computing & Optimization*, pp. 16–27. Cham: Springer Nature Switzerland, 2023.
34. ISO 21384-3:2023. *Unmanned Aircraft Systems — Part 3: Operational Procedures*. International Organization for Standardization.

Index

Note: **Bold** page numbers refer to tables and *italic* page numbers refer to figures.

Accelerometers, 30, 32, 40, 105, 107, 109, 111, 114, 118, 121, 125
  calibration of, 100, 102
  and gyro, 116
  quantum, 48
Acoustic/ultrasonic sensors, 32
Advanced Encryption Standard (AES), 221
Advanced lithium-ion batteries, 17
Aerial photogrammetry
  drone-based, 314
  low-altitude, 315
Aerial photography, for real estate marketing, *318*, 318–319
"Aerial Targets", 5
Aeroqual PID sensors, **271**, *273*, **274**
AES-128, 176, 221
Age of covert information (AoCI), 174
Agribusiness Students Female (ASF)-29, awareness of, **263–264**
Agribusiness Students Male (ASM)-36, awareness of, **261–262**
Agriculture, UAVs in, 19–20, 237–238
  advantages of UAVs, 243
  challenges of UAVs, 242–244
  cost-effectiveness, 240
    accuracy and use of data, 240
  efficiency and productivity, 240
  environmental impact, 240
  interpretation of awareness of UAV challenges, 248
    additional value of present research, 249–250
    agribusiness students (male and female, ASM and ASF), 249
    recommendations to study findings, 249
    rural farmers (female and male, RFM and RFF), 248
    urban farmers (male and female, UFM and UFF), 248
  interpretation of data on awareness of UAV benefits, 247, *248*
  limitations of study, 250
  method of survey and demographics of participants, 245, **246–247**
  need for current study, 241
  perks of using UAVs for farming, 238–239
  present trends of UAV in India, 242–243
  rationality of study/need of study, 244
  regulatory limitations, 241
  research design, 244–245
  research questions, 241
  safety issues, 241
  sample size determination, 245, **246**
  scope for future study, 250
  scope of study, 245
  technical challenges, 242
AI-augmented human–UAV collaboration, 379
AI-based predictive failure modeling, 341–342
AI-driven machine vision analytics, 367
AI-driven UAV
  coordination for multi-agent emergency response, 359
  system interoperability, 371
  traffic management systems, 355

AI-enabled autonomous UAV collision avoidance, 366–367
AI-guided UAV swarms, 354
AI-powered UAVs
emergency logistics systems, 359
with multi-agent swarm intelligence systems, 377
Airborne Command Operations (ACOs), 267
Aircraft Act 1934, 68
Aircraft Rules 1937, 68
Air India Ltd. v. Cochin International Airport Ltd. (n.d.) case, 68
AirSim (Aerial Informatics and Robotics Simulation), 86
Air traffic control (ATC) systems, 369
All-In-One (AIO), 110
ALOHA, 177
American Telephone and Telegraph (AT&T) Company, 314
Antović and Mirković v. Montenegro (n.d.), 70
Application Program Interfaces (APIs), REST, 182–183
ArduPilot, 84, 88–90, 124
firmware, 104–106
open-source autopilot, 144
open-source firmware, 108
Windows GCS for, 204
ArduPilot Mega (APM), 103, *104*
calibration, 105
challenges, 104
cons, 105–106
pros, 105
Artificial intelligence (AI)
with machine learning algorithms, 3
and swarm intelligence for UAVs, 361–364, *362*
Asymmetric cryptography, 218
MAVLink, 210–212
Authentication, 19, 176, 225
asymmetric, 223
and authorization, 146–147
blockchain-enabled, 372
blockchain-enabled UAV, 369
blockchain-encrypted UAV, 378
blockchain-secured UAV, 356
certificate-based, 212
credentials, 222
encryption and, 151, 182
inadequate, 208
and key establishment in drone networks, 218
message, 210, 212
ML-DSA for drone, 231
protocols, 137
UAV, 370
user mutual, 208
Automated warehouse mapping and stock auditing, 345
Autonomous drones, in conflict zones, 24
Autonomously guided vehicles (AGVs), 337
Autonomous Vehicle for Environmental Research (AVE-R) platform, 188–189, *189*
Awareness
Agribusiness Students Female (ASF)-29, **263–264**
Agribusiness Students Male (ASM)-36, **261–262**
Rural Farmers Female (RFF)-16, **255–256**
Rural Farmers Male (RFM)-34, **253–254**
Urban Farmers Female (UFF)-24, **259–260**
Urban Farmers Male (UFM)-43, **257–258**

Barometer, 31, 32, 38, 39, 42–44, **97**, 113
Barometric altimeter, 42–43
Beyond-visual-line-of-sight (BVLOS), 13, 368
flights, long-endurance, 9
operations, 3, 47, 66
Bias drift, 40
BIM. *See* Building information modelling (BIM)
Bio-inspired drone design, 11
Blended-wing body (BWB) drones, 12
Blockchain-backed logging system, 278
Blockchain-enabled autonomous UAV logistics, 379
Boggs v. Merideth (n.d.), 70
Brushless DC (BLDC) motors, 17
Building information modelling (BIM), 309
BVLOS. *See* beyond-visual-line-of-sight (BVLOS)
BWB drones, 12, 15

Calibration methods, IMU errors, 40
California Consumer Privacy Act (CCPA), 369

California Wildfires 2022, 299
Cameras
  infrared (thermal), 34–35
  long-wave infrared, 34
  mid-wave infrared, 34
  RGB, 33–34
  stereo and depth, 35–36
CAN. *See* Controller area network (CAN)
Carbon fiber, 17
Carrier-Sense Multiple Access with Collision Avoidance (CSMA/CA), 177
Checksum techniques (CRC), 208
Chemical sensors, 44
Chicago Convention on International Civil Aviation, 56
Cinematography, UAVs applications in, 307
CISA. *See* Cybersecurity and Infrastructure Security Agency (CISA)
The Civil Aviation Safety Authority v. Marsh 2014, 70
Classical encryption, 219, 221
Cloud-based intelligence, *172*
Cluster Head (CH), 171
  UAV transmits, 185–186
CNN. *See* Convolutional neural network (CNN)
CoAP. *See* Constrained Application Protocol (CoAP)
Code-based cryptography, 225
Cold chain logistics
  in industrial supply chains, 346–347
  with UAVs, 356–357
Combat drones, 307
Commercial drone, 6, 98
Commercial flight controller, 122–126
Communication and control, 18, *18*
  BVLOS operation, 18–19
  LOS (line of sight) operation, 18–19
  redundancy and secure protocols, 19
Communication interfaces and synchronization, 46–47
Communication protocols, 8, 104, 106, 130, 133, 369
  decentralized, 363
  integration with emerging, 214
  lightweight, 197
  packet-based, 197
  UAV swarm, 176
Compass (magnetometer), 97
Complementary filter, 40
Complementary Metal-Oxide-Semiconductor (CMOS) sensor, 33
Complex Event Processor, 175
Computer vision systems, MAVLink in, 207
Constrained Application Protocol (CoAP), 173, *183*, 183–185, *185*
Consumer drones, evolution of, 6
Controller area network (CAN), 46
Convolutional neural network (CNN), 271, 323, 364
CopterSim, 83, 91–93
Counter-drone technology, 10–11, 242
CPSs. *See* Cyber-Physical Systems (CPSs)
Criminal trespass, 67
Cryptographic algorithms, for drone networks, 212, 227
Cryptographic protocols, 217, 370
Cryptography
  code-based, 225
  lattice-based, 224–226, *226*
  multivariate, 225
Cube. *See* Pixhawk 2.1 (Cube)
Cyber-physical systems (CPSs), 96, 171, 340, 372
Cybersecurity
  issues in UAV airspaces, 356
  and UAV data protection, 369–370
Cybersecurity and Infrastructure Security Agency (CISA), 222

DART. *See* Dynamic Animation and Robotics Toolkit (DART)
Data
  compression and transmission, 47
  transmission rates, 177
Data Distribution Service (DDS) protocol, 137
Datagram Transport-Layer Security (DTLS), 173, 212
Decentralization of AI-driven UAV logistics and supply chains, 379
Deep neural networks (DNNs), 45, 364
Deep Q Network (DQN), 267
Deep Reinforcement Learning (RL), 86, 267, 362
Defence and security, UAVs applications in, 307

Defense Advanced Research Projects Agency (DARPA), 130
DEMs. *See* Digital Elevation Models (DEMs)
Depth fusion, 36
Detect-and-avoid (DAA) systems, 368
DGCA. *See* Directorate General of Civil Aviation (DGCA)
DGCA's Digital Sky, 64
Differential GPS (DGPS), 41
Diffie–Hellman, elliptic curve, 217, 222, 223
Diffie–Hellman key exchange scheme, 219
Digital Elevation Models (DEMs), 320
Digital Personal Data Protection Act, 73
Digital surface models (DSM), 310
Digital Terrain Models (DTMs), 320
Directorate General of Civil Aviation (DGCA), 63, 68, 241
Direct time-of-flight (dToF) method, 36
Disaster management
  application of UAV swarms in, 286
    damage evaluation and structural assessment, 286–287
    search-and-rescue operations, 286
  case studies and real-world deployments, 298
    Earthquake response, 298–299
    flood management and predictive modeling, Southeast Asia Floods 2021, 299–300
    industrial disaster management, China 2021, 300–301
    telecommunication restoration in disaster zones, Puerto Rico Earthquake 2020, 300
    wildfire monitoring and containment, California Wildfires 2022, 299
  challenges in implementing UAV sensor systems, 291–292
    data overload and bandwidth constraints, 293
    environmental constraints and weather resistance, 293–294
    high power consumption and limited mission duration, 292–293
    integration with disaster response frameworks, 294
    real-world relevance, 294–295
    sensor alignment and calibration in dynamic environments, 292
  communication system technologies, 281
    LoRa (long range), 281–282
    LTE (4G), 282
    mesh networks (IEEE 802.11s), 282
    satellite communication, 282
  data-driven review of current disaster response inefficiencies, 266
  data processing and AI-driven insights, 288–289
    AI-driven sensor fusion for enhanced detection, 289
    mechanism of fusion, 289
    performance metrics, 289
    real-time data processing with edge computing, 290
  edge computing platform, 274
  energy optimization strategies, 282–283
  energy systems, 275
    collaborative power sharing, 275
    solar-powered UAVs, 275, 275
  flood mapping and management, 287–288
  future trends and innovations in UAV technology for, 295
    AR and VR for enhanced visualization, 297–298
    predictive AI models for disaster forecasting, 296
    quantum sensors for ultra-precise measurements, 295
    renewable energy and hybrid power systems, 297
    6G communication and terahertz technologies, 296–297
  hazard monitoring and threat identification, 287
  multi-sensor UAV swarms, 288
  operational challenges during disaster deployments, 284–285
    ethical and data privacy issues, 285
    flight endurance and power limitations, 285–286
    regulatory constraints and airspace restrictions, 285

operational workflow of UAV swarms, 276
continuous improvement protocols, 278, **278**
data collection, 277, **277**
deployment and navigation, 276–277, **276**
emergency protocols and risk mitigation, 280, **280**
mission planning and initialization, 276, **276**
post-mission analysis and blockchain logging, 278
review and adaptive strategies, 280, **281**, *281*
task allocation and resource management, 277–278, **277**
training and skills development, 279, **279**
vendor support and maintenance, 278–279, **279**
predictive maintenance models using machine learning, 290–291
resource delivery and telecommunication restoration, 288
role of UAV technologies, 265–266
sensor technologies, 270–271, **271**, *272–273*, **274**
swarm intelligence and sensor integration, 266–267
fixed-wing UAVs, 268, *268*, **270**
hybrid UAVs, 269, *269*, **270**
rotary-wing UAVs, 269, *269*, **270**
system architecture overview, 283
command and control flow layer, 284
communication and network coordination layer, 284
edge processing and data acquisition layer, 284
sensor integration layer, 283
UAV
platforms, *268*, 268–269, *269*, **270**
sensor classification and disaster applications, **271**, *272*, *273*
swarm architecture, 268
visualization and decision-making tools, 291
Disaster response, 8
Discrete logarithm problem (DLP), 219
DJI Matrice 300 RTK, 269, *269*
DJI Matrice 600, 123
calibration, 123
challenges, 123
disadvantages, 123
pros, 123
DJI NAZA A3, 121
calibration, 121
challenges, 121
cons, 121–122
pros, 121
DJI NAZA H, 122
calibration, 122
challenges, 122
cons, 122
pros, 122
DJI NAZA M Lite, 108, 110
calibration, 109
challenges, 109
cons, 109–110
pros, 109
DJI NAZA N3, 120
calibration, 120
challenges, 120
cons, 121
pros, 120–121
DJI Phantom in 2012, 5–6
DLP. *See* Discrete logarithm problem (DLP)
DNNs. *See* Deep neural networks (DNNs)
Doppler effect, 31, 38
Drone, *See also* Unmanned Aerial System (UAS); Unmanned aerial vehicles (UAVs); *specific types*
advent of ruggedized, 7
authentication, ML-DSA for, 231
based logistics systems, 10
benefits of, 21
capabilities of, 29
classification
autonomy level, 13, *14*
design, 14–15
function, 13–14
type and size, 11–12, *12*, **13**
composite frames of, 17
delivery networks, 10
design, bio-inspired, 11
emerging technologies and trends
autonomy and AI, 22

Disaster management (*cont.*)
delivery and logistics, 23
extended flight and energy sources, 22
integrated Internet of Things drones, 23
smart communication and UTM, 22–23
swarm intelligence, 22
enabled NDVI-based fertilizer allocation, 20
environmental impact, 24
for environmental monitoring, 14
ethical and legal regulatory, 24
future prospects, 24
issues on airspace, 23
leading industry players, 25
market size 2020–2030, *4*
MAVLink in, 200
operators, 60
privacy and security, 23
properties of, 15–16
reduce carbon footprints, 24
regulatory environment
global developments (2020–2025), 21
international frameworks, 21
representative application domains, 19–20
ROS-enabled, 130
sensors, 48–49
structural materials, 17–18
training, 242
Drone communication, 217
cryptographic challenges arise in, 218
harvest now, decrypt later attacks on, 221–224
lattice-based cryptography for, 225–226, *226*
closest vector problem, 228
important lattice problems, 227
learning with errors, 227
mathematical basis of lattices, 226–227
shortest independent vectors problem, 228
shortest vector problem, 228
lattice-based key encapsulation, 228–229
MAVLink in advancing, 214
networks, 217
quantum-resistant approaches for, 224–225
quantum-safe, future direction for, 231–232
Drone communication security
NIST standardization and the future of, 229
quantum computing threats to, 220–221
Drone communication system, 231
Drone evolution
civil and commercial drone expansion, 5–6
contemporary trends (2020–2025), 6–7, *7*
bio-inspired drone design, 11
drone delivery networks, 10
5G-enabled drones, 9–10
integration of AI and machine learning, 7–8
long-endurance and BVLOS flights, 9
swarm drones, 8–9
early developments, 5, *6*
Drone key exchange, ML-KEM for, 230–231
Drone networks, 217
cryptographic algorithms for, 227
Drones Air Space Management Cell (DASMC), 242
Drone security
architecture, 229
post-quantum, 232
communication, 220
cryptographic landscape in, 218–220
protocols, 217
SLH-DSA for conservative, 231
DroneSim Pro, 83, 91
Drone swarm network architecture, 220, *220*
Drone system architectures
flight controller, 16, *16*
propulsion and energy systems, motors and propellers, 16–17
Drone technology, 5–7, 217
DSM. *See* Digital surface models (DSM)
DTMs. *See* Digital Terrain Models (DTMs)
Ducted-fan drones, 12, 15
Dynamic Animation and Robotics Toolkit (DART), 84

Earthquake response, UAV, 298–299
ECDH. *See* Elliptic Curve Diffie–Hellman (ECDH)
ECDLP. *See* Elliptic Curve Discrete Logarithm Problem (ECDLP)
ECDSA. *See* Elliptic Curve Digital Signature Algorithm (ECDSA)
Edge computing and real-time data processing, 364–365
EKFs. *See* Extended Kalman filters (EKFs)
Electric *vs.* hybrid UAV propulsion efficiency, 361
Electronic speed controller (ESC), **97**, 107
Elliptic Curve Cryptography (ECC), 217, 219
Elliptic Curve Diffie–Hellman (ECDH), 230
Elliptic Curve Digital Signature Algorithm (ECDSA), 231
Elliptic Curve Discrete Logarithm Problem (ECDLP), 219
Emergency and disaster response logistics, 357–359
Emergency Services and Disaster Response, 318
Encryption, 212, 222–225
  absence of default, 208
  algorithms, 137
  authenticated, 229
  and authentication, 151, 182
  classical, 219, 221
  conventional methods of, 374
  for data confidentiality, 176
  and decryption of data, 209
  end-to-end, 369
  lattice-based, 227
  old-fashioned, 369
  RSA, 219
  symmetric, 210, 212, 221, 223, 229, 230
  transport-level, 215
End-to-end encryption, 369
Energy consumption, 174
Energy system, advances in, 3
Entertainment, UAVs applications in, 307
Environmental and mission-specific sensors, 43–44
Environmental monitoring, drones for, 14
Environmental sensors, 30, 31, 43
Ethernet, 46–47, 198, 201
European Aviation Safety Agency, 58
European Court of Human Rights (ECHR), 70
European Union Aviation Safety Agency (EASA), 368
Extended Kalman filters (EKFs), 44–45

FC. *See* Flight controller (FC)
Federal Aviation Administration (FAA), 54, 315, 368
  for pilot training, 88
Federal Trade Commission, 60
field of view (FOV), 33
Filters
  complementary, 40
  Kalman. *See* Kalman filters
  particle, 45
Fire behavior observation, 271
First-person view (FPV)
  drones, 307
  racing drones, 110
5G, 7
  5G-enabled drones, 9–10
Fixed-pitch propellers, 17
Fixed-wing drones, 12, 15, 242–243, 268, *268*, **270**
Flash LiDAR, 37
Flight controller (FC), 40, 96–97
  ArduPilot Mega, 103–106, *104*
  classification of different, 98
  commercial, 122–126
  core components of, **97**
  core functions of, 98
  DJI NAZA M Lite, 108–110, *110*
  in filming category, 98
  for hobbyists/DIY builders, 98
  K.K.2.1.5, 98–101, *99*, *100*
  Pixhawk, 106–108, *107*
  racing, 98, 110–118, *See also* Racing flight controller
  role in UAV systems, 97
  Thunder QQ Super, 101–103
Flight endurance, 9
FlightGear, 90–91
Flood control and prediction, 270
Flood management and predictive modeling, Southeast Asia Floods 2021, 299–300
Flywoo GOKU F405 HD 1-2S 12A AIO V2 FC, 117

Flywoo GOKU (*cont.*)
  calibration, 118
  challenges, 116
  cons, 118
  pros, 117
Flywoo GOKU GN745 AIO V3, 114–115
  calibration, 115
  challenges, 114–115
  cons, 115
  pros, 115
FMCW. *See* Frequency-modulated continuous wave (FMCW)
Forward Error Correction (FEC) level, 186, 192, *192*
FPGA-accelerated UAV edge processing, 365
Frame rate, 33, 48
Frequency-modulated continuous wave (FMCW), 38
Fuel cells, hydrogen, 9, 17

Galois/Counter Mode (AES-256-GCM), 210
Gas leak detection, 44
Gazebo, 80, 82, 83, 85, 87, 130, 133, 147–148, 151
  interface to ROS, 143–144
  simulation, 149
  3D simulation program, 143
  using ROS with, 147–151
Gazebo-compatible simulation environment, 144
Gazebo+PX4, 83–84
  Gazebo+PX4+ROS integration, 85
  key features, 84–85
  workflow, 85
GCPs. *See* Ground control points (GCPs)
GCS. *See* Ground control station (GCS); Ground control system (GCS)
General Data Protection Regulation (GDPR), 369
General Number Field Sieve (GNFS), 218–219
Gigabit Ethernet, 46–47
Gilbert–Elliott channel model, 185
Global positioning system (GPS), 32, 96
  denied navigation, 351
  module, 5, 97
  signal, 41, 147
GNDVI. *See* Green normalized difference vegetation index (GNDVI)
GNFS. *See* General Number Field Sieve (GNFS)
GNSS (GPS, GLONASS, Galileo, BeiDou), 41–42
GNSS-denied navigation, 41–42
GOKU GN745, 114
GPS. *See* Global positioning system (GPS)
Graphical User Interface (GUI), 102
Green normalized difference vegetation index (GNDVI), 43–44
Green UAV technologies, 335
Ground control points (GCPs), 311
Ground control station (GCS), 197, 323–324
  MAVLink and, 204–205
Ground control system (GCS), 81
Group-based Social Navigation (GSON) framework, 128
Grover's algorithm, 223
Grover's quantum search algorithm, 221
Gyroscopes, 30, 32, 40, 48, 105, 107, 109, 118, 119, 121, 125

HappyModel X12AIO, 115
  calibration, 116
  challenges, 116
  cons, 116
  pros, 116
Hardware-in-the-loop (HITL) simulators, 80–83
Harvest now, decrypt later (HNDL)
  attacks on drone communication, 221–224
  threat, 224
    for communication with drones, 223
Hash-based Message Authentication Code (HMAC), 210
Hash-based signatures, 224–225
HEARTBEAT, 198
Hexacopters, 15, 101
High-speed logistics and material handling, 346
HITL simulators. *See* Hardware-in-the-loop (HITL) simulators

HMAC. *See* Hash-based Message Authentication Code (HMAC)
Holybro Pixhawk 4, 118
  calibration, 118–119
  challenges, 118
  disadvantages, 119
  pros, 119
Hotter objects, 34
Human–AI collaborative UAV systems, 335
Human–UAV cooperative assembly line operations, 348
Humidity sensors, 31, 43
Hybrid (fuel-electric) battery, **13**
Hybrid drones, 17, 269, *269*, **270**
Hydrogen fuel cell, 3, 9, 17, 22, 283, 285, 353, 360
  hydrogen fuel cell–powered UAVs, 360–361
Hyperspectral sensors, 3, 43, 158
Hypertext Transfer Protocol (HTTP), 182, 183

ICAO. *See* International Civil Aviation Organisation (ICAO)
IDEs. *See* Integrated development environments (IDEs)
IETF. *See* Internet Engineering Task Force (IETF)
iFlight Beast F7 AIO 55A, 113
  calibration, 114
  challenges, 113
  cons, 114
  pros, 114
Image-processing algorithms, 33, 158
IMU. *See* Inertial measurement unit (IMU)
Indian Telegraph Act of 1885, 65
Industrial AI-optimized drones, 362
Industrial disaster management, China 2021, 300–301
Industrial risk mitigation and fire response, 358
Industrial transportation and logistics, 352–353
Industrial UAVs, 333
  in additive and hybrid manufacturing, 339
  advanced AI and swarm intelligence, 377
  advanced UAV navigation and autonomy, 366–367
  AI and swarm intelligence for UAVs, 361–364, *362*
  AI-based predictive failure modeling, 341–342
  AI-driven UAV coordination for multi-agent emergency response, 359
  AI-driven UAV system interoperability, 371
  AI-enabled autonomous UAV collision avoidance, 366–367
  challenges and solution in industrial UAV technology, 367–371
  for cold chain logistics in industrial supply chains, 346–347
  cold chain logistics with UAVs, 356–357
  collaboration in smart factories, 347–348
  cybersecurity and UAV data protection, 369–370
  edge computing and real-time data processing, 364–365
  electric *vs.* hybrid UAV propulsion efficiency, 361
  in emergency and disaster response logistics, 357–359
  enabling technologies for industrial UAV operations, 351
    GPS-denied navigation, 351
    IoT integration and edge connectivity, 352
    LiDAR-based spatial awareness, 351
    SLAM, 352
  evolution of UAVs in industrial applications, 333–334, *334*
  FPGA-accelerated UAV edge processing, 365
  future outlook of UAVs in Industry 5.0, 379
  future trends and research directions, 371–376
  in high-speed logistics and material handling, 346
  human–UAV cooperative assembly line operations, 348
  hydrogen fuel cell–powered UAVs, 360–361
  implementation of UAVs in manufacturing and logistics, 376–377

Industrial UAVs (*cont.*)
  industrial integration and scalability challenges, 370–371
  in industrial manufacturing, 337
  for industrial risk mitigation and fire response, 358
  in industrial transportation and logistics, 352–353
  integration in smart manufacturing execution systems (MESs), 342
  key performance indicators (KPIs), 336
  last-mile delivery systems with UAV swarms, 354
  latency-optimized UAV communication networks, 365
  lightweight high-strength composites for UAV manufacturing, 376
  in manufacturing and logistics, 377–378, *378*
  market growth and adoption trends, 335–336
  multi-agent UAV coordination for industrial applications, 363
  nanotechnology-enabled self-repairing UAV structures, 375, *375*
  neural network–enabled predictive path planning, 364
  next-generation UAV energy systems, 377–378
  for predictive maintenance, 338, *339*
  for process optimization and waste reduction, 343
  propulsion and energy systems, 360
  QNNs in UAV AI systems, 373, *374*
  quantum computing
    and secure UAV operations, 378
    for UAV optimization, 373
  quantum cryptographic
    security for UAV networks, 369–370
    UAV communication security, 374
  quantum sensors and next-generation UAV perception systems, 376
  for real-time anomaly detection in smart factories, 349–350
  regulatory and compliance challenges, 368
    privacy and data security regulations, 369
    UAV flight restrictions and legal barriers, 368–369
  reinforcement learning–based UAV optimization, 362–363
  role of UAVs in Industry 4.0 and 5.0, 334–335
  self-healing UAV materials and advanced composites, 374–376
  6G-enabled UAV communication and edge computing, 378
  6G-enabled UAV networks for industrial applications, *372*, 372–373
  SLAM-based UAV navigation in GPS-denied environments, 366
  solar-powered UAVs for extended operations, 361
  in supply chain and inventory management, 344
    automated warehouse mapping and stock auditing, 345
    just-in-time logistics optimization, 344–345
  technological impact, 377–378
  technological innovations in UAVs, 360–367
  for telecommunications and power restoration during crises, 358–359
  traffic management systems, 354, *355*
    5G and 6G in UAV fleet communication, 355
    AI-powered UAV air traffic optimization, 355
    blockchain-based secure UAV air traffic control, 356, *356*
  UAV-assisted disaster relief and supply drops, 357–358
  UAV-assisted ergonomic optimization in industrial workflows, 349
  UAV-assisted smart manufacturing, 337
    quality control and defect detection using AI, 338
    real-time monitoring and automation, 337–338
    role of UAVs in precision machining, 339

UAV-based energy management in smart factories, 342–343
UAV-based logistics and freight transport, 353
high-capacity UAVs for heavy cargo transport, 353
UAV-based quantum LiDAR and imaging systems, 376
UAV-based RFID and IoT synchronization, 345–346
UAV-based supply chain automation, 347
UAV-digital twin integration, future implications, 343
UAV-driven real-time factory simulation, 340–341
UAV-enabled AI and machine learning integration for smart factory intelligence, 350
UAV-enabled digital twin integration, 340, *340*
UAV-integrated supply chain optimization, 354
UAV-to-IoT infrastructure integration, 371
vision systems for industrial inspection, 367
vulnerabilities and countermeasures, 370
Industry 4.0, 183, 309, 333
role of UAVs in, 334–335
Industry 5.0
future outlook of UAVs in, 379
UAV applications, 335
Inertial measurement unit (IMU), 39–41, 97
in attitude estimation, 40
MEMS-based, 40
role of, 40
Inertial/mechanical sensors, 32
Inertial navigation systems (INS), 276
Infrared (thermal) cameras, 34–35
Infrastructure and traffic management, 317
Input/output ports, 97
Integrated development environments (IDEs), 144
Intelligence, Surveillance, and Reconnaissance (ISR), 222
I2C (Inter-Integrated Circuit), 46
Inter-ministerial committee, 242
International Civil Aviation Organisation (ICAO), 54, 55, 368
Convention on International Civil Aviation, 56
Remotely Piloted Aircraft Systems (RPAS), 55, 56
Remotely Piloted Aircraft Systems Panel, 54
role of, 55–58
SARPs, 56
strategies, 57
UN ICAO, 55
Internet Engineering Task Force (IETF), 173, 183
Internet of Things (IoT)
drones, integration with, 23
integration and edge connectivity, 352
interaction with IoT devices, 182
networking protocols, 171–172
Internet of Unmanned Aerial Vehicle (I-UAV), 174
networks, 171
scenarios, 171–172
Internet Protocol version 6 (IPv6), 173, 178–179, *180*
IPv6 over Low-Power Wireless Personal Area Networks (6LoWPAN), 173

JMAVSim, 80, 87–88
Justice K.S. Puttaswamy v. Union of India, 67, 70
Just-in-time (JIT) logistics optimization, 344–345

K. Krishnan v. State of Tamil Nadu (n.d.)., 68
Kalman filters, 40, 44, 271
extended Kalman filters (EKFs), 44–45
unscented, 44–45
Kapanga firebird
emulation topology, 189, *190*
user interface, 188, *188*
Key performance indicators (KPIs), 336
K.K. Multicopter, 98
K.K.2.1.5 FC, 98–99, *99*
calibration, 100
challenges, 99–100
cons, 101
pros, 100–101
Knife Edge Software, 89

Last-mile delivery systems with UAV swarms, 354
Latency-optimized UAV communication networks, 365
Lattice-based cryptography, 224
  for drone communications, 225–226, *226*
Law enforcement organisations, 73
LiDAR. *See* Light detection and ranging (LiDAR)
Light detection and ranging (LiDAR), 32, 48, 310
  based spatial awareness, 351
  flash, 37
  MEMS, 37
  OPA (optical phased-array), 37
  range and proximity sensors, 36–37
  sensor, 36–37
  solid-state, 37
Lithium-ion (Li-ion) battery, **13**, 17
Lithium polymer (LiPo) battery, **13**, 17
LLNs. *See* Low-Power Lossy Networks (LLNs)
Long-endurance and BVLOS flights, 9
Long-Range Wide-Area Network (LoRaWAN)
  link layer, 177–178
  Media Access Control, 177
Long-wave infrared (LWIR) cameras, 34
LoRa, 176
  physical layer, 176–177
  technology for Search and Rescue (SaR) missions, 175
LoRaWAN. *See* Long-Range Wide-Area Network (LoRaWAN)
Low-altitude aerial photogrammetry, 315
6LoWPAN topologies, 180–182, *181*
Low-Power Lossy Networks (LLNs), 180
Low-Power Wide-Area Network (LPWAN) topology, 173, 176

Machine learning algorithms, artificial intelligence with, 3
Magnetometer, 40
Manufacturing execution systems (MESs), integration in smart UAVs, 342
MAVLink. *See* MAVLink (Micro Air Vehicle Link) protocol
MAVLink 1, 197, 199
MAVLink 2, 197, 199, 204, 208, 210
MAVLink C Library, 201
MAVLink (Micro Air Vehicle Link) protocol, 197, 199
  in advancing drone communication, 214
  communication system, 200–201
    advanced projects, 202–203
    hardware interfaces, 201
    Pixhawk and Mission Planner/QGroundcontrol, 201–202
    software libraries, 201
  comparative analysis of cryptographic solutions, 212–213
  in computer vision systems, 207
  in drones and robotics, 200
  ecosystem, security and regulation, 215
  5G and edge AI, enabling real-time intelligence, 214–215
  and GCS, 204–205
  integration with emerging communication protocols and robotics frameworks, 214
  message handling, 203–204
  MicroRTPS, 214
  original purpose of, 205–206
  parameter protocol in, 198
  routing, 207
  security, 207–208
  security limitations of, 208–209
    asymmetric cryptography, 210–212
    post-quantum cryptography, 212
    symmetric cryptography, 209–210
  swarm robotics with, 206–207
  timestamps, 47
  use cases, 205–207
MAVProxy, 201, 202, 204, 207
MAVROS, 214
Maximum Transmission Units (MTUs), 182
Media Access Control (MAC), LoRaWAN, 177
Medium drones, 12
MEMS-based IMUs, 40
MEMS (Micro-Electro-Mechanical Systems) LiDAR, 37
Mesh networking, 7

Message-based MAVLink protocol, 203–204
Microcontroller unit (MCU), 97
MicroRTPS, MAVLink, 214
Microsoft AirSim, 82
Microsoft Kinect, 35
Mid-wave infrared (MWIR) cameras, 34
Mission Planner, 106, 204, **205**
  firmware, 104, 105
  and QGroundControl, 107, 198, 201–202
  software, 104, 105, 118–119
Mist-based intelligence, *172*
ML-DSA, 229
  for drone authentication, 231
ML-KEM, 229
  for drone key exchange, 230–231
Modern drones, 22
Modernisation and Reform Act of 2012, 60
Motor Vehicles Act, 68
MTUs. *See* Maximum Transmission Units (MTUs)
Multi-Agent Reinforcement Learning (MARL), 267
Multi-agent swarm intelligence system, AI-powered UAVs with, 377
Multi-agent UAV coordination, 363
Multirotor drones, 15
Multispectral sensors, 14, 30, 43, 48, 136, 238, 239, 287, 293, 299
Multivariate cryptography, 225

Nanotechnology-enabled self-repairing UAV structures, 375, *375*
NAT. *See* Network Address Translation (NAT)
National Security Agency (NSA), 221–222
Navigation and positioning sensors
  barometric altimeter, 42–43
  GNSS (GPS, GLONASS, Galileo, BeiDou), 41–42
  inertial measurement unit (IMU), 39–41
Navigation sensors, 29, 30
Navigation system in UAVs, 357–358
NDVI. *See* Normalized difference vegetation index (NDVI)
Neighbor Discovery (ND), 182
Network Address Translation (NAT), 178
Network layer, 178
  6LoWPAN, 180–182, *181*
  IPv6, 178–179, *180*
Neural network, quantum, 378
  neural network-assisted UAV path planning, 364
  neural network–based planning models, 276
  neural network–enabled predictive path planning, 364
Next-generation UAV energy systems, 377–378
No-fly zones, 64, 68, 73
Normalized difference vegetation index (NDVI), 43
NSA. *See* National Security Agency (NSA)
NVIDIA Jetson Xavier NX, 271, *274*

Object-Based Image Analysis (OBIA), 314
Octocopters, 12, 15, 101
ODE. *See* Open Dynamics Engine (ODE)
On-screen display (OSD), 98
OPA (optical phased-array) LiDAR, 37
Open Dynamics Engine (ODE), 83–84
Open Source Robotics Foundation, 83
Optical sensors, 32, 38, 301, 320, 339

Pacific Gas and Electric Company (PG&E), 315
Particle filters, 45
Particle Swarm Optimization (PSO), 267
Path Loss (PL), 175
Payload sensors, 30, 32
People's Union for Civil Liberties v. Union of India (n.d.), 70
Perception sensors, 29, 31
Peter Shor's quantum algorithm, 221
Phase shift method, 36
Photogrammetry, 320
  low-altitude aerial, 315
pH sensors, 44
Pixhawk, 106, *107*
  calibration, 107–108
  challenges, 106–107
  cons, 108
  pros, 108
Pixhawk 2.1 (Cube), 123
  calibration, 124
  challenges, 124
  disadvantages, 124
  pros, 124

Post-Processing Kinematic (PPK), 320
Post-quantum cryptography (PQC), 218
  algorithms, 229, *230*
  MAVLink, 212
  NIST standardization of, 229
Potential Field Algorithm and RL, 276
Power management circuitry, 97
Power transmission, telecommunication and, 358–359
PQC. *See* MAVLink; Post-quantum cryptography (PQC)
Precision time protocol (PTP), 47
Prediction success, 190
Property Technology (PropTech), 309
Proportional–Integral–Derivative (PID), 198
PTP. *See* Precision time protocol (PTP)
Public key cryptography. *See* Asymmetric cryptography
Puerto Rico Earthquake 2020, 300
PX4, 80, 82, 118, 124, 134, 144, 204
  Gazebo+PX4, 83–84
    Gazebo+PX4+ROS integration, 85
    key features, 84–85
    workflow, 85
  lightweight UAV simulation for, 87–88
Pymav link (Python), 201

QGroundControl, 204
QKD protocols. *See* Quantum-secured key distribution (QKD) protocols
Quadcopters, 15
Quantum AI–based UAV logistics, 335
Quantum algorithms, 217–219
Quantum computers, 220
Quantum computing
  emergence of, 232
  evolution of, 223
  and secure UAV operations, 378
  threats to drone communication security, 220–221
  for UAV optimization, 373
Quantum computing-driven by UAV decision systems, 379
Quantum cryptographic security for UAV networks, 369–370
Quantum cryptographic UAV communication security, 374
Quantum neural networks (QNNs), 378
  in UAV AI systems, 373, *374*
Quantum-resistant approaches, for drone communications, 224–225
Quantum-resistant cryptography. *See* Post-quantum cryptography (PQC)
Quantum-safe cryptography. *See* Post-quantum cryptography (PQC)
Quantum-safe drone communication, future direction for, 231–232
Quantum-secured key distribution (QKD) protocols, 374
Quantum sensors and next-generation UAV perception systems, 376

R. Rajagopal v. State of Tamil Nadu (n.d.), 70
Racing flight controller, 98
Radar and mmWave sensors, 38
Radiation sensors, 31, 43
Radio frequency (RF)
  based sensors, 32
  detection, 10
  hardware, 174
"Radioplane OQ-2", 5
Radio waves, 31, 32, 36, 38
Ramlal v. Mustafabad Oil Mills 1968, 67
Range and proximity sensors
  LiDAR (light detection and ranging), 36–37
  radar and mmWave sensors, 38
  ultrasonic sensors, 39
Real estate, 304
  applications of UAVs, 306, *306*
    agriculture, 306–307
    cinematography, 307
    defence and security, 307
    entertainment, 307–308
    infrastructure, 308
    underwater operations, 308
  challenges in implementing drone technology, 321
    data processing challenges, 322
    operational challenges, 322
    privacy and surveillance concerns, 323
    real-world challenges, 323
    technical challenges, 321–322
    workforce readiness, 321

classification of UAVs, 304–305
based on aerodynamics, 305–306, *305*
based on size, 305
diverse roles of drones in, 308–309, *309*
aerial photography and videography for real estate marketing, *318*, 318–319
land mapping and topographic surveying, 310–311, *311–313*, **313**
property inspection and structural analysis, 312–316, *315–317*
smart cities and sustainable development, 317–318
potential future prospects for drone technology in, 323–324
techniques and tools
drone swarm technology, 319–320
geospatial and mapping tools and techniques, 320
RealFlight, 83, 89
Real-time kinematic (RTK), 276, 320
Red, Green and Blue (RGB) optical cameras, 271
Regulatory compliance, 152, 215, 356
Reinforcement learning–based UAV optimization, 362–363
Remotely piloted aircraft system (RPAS), 3, 55
Regulatory Framework, 55
Representational State Transfer (REST), 173, 182, *184*
Application Program Interfaces (APIs), 182–183
transactions, 173
Return-to-home (RTH) function, 98
RGB cameras, 33–35
Rivest-Shamir–Adleman (RSA), 217, 219
Robot arms, 337, 338, 348
Robotics
development process, 128
frameworks, integration with emerging communication protocols and, 214
MAVLink in, 200
Robot Operating System (ROS), 128–129, 200
AI implementation, 152
architecture, 130–131, *132*
challenges and considerations in ROS-based UAS development, 144–147
communication and data protection security, 145
comparison with non-ROS alternatives, 134–135
complexity of software and learning curve, 146
compliance and regulatory issues, 146
enabled drones, 130
future trends in, 151–152
with Gazebo for UAV and robot simulation, 147–151
integration of UAS with, 133–134
lack of real-time processing capabilities, 145
Navigation Stack, 130
network protection, 146–147
prominent characteristics concerned with, 129–130
real-world case studies, 130
resource consumption and minimal system power, 145
resource restrictions, 147
reviewing hardware combination problems, 145
ROS 1, 128
to ROS 2 evolution, 129
ROS 2, 128
developments, 151
sensor and actuator diversities, 147
special interfaces and middleware, 147
standardization efforts, 152
system defect and reliability problems, 145–146
system real-time management, 146
tools and resources in the development of UAS, 143–144
UAS devices, 146
Robot simulation, ROS with Gazebo for UAV and, 147–151
Robot System, 132
Rogue drones, threat of, 10
ROS. *See* Robot Operating System (ROS)
Rotary-wing UAVs, 269, *269*, **270**
Rotor UAV, 11

RPAS. *See* Remotely piloted aircraft system (RPAS)
RSA encryption, 219
RTK (real-time kinematic) positioning, 41
Rural Farmers Female (RFF)-16, awareness of, **255–256**
Rural Farmers Male (RFM)-34, awareness of, **253–254**
Rust/JavaScript/Swift Bindings, 201
R v. Rogers (2020), 70
Ryan Firebee, 5

SAA. *See* Stateless Address Autoconfiguration (SAA)
Salinity sensors, 44
Search and Rescue (SaR), 270
 missions, LoRa technology, 175
 operations, 175
Secure communication, 219, 220
 essential, 137
 frameworks, 137
 of mission-critical information, 226
 PQC for, 48
 protocols, 130
Secure Digital (SD) card, 108
Self-healing UAVs
 with adaptive flight materials, 379
 materials and advanced composites, 374–376
Semi-autonomous drone, 314
Sensor, *See also specific types*
 acoustic/ultrasonic, 32
 chemical, 44
 CMOS, 33
 environmental, 30, 31, 43
 environmental and mission-specific, 43–44
 fusion, 40, 42, 46–47
  and data processing, 44–46
  and swarm intelligence, 267
 humidity, 43
 hyperspectral, 43
 inertial/mechanical, 32
 light detection and ranging (LiDAR), 36–37
 multispectral, 43
 navigation, 29
 optical, 32
 payload, 30, 32
 perception, 29, 31
 pH, 44
 radar and mmWave, 38
 radiation, 43
 radio frequency-based, 32
 salinity, 44
 smoke, 44
 systems equip drones, 29
 temperature, 43
 time-of-flight, 35–36
 turbidity, 44
 ultrasonic, 39
Sensor taxonomy in drone systems, classification
 based on function, 30–32
 based on physical principle, 32
Sensor technology, 270–271
 challenges and future directions, 47–48
 in drone operations, 29–49
Serial peripheral interface (SPI), 46
Shor's algorithm, 222
Shreya Singhal v. Union of India, 67, 68
Signal-to-Noise Ratios (SNRs), 171, 178
Simulation environment, 80, 85
 Gazebo-compatible, 144
Simulators for UAVs, 79
 AirSim, 86–87
 comparison of UAV simulators, **92**
 CopterSim, 91–93
 DroneSim Pro, 91
 FlightGear, 90–91
 Gazebo+PX4, 83–85
 hardware-in-the-loop simulators, 80–83
 JMAVSim, 87–88
 lightweight UAV simulation for PX4, 87
 RealFlight, 89
 software-in-the-loop simulators, 79–80, *80*
 types, 79, *80*
 variants, 83, *84*
 Velocidrone, 89–90
 X-Plane, 88–89
Simultaneous localization and mapping (SLAM), 7, 8, 277, 352
 based UAV navigation in GPS-denied environments, 366
 LiDAR sensor, 37
 RGB cameras, 34
Single-rotor drones, 11–12
Single-rotor helicopters, 15

SITL simulators. *See* Software-in-the-loop (SITL) simulators
6G-enabled UAV communication and edge computing, 378
6G-enabled UAV networks for industrial applications, *372*, 372–373
SLAM. *See* Simultaneous localization and mapping (SLAM)
SLH-DSA, 229
  for conservative drone security, 231
Small drones, 12
Smart communication and UTM, 22–23
Smoke sensors, 44
SNRs. *See* Signal-to-Noise Ratios (SNRs)
Software in the loop (SITL) simulation, 204
Software-in-the-loop (SITL) simulators, 79–80, *80*
Soil health, quantification of, 239
Solar-powered designs, 9
Solar-powered UAVs, 22, 275, *275*
  for extended operations, 361
Solid-state LiDAR, 37
Sophisticated image-processing techniques, 320
Southeast Asia Floods 2021, 299–300
SpeedyBee F7 V3, 119
  calibration, 119
  disadvantages, 120
  pros, 119–120
SpeedyBee F405, 112
  calibration, 112–113
  challenges, 112
  cons, 113
  pros, 113
SpeedyBee F405 V4, 111
  calibration, 111
  challenges, 111
  cons, 112
  pros, 111–112
SPI. *See* Serial peripheral interface (SPI)
Standards and Recommended Practices (SARPs), 56
Stateless Address Autoconfiguration (SAA), 182
Stereo
  and depth cameras, 35–36
  vision principle, 35
Structural health monitoring (SHM), 324
Structure from Motion (SfM) algorithms, 161
Surveillance and security drones, 13
Sustainability and environmental monitoring, 317–318
Swarm behavior of UAV systems, 363
Swarm drones, 8–9
  flight control, 138
Swarm intelligence
  AI-powered UAVs with multi-agent, 377
  sensor fusion and, 267
Swarm robotics with MAVLink, 206–207
Symmetric cryptography, MAVLink, 209–210
Symmetric encryption, 210, 212, 221, 223, 229, 230
Synchronization
  and calibration, 46
  communication interfaces and, 46–47

Taylor v. Federal Aviation Administration (n.d.), 70
Telecommunication
  and power transmission, 358–359
  restoration in disaster zones, 300
Telemetry system, UAV, 174
Temperature sensors, 39, 43
Thermal cameras, 34–35
Thunder QQ Super FC, 101–102
  calibration, 102
  challenges, 102
  cons, 103
  pros, 103
Tilt-rotor drones, 12, 15
Time-of-flight (ToF)
  principle for ultrasonic sensors, 39
  sensors, 35–36
Time synchronization, 46–47
Toxic release response, 270
Traditional AI-based UAV decision models, 373
Traditional drones, 11
Traffic management systems, industrial UAVs, 354, *355*
  AI-powered UAV air traffic optimization, 355
  blockchain-based secure UAV air traffic control, 356, *356*
  5G and 6G in UAV fleet communication, 355

Transmission Control Protocol (TCP), 182, 198
Transponder, 175
Transport and application layers, 182–185, *183–185*
Transport-level encryption, 215
Triangulation, 35
Turbidity sensors, 44

UART. *See* Universal asynchronous receiver-transmitter (UART)
UAS. *See* Unmanned Aerial System (UAS)
UAVCAN. *See* Unmanned Aerial Vehicle Controller Area Network (UAVCAN)
UAVs. *See* Unmanned aerial vehicles (UAVs)
UDP. *See* User datagram protocol (UDP)
Ultrasonic sensors, 39
Underwater operations, UAVs applications in, 308
UN ICAO, 55
Universal asynchronous receiver-transmitter (UART), 46, 100, 201
Unmanned Aerial System (UAS), *See also* Drone; Unmanned aerial vehicles (UAVs)
  applications, 135
    autonomous navigation and path planning, 136
    multiple unmanned aircraft systems, 136–137
    sensor integration and data processing, 136
  case studies and applications, 137, *138*, **139–142**
  challenges and considerations in ROS-based UAS development, 144–147
  development environments, 144
  devices, 146
  flight controllers, 144
  future trends in, 151–152
  integration of UAS with ROS, 133–134
  ROS tools and resources in the development of, 143–144
  simulation tools of, 143–144
  technology, 238, 240
Unmanned Aerial Vehicle Controller Area Network (UAVCAN), 47
Unmanned aerial vehicles (UAVs), 3, 4, 7, 10, 21, 52–53, 237, *See also* Drone; Unmanned Aerial System (UAS)
  advanced image-processing applications, 162
    agricultural and environmental monitoring, 162–163, **164**
    urban infrastructure and security applications, 163, **165**
  advances in, 156
  advantages of, 243
  agricultural, 19–20
  in agriculture. *See* Agriculture, UAVs in
  applications of, 65, 306, *306*
    agriculture, 306–307
    cinematography, 307
    defence and security, 307
    entertainment, 307–308
    infrastructure, 308
    underwater operations, 308
  assisted and traditional farming techniques, 239
  AVE-R, 188–189, *189*
  challenges of, 243–244
  CH to UAV message flow, 185–186, *186*
  classification and characteristics, 157, **163**
    operation-based classification system, 157–158, *159*
    size-based classification, 157, *158*
  dangers of UAV operations, 56
  examples of, *158*
  experimental framework, 188–192
  flight controller role in, 97
  fully autonomous level, 13
  historical development, regulations, 53–55
  ICAO regulation, 56
  image processing important to UAV systems, 156, **164**
  image-processing methodology framework, 158, **164**
    advanced classification systems, 159–160
    data acquisition and preprocessing, 158–159

segmentation and analysis framework, 159, *160*, *161*
three-dimensional reconstruction methodologies, 161, *162*, **165**
international legal frameworks governing, 55
European Union's approach, 58–59
role of ICAO, 55–58
role of UN and multilateral efforts, 61–63
United States framework, 59–61
and liability issues, 67–69
link layer, 177–178
low-latency alternative, 171
market, 3
mist computing infrastructure, 185–187, *186*
national regulations and diverging approaches, 63–66
operation of, 29
performance comparison and analysis, 163, **165**
physical layer, 176–177
pilots of, 11
predictability as a function, 190, *191*
rapid proliferation of, 74
regulations
futures, 72–73
international harmonisation, 75
legal precedents and judicial interpretations, 69–72
role of RGB cameras in, 34
ROS with Gazebo for UAV and robot simulation, 147–151
rotor, 11
sensor systems, 48
categories, 29–30
simulator for. *See* Simulators for UAVs
small, 16
small- to mid-scale, 17
solar-powered, 22
swarms of, 174–175
technological evolution, 54
technology with modern image-processing capabilities, 166–167
telemetry system, 174
transport and application layers, 182–185, *183–185*
types, *4*
Unmanned Surface Vehicle (USV), 52
Urban Farmers Female (UFF)-24, awareness of, **259–260**
Urban Farmers Male (UFM)-43, awareness of, **257–258**
Urban planning and development, 320
U.S. Department of Defense, 130
US Air Force, 60
User datagram protocol (UDP), 173, 198
network interface, 88
transport layer, 183
UTM, smart communication and, 22–23

Variable-pitch propellers, 17
Vector V2 by Eagle Tree systems, 124
calibration, 124
challenges, 124
disadvantages, 125
pros, 125
Velocidrone, 83, 89–90
Velodyne VLP-16 Puck, **271**, *273*
Vertical take-off and landing (VTOL) UAVs, 9, 276
Videography, for real estate marketing, *318*, 318–319
Vision-based navigation systems, 7
Vision-based sensing
infrared (thermal) cameras, 34–35
RGB cameras, 33–35
stereo and depth cameras, 35–36
Visual line-of-sight (VLOS), 13, 368

Warehouse management system (WMS), 352, 354
Wearable sensor technology, 348
Wildfire monitoring and containment, California Wildfires 2022, 299
Windows GCS for ArduPilot, 204
WingtraOne VTOL, 269, *269*
WPANs
categories of, 181
devices within, 182

X-Plane, 82, 88–89

For Product Safety Concerns and Information please contact our EU representative GPSR@taylorandfrancis.com Taylor & Francis Verlag GmbH, Kaufingerstraße 24, 80331 München, Germany

Batch number: 10761548

Printed by Printforce, the Netherlands